Sustainable Carbon Capture

Sustainable Carbon Capture

Technologies and Applications

Edited by
Humbul Suleman, Philip Loldrup Fosbøl,
Rizwan Nasir, and Mariam Ameen

CRC Press
Taylor & Francis Group
Boca Raton   London   New York

CRC Press is an imprint of the
Taylor & Francis Group, an **informa** business

MATLAB® is a trademark of The MathWorks, Inc. and is used with permission. The MathWorks does not warrant the accuracy of the text or exercises in this book. This book's use or discussion of MATLAB® software or related products does not constitute endorsement or sponsorship by The MathWorks of a particular pedagogical approach or particular use of the MATLAB® software.

First edition published 2022
by CRC Press
6000 Broken Sound Parkway NW, Suite 300, Boca Raton, FL 33487–2742

and by CRC Press
2 Park Square, Milton Park, Abingdon, Oxon, OX14 4RN

CRC Press is an imprint of Taylor & Francis Group, LLC

© 2022 selection and editorial matter, Humbul Suleman, Philip Loldrup Fosbøl, Rizwan Nasir, Mariam Ameen; individual chapters, the contributors

Library of Congress Cataloging-in-Publication Data
Names: Suleman, Humbul, editor. | Fosbøl, Philip Loldrup, editor. | Nasir, Rizwan, editor. | Ameen, Mariam, editor.
Title: Sustainable carbon capture : technologies and applications / edited by Humbul Suleman, Philip Loldrup Fosbøl, Rizwan Nasir, Mariam Ameen.
Description: First edition. | Boca Raton : CRC Press, 2022. | Includes bibliographical references and index. | Summary: "A comprehensive resource on different aspects of sustainable carbon capture technologies including recent process developments, environmentally friendly methods, and roadmaps for implementations. It discusses also the socio-economic and policy aspects of carbon capture and the challenges, opportunities, and incentives for change with a focus on industry, policy, and governmental sector. This book provides guidelines for sustainable and responsible carbon capture and addresses current and future global energy, environment, and climate concerns"— Provided by publisher.
Identifiers: LCCN 2021040320 (print) | LCCN 2021040321 (ebook) | ISBN 9780367755140 (hardback) | ISBN 9780367755157 (paperback) | ISBN 9781003162780 (ebook)
Subjects: LCSH: Carbon sequestration—Technological innovations. | Green technology. | Sustainable engineering.
Classification: LCC TD885.5.C3 S87 2022 (print) | LCC TD885.5.C3 (ebook) | DDC 628.5/32—dc23/eng/20211116
LC record available at https://lccn.loc.gov/2021040320
LC ebook record available at https://lccn.loc.gov/2021040321

ISBN: 978-0-367-75514-0 (hbk)
ISBN: 978-0-367-75515-7 (pbk)
ISBN: 978-1-003-16278-0 (ebk)

DOI: 10.1201/9781003162780

Typeset in Times
by Apex CoVantage, LLC

Dedicated to

A greener world
and
prevention of climate change skepticism

Contents

Preface

Carbon dioxide emissions from fossil fuels and industries are warming-up the world. These CO_2 emissions, along with other greenhouse gas emissions (like methane, NO_X, and CFCs), must be cut down to limit the global temperature increase below 1.5°C as per the Paris Agreement and in line with the UN's Sustainable Development Goals.

Carbon capture, along with storage and utilization technologies, is the frontline technology concept for capturing 80% of global CO_2 emissions, with an opportunity of achieving carbon-negative emissions in the long run. Although the technology is quite mature, sustainable capture of carbon dioxide is still a challenge. Recent developments point to a new suite of technologies that can make carbon capture environmentally friendly and economically feasible, which forms the motivation for this book.

This book provides the latest information on the developments in carbon capture for climate change mitigation, alongside bridging a knowledge gap between the contribution of new materials and the latest technologies towards sustainability. Also, it discusses newfound applications of carbon capture in the field of biofuels, hydrogen, and carbon-neutral fuels.

The opening chapter of the book, by our editor Foshøl, Suleman and co-authors, introduces the history of carbon capture applications prior to decarbonization protocols, current state-of-the-art technologies, and the sustainability level of carbon capture. Further, this chapter describes the role and infrastructure of carbon capture in the forthcoming years. Chapter 2 of the book, written by Puxty and co-authors, addresses recent developments in reactive chemical absorption of CO_2 by organic molecules. They discuss the research developments in carbon dioxide absorption, especially for the aqueous amines and their derivatives. The chapter also examines the potential of removing water from these absorbents, and their beneficial biphasic features, developing an alternate chemistry (to aqueous amines) for carbon dioxide absorption.

Ionic liquids (ILs) can potentially increase the efficiency of available CO_2 capture technologies. In Chapter 3, Economou and co-authors present details of the current and prospective carbon capture methods involving ILs, followed by an in-depth examination of separator performance. Furthermore, discussion of computational methods provides insight into the chemical structure of ILs, macroscopic behavior, and reliable property prediction.

Gas hydrates are ice-like solids formed at low temperature and high pressure. They are a significant concern in the oil and gas sector due to flow assurance issues. Contrarily, they can also be used as separation reagents, wherein they remove CO_2 from mixed flue gas. Von Solms authored Chapter 4 on the use of gas hydrates for carbon capture. To help those interested in gas hydrates, he has given a summary of the thermodynamic characteristics of gas hydrates. He has also covered previous and current efforts to use gas hydrates for sustainable carbon capture.

The metal-organic framework (MOF) is a novel family of crystallized porous media that has recently gained research importance. Up-to-date advances allow transforming ambient carbon dioxide into usable compounds, described by Basu, Khan and co-authors in Chapter 5. Their contribution also includes the use of MOF as fillers in CO_2 separation membranes. The latter are then discussed separately in Chapter 6 by Mannan's group in which researchers evaluate membrane innovations in carbon capture and set forth carbon separation pathways using new polymeric materials. Maqsood and colleagues provide an exhaustive account of cryogenic CO_2 capture in Chapter 7, focusing on inherent problems, their possible solutions, and prospects of cryogenic separation-based technologies.

Syngas production from carbon dioxide is a promising CO_2 conversion technology. These carbon-neutral fuels can replace fossil fuels in industrial and automobile applications. Since the process and associated technologies are in early developmental stages, the catalysts are not 100% efficient. Hence, the product has substantial quantities of residual carbon dioxide feed, which must be separated using carbon capture. Chapter 8, authored by Azizan and his research group, discusses the issues related to alternative fuel production from carbon dioxide and the use of carbon capture as a potential solution in improving product quality. Similarly, the blue hydrogen sector is another potential user of carbon capture, as the hydrogen produced from the natural gas requires the process CO_2 to be captured. Blue hydrogen is an essential part of the world's transition to a sustainable hydrogen economy. Mature technology and infrastructure exist for large-scale production and transportation of blue hydrogen. In Chapter 9, Abdus Salam's group provides comprehensive information about the various techniques of producing blue hydrogen (involving carbon capture) and possibilities of supplementing it via bio-hydrogen pathways.

The engineering design of carbon capture systems has witnessed significant advances in the last two decades. The advancement of process design, simulation, and process control with respect to carbon capture is described in Chapter 10 by Zabiri, Maulud and co-authors. Different thermodynamic approaches, kinetic models, and process design methodologies are explained.

Four case studies related to sustainable carbon capture are reported in a joint chapter (Chapter 11) from Goetheer, Neerup, and co-authors about carbon capture applications in the liquefied natural gas (LNG) sector, shipping vessels, biogas upgrading, and conventional chemical industries (cement, iron/steel).

A net-zero economy calls for revolutionary changes in demographics, economies, and societies. It is doubtful that any new carbon capture policy will be put in place until its effects on humans, society, and the economy are established, which forms the content of Chapter 12. MacDowell and co-authors put forward a convincing view on the developed models for determining the socio-economic impacts of carbon capture on common human practices. The chapter also devises pathways for the energy system change in the near future.

The book concludes with Chapter 13 from our editors Nasir, Ameen, and co-authors. The chapter explores emerging trends, advancements in the allied technologies and provides an overview of the research hotspots for carbon capture. The chapter entails a detailed account of possible future scale-up technologies, including direct air carbon capture, enzymatic carbon capture, and mixed chemical looping.

The world today stands at a critical juncture, when future research for the next 30 years (i.e., until 2050) is motivated by the decarbonization goals set by the Paris Agreement. Training our aspiring researchers, young engineers, and general academia about the acquired sustainability in the field of carbon capture is undeniably crucial. This book is a humble effort to disseminate the scientific knowledge behind carbon capture and educate people about its future role in reducing new CO_2 emissions. The editors believe it will help to understand the technology transition in the field, instigating new research to develop concurrent new strategies in applied and allied sciences for a green and sustainable world.

The Editors
Humbul Suleman
Philip Loldrup Fosbøl
Rizwan Nasir
Mariam Ameen

Acknowledgments

The editors would like to appreciate all the contributors who have led to the successful completion of this book. We also thank all the reviewers who refined the technical contents of this book.

The editors would like to extend a special thanks to Randi Neerup, Center for Energy Resources Engineering, Department of Chemical and Biochemical Engineering, Technical University of Denmark, for her help in proofreading many of our chapters and helping us gather materials/resources for this manuscript. Words of gratitude are also due to Deval Divyangkumar Patel, School of Computing, Engineering and Digital Technologies, Teesside University, UK, who spent his time formatting some of the chapters for our book.

Editor Biographies

Humbul Suleman is a senior lecturer in chemical engineering at Teesside University, United Kingdom. He is also a member of the Hydrogen Economy and Decarbonisation Technologies group and Centre for Sustainability Engineering at the university. Externally, he is an active member of the UK Carbon Capture and Storage Research Centre (UKCCSRC) and has delivered invited lectures on carbon capture in many universities. He is involved in research for sustainable carbon capture using advanced biological compounds and ionic liquids for hybrid carbon capture systems. Currently, he is a principal and co-investigator on various funded studies for sustainable carbon capture totaling £1.1 million. Previously, he has worked on six research projects related to sustainable carbon capture in various capacities. He has contributed multiple peer-reviewed publications, media articles and book chapters in the field. Apart from his carbon capture research and work with UKCCSRC, he is also involved in knowledge-sharing and professional development activities with the Institution of Chemical Engineers UK (IChemE), Marie-Curie Alumni Association (MCAA), and The Research Council of Oman (TRC).

Philip Loldrup Fosbøl is an associate professor at the Center for Energy Resources Engineering, Department of Chemical and Biochemical Engineering at the Technical University of Denmark, where he currently leads the Carbon Capture and Bioenergy research group with three PDRAs and seven PhD students. His students are focused on technology development for sustainable carbon capture, storage, and utilization at the molecular, process, and network scales. The research entails lab analysis covering thermodynamics, energetic properties, kinetics, and also intermediate bench scale and larger unit operation pilot-scale facilities.

He is currently PI of four large Danish-funded CCS and gas-cleaning projects with a total budget of €9 million. An additional collaboration project includes the ongoing large-scale 3D EU project focusing on CCS for the steel sector. Denmark is currently entering into a very active CCS strategy, and he has acted as the main advisor for the political negotiations to reach the ambitious sustainability goal for the Danish state to reduce emission by 70% before 2030. He has been active in social and print media to promote low carbon energy. His current research focus is CO_2 capture, biogas upgrading, CO_2 conversion, and electrochemical gas cleaning applications. The technology development is performed through a close collaboration to industrial partners, and a consortium of 20 industries supports the research.

Rizwan Nasir is an assistant professor in chemical engineering at the University of Jeddah, Saudi Arabia. He is a member of the European Membrane Society (EMS), International Association of Engineers (IAENG), and Society of Chemical Industry (SCI). He has accomplished research expertise in membrane technology, materials development, characterization, and potential applications in separation processes related to energy and the environment. He is a author of several high-indexed publications, book chapters and conference proceedings. Also, he has presented his research articles at various national and international conferences and exhibitions and received recognition in the form of awards, medals, and appreciation. Moreover, he is an invited reviewer for many high-quality journals in the field of carbon capture and membranes.

He is currently the principal investigator on the Screening of Forward Osmosis (FO) Membranes for Upstream Produced Water Treatment before Water Injection Practice and the co-principal investigator on the Sustainable Green Approach for Biodiesel Synthesis from Algae and the Synthesizing of COF (Covalent Organic Framework)/Polymer composite membranes for N_2, CO_2, and CH_4 Separation.

Mariam Ameen is a senior postdoctoral research associate in the Higher Institution Center of Excellence (HICoE), Centre for Biofuels and Biochemical Research (CBBR) at the Department of Chemical Engineering, Universiti Teknologi PETRONAS, Malaysia. She is a member of RSC | ACS | WSSET | SCI | ISEST| IChemE. She is a forward-thinker, highly motivated, and solution-oriented chemist/research scientist backed by the experience of over six years in collaborative (industry and academic) research in applied heterogeneous catalysis, sonochemistry, materials characterization, and reaction mechanisms leading to the successful investigation of novel materials. She is engaged in designing of various catalyst types, characterization, and testing at industry-relevant conditions for direct biomass conversion into biofuels (e.g., green diesel, biogasoline, jet fuel, hydrogen and biogas production on high temperature and pressure at the lab and pilot scales). Mariam is an emphatic and valiant science communicator making complex scientific concepts accessible to audiences of various backgrounds, an excellent team player, and a strong builder of prolific collaborations in various aspects of basic and applied research. She has contributed to more than eight research projects in collaboration with industry and academia and several peer-reviewed scientific publications in high impact journals. She has one Malaysian patent filed, has written one book chapter, and has been an invited reviewer for many reputed journals. She is a member of scientific societies and presenter in conferences, seminars, webinars, and meetings on scientific research. She is currently involved in different projects related to biofuel technology, such as biogas from waste materials, oxygen carrier, aqueous-phase reforming for hydrogen and value-added chemical productions, bio-lubricant synthesis, bio-gasoline, and green diesel production. She believes that research and knowledge move forward by sharing with others.

Contributors

Abdulgadir, Alamin Idris
Department of Engineering and
Chemical Sciences
Karlstad University
Sweden

Maulud, Abdulhalim Shah
Department of Chemical Engineering
Universiti Teknologi Petronas
Malaysia

Abdulrahman, Aymn
Department of Chemical Engineering
University of Jeddah
Jeddah, Saudi Arabia

Affian, Muhammad Afif Asyraf
Department of Chemical Engineering
Universiti Teknologi Petronas
Malaysia

Ali, Abulhassan
Department of Chemical Engineering
University of Jeddah
Jeddah, Saudi Arabia

Ameen, Mariam
HICoE, Center for Biofuel and
Biochemical Research (CBBR)
Institute of Self-Sustainable Building
(ISB)
Department of Chemical
Engineering
Universiti Teknologi Petronas
Malaysia

Azizan, Mohammad Tazli
Faculty of Chemical Engineering
Technology
Universiti Malaysia Perlis
Malaysia

Basu, Subhankar
Department of Applied Sciences and
Humanities
National Institute of Foundry and Forge
Technology (NIFFT)
India

Bennett, Robert
The Commonwealth Scientific and
Industrial Research Organisation
(CSIRO) Energy
Australia

Economou, Ioannis G.
Institute of Nanoscience and
Nanotechnology
National Center for Scientific
Research
Greece
Chemical Engineering Program
Texas A&M University
Qatar

Fatima, Anmol
Department of Chemical
Engineering
Universiti Teknologi Petronas
Malaysia

Fosbøl, Philip Loldrup
Center for Energy Resources
Engineering (CERE)
Department of Chemical and
Biochemical Engineering
Technical University of Denmark
Denmark

Goetheer, Earl L.V.
TNO
The Netherlands

Gong, Wentao
Center for Energy Resources
Engineering (CERE)
Department of Chemical and
Biochemical Engineering
Technical University of Denmark
Lyngby, Denmark

Hauschild, Michael Z.
Department of Technology,
Management and Economics
Technical University of Denmark
Lyngby, Denmark

Isa, Faezah
Department of Chemical Engineering
Universiti Teknologi Petronas
Malaysia

Islam, Md. Tauhidul
Hydrogen Energy Laboratory
Bangladesh Council of Scientific and
Industrial Research (BCSIR)
Bangladesh

Jamil, Asif
Department of Chemical Polymer
and Composite Material
Engineering
University of Engineering and
Technology
Pakistan

Jørsboe, Jens Kristian
Center for Energy Resources
Engineering (CERE)
Department of Chemical and
Biochemical Engineering
Technical University of Denmark
Denmark

Khan, Asim Laeeq
Department of Chemical Engineering
COMSATS University
Pakistan

Kumari, Ranjana
Department of Applied Sciences and
Humanities
National Institute of Foundry and Forge
Technology (NIFFT)
India

Lakra, Reshma
Department of Applied Sciences and
Humanities
National Institute of Foundry and Forge
Technology (NIFFT)
India

Latif, Muhammad
Institute of Energy and Environmental
Engineering
University of the Punjab
Pakistan

Mac Dowell, Niall
Centre for Environmental Policy
Imperial College London
United Kingdom

Maeder, Marcel
School of Environmental and Life
Sciences
The University of Newcastle
Australia

Mannan, Hafiz Abdul
Institute of Polymer and Textile
Engineering
University of the Punjab
Pakistan

Maqsood, Khuram
Department of Chemical Engineering
University of Jeddah
Jeddah, Saudi Arabia

Mohshim, Dzeti Farhah
Department of Petroleum Engineering
Universiti Teknologi Petronas
Malaysia

Monteiro, Juliana G.M-S.
TNO
The Netherlands

Mukhtar, Hilmi
Department of Chemical
Engineering
Universiti Teknologi Petronas
Malaysia

Nasir, Rizwan
Department of Chemical Engineering
University of Jeddah
Saudi Arabia

Naveed, Hira
Department of Chemical
Engineering
COMSATS University
Pakistan

Neerup, Randi
Center for Energy Resources
Engineering (CERE)
Department of Chemical and
Biochemical Engineering
Technical University of Denmark
Denmark

Papri, Nasrin
Hydrogen Energy Laboratory
Bangladesh Council of Scientific and
Industrial Research (BCSIR)
Bangladesh

Patrizio, Piera
Centre for Environmental Policy
Imperial College London
United Kingdom

Puxty, Graeme
The Commonwealth Scientific and
Industrial Research Organisation
(CSIRO) Energy
Australia

Qadir, Danial
Centre for Sustainable Engineering
Teesside University
United Kingdom

Romanos, George E.
Institute of Nanoscience and
Nanotechnology
National Center for Scientific
Research
Greece

Ros, Jasper A.
TNO
The Netherlands

Salam, Md. Abdus
Hydrogen Energy Laboratory
Bangladesh Council of Scientific and
Industrial Research (BCSIR)
Bangladesh

Shahbudin, Muhammad Izham
Department of Chemical Engineering
Universiti Teknologi Petronas
Malaysia

Shaheen, Haleefa
Department of Chemical Engineering
COMSATS University
Pakistan

Sharif, Rabia
Department of Chemical and Polymer
Engineering
University of Engineering and
Technology Lahore (Faisalabad
Campus)
Pakistan

Suleman, Humbul
School of Computing, Engineering and
Digital Technologies
Teesside University
United Kingdom

Sultan, Tahir
Department of Chemical Engineering
Universiti Teknologi Petronas
Malaysia

Svendsen, Hallvard F.
Department of Chemical Engineering
Norwegian University of Science and
Technology
Norway

Syuhada, Ain
Department of Chemical Engineering
Universiti Teknologi Petronas
Malaysia

Vergadou, Niki
Institute of Nanoscience and
Nanotechnology
National Center for Scientific Research
Greece

Villadsen, Sebastian Nis Bay
Center for Energy Resources
Engineering (CERE)
Department of Chemical and
Biochemical Engineering
Technical University of Denmark
Denmark

Vinjarapu, Sai Hema Bhavya
Center for Energy Resources
Engineering (CERE)
Department of Chemical and
Biochemical Engineering
Technical University of Denmark
Denmark

von Solms, Nicolas
Center for Energy Resources
Engineering (CERE)
Department of Chemical and
Biochemical Engineering
Technical University of Denmark
Denmark

Xie, Judy Jingwei
Centre for Environmental Policy
Imperial College London
United Kingdom

Zabiri, Haslinda
Department of Chemical Engineering
Universiti Teknologi Petronas
Malaysia

1 Introduction to Sustainable Carbon Capture

Philip Loldrup Fosbøl, Humbul Suleman,
Hallvard F. Svendsen, and Michael Z. Hauschild

CONTENTS

The emission of carbon dioxide (CO_2), a greenhouse gas, has led to a rise in average atmospheric temperature. As of May 2021, atmospheric CO_2 levels are at 419 ppm despite a halt in the world economy due to a raging pandemic [1]. To maintain the rise in global temperature below the 1.5 °C benchmark (as stipulated by the Paris Agreement [2]), worldwide CO_2 emissions must be reduced. Many methodologies have been suggested, such as process change (e.g., converting to power generation based on renewables instead of fossil fuels), design change (e.g., adapting to green building standards), or efficiency improvement (e.g., better use of energy) [3]. Nevertheless, a major portion of CO_2 emissions comes from fossil fuels and the process industry [4]. Even if the use of fossil fuels is brought to zero (which is not possible in the near future), emissions from industries like cement, steel, paper, pharmaceuticals, lime, and others will linger until we find an alternative method

DOI: 10.1201/9781003162780-1

to eliminate them from the start. These emissions usually originate at stationary points and in large quantities. To mitigate such emissions, carbon capture utilization and storage (CCUS) is a promising technology concept. It is a mix of three separate technology paradigms that work in tandem to reduce carbon dioxide emissions from large point sources. Carbon capture involves technologies that separate carbon dioxide from other gases, either pre-combustion (e.g., sweetening of natural gas) or post-combustion (e.g., cleaning of flue gas). Carbon utilization is a set of newly evolving technologies that convert the captured CO_2 into useful chemicals, value-added products, and/or process commodities for direct use. Carbon storage is a multitude of technical methods to store carbon dioxide in underground coal seams, exhausted reservoirs, impervious rock layers, and aquifers [5]. Figure 1.1 presents a schematic view of the CCUS concept.

Carbon capture enjoys an overall better technology readiness than the carbon utilization (sometimes referred to as carbon usage) and carbon storage sectors. Yet the technology requires rapid technological improvements (via intensified research activity), social acceptance, commercialization, and wide-scale implementation to reach the sustainable and environmental goals of decarbonization.

This book focuses on the technological developments, current status, and future applications of carbon capture for reducing global CO_2 emissions sustainably. The collection includes 13 chapters focusing on various technologies (Chapters 2 to 7), upcoming applications (Chapters 8 and 9), recent developments in process design (Chapter 10), success stories (Chapter 11), and social policy around carbon capture

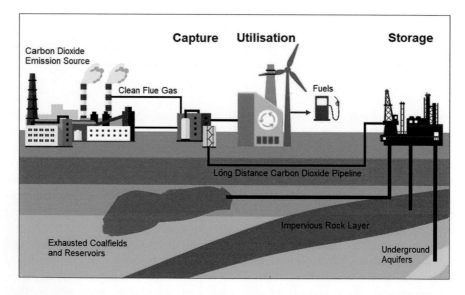

FIGURE 1.1   Schematic view of carbon capture, utilization, and storage technology concept.

(Chapter 12). The book closes with a chapter on emerging trends within carbon capture technologies, which outlines upcoming technologies we can potentially expect to see full scale in the future (Chapter 13).

1.1 INTRODUCTION TO CARBON SEPARATION TECHNOLOGIES

The carbon separation technologies are roughly divided into four sub-areas, namely, absorption, adsorption, membranes, and cryogenics [6–7], as shown in Figure 1.2.

Absorption is the dominant technology in the field, which involves use of physical and chemical solvents to absorb carbon dioxide from a gaseous mixture. Driven by a need to reduce energy consumption, avoid solvent degradation, and minimize solvent loss, many classes of chemical solvents have been tested and have found selective commercialization (e.g., alkanolamines, amino acids, potassium carbonate, ammonia, ionic liquids) [8]. Selexol is a prime example of physical solvents [9]. A divergence of the concept is the use of gas hydrates to absorb/adsorb carbon

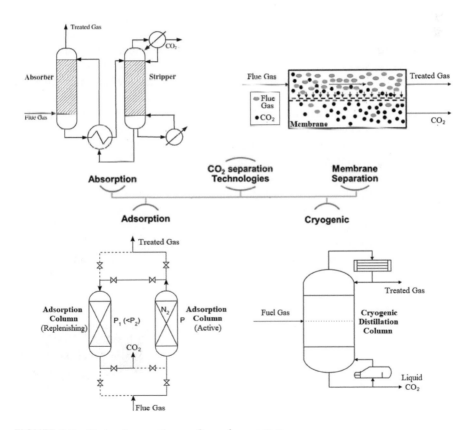

FIGURE 1.2 Technology pathways for carbon capture.

dioxide [10]. However, the technology platforms available today are not developed enough to effectively utilize the gas hydrate concept.

Adsorption is another promising technology for CO_2 capture, with substantial reduction in energy and costs as compared to the conventional absorption technology. Zeolites and activated carbon have remained mainstay adsorbents for this technology [11]. Several modifications and functionalization of the adsorbents have been researched. The past decade has seen a steep rise in the use of adsorbents that are cycled through sorption and desorption cycles for CO_2 removal from ultra-dilute gases such as air. The captured CO_2 reacts with metal cations (usually with a hydroxide ion as an anion) to form carbonate minerals, a concept called CO_2 mineralization. There are several alkali/alkaline minerals capable of this feat (e.g. calcium, sodium, potassium hydroxides), but the one gaining the most attention is olivine [12], a magnesium-containing crystal found in igneous rocks, which is converted to magnesium hydroxide (brucite) by a complex series of reactions. One pound of olivine can absorb as much as one pound of carbon dioxide from the air, under ideal circumstances [13]. A disruptive concept is to use the metal organic frameworks for CO_2 capture [14].

Membranes offer a cost-effective carbon capture option and have seen massive development but limited application in CO_2 capture in the last two decades [15–16]. They offer solvent-free separation, but durability and operational issues still require further development.

Cryogenic CO_2 separation is a well-known area that has revived after a need to monetize natural gas containing high levels of CO_2 (as natural gas fields with low CO_2 content deplete). The commercial applications are restricted by the high costs of energy, mostly required to condense all or a portion of the gaseous mixture. Novel processing concepts in cryogenic distillation and a new cheap (but strong) class of materials offer new possibilities for this technology.

Calcium looping, or the regenerative calcium cycle, is another carbon capture technology. It is the next step to carbonate looping. Calcium oxide (CaO), a sorbent, directly reacts with carbon dioxide (CO_2) to form calcium carbonate ($CaCO_3$), hence separating carbon dioxide from a mixture of gases. The formed carbonate is then calcined in another section of the plant to retrieve the reacted CO_2. As a dry process involving high temperatures (550–1150 °C), calcium looping is advocated as a feasible option for cement and hydrogen production [17].

1.2   HISTORY OF CARBON CAPTURE APPLICATIONS BEFORE DECARBONIZATION PROTOCOLS

The history of carbon capture can be traced back to the 1920s, when it was first used to remove carbon dioxide from the natural gas to make the latter marketable. The absorption method using chemical solvents (alkanolamines) was the preferred method. For the next 50 years, absorption was the dominant technology for carbon dioxide separation, and this is one of the reasons that it is the most mature carbon capture technology among its competitors. In this period, the major improvement for the process was its upgrading to high-pressure systems and testing of sister alkanolamines (chemical solvents), with the need for reducing energy penalty and reducing solvent loss [18–21].

In the early 1970s, carbon capture enjoyed revived interest thanks to carbon dioxide's commercial use in enhanced oil recovery. For example, in 1972, the Terrell gas processing plant in Texas (US) was the first commercial carbon capture facility installed, with a purpose of using captured CO_2 for monetary benefit, rather than a process detriment as before. The plant captured carbon dioxide from natural gas and transported it to a West Texas oilfield for enhanced oil recovery. Many natural gas processing facilities in the United States followed this pattern. In 1982, the Koch Nitrogen Enid Fertilizer Plant in Oklahoma (US) was the first chemical processing plant to join the group; captured carbon dioxide was transported to an oilfield in Southern Oklahoma for enhanced oil recovery. In 1986, ExxonMobil, Chevron, and Anadarko Petroleum developed a joint carbon capture facility in Wyoming. Known as the Exxon Shute Creek Gas Processing Facility, the carbon capture plant had an annual capture capacity of 7 million tons. Obviously, the captured gas was used for enhanced oil recovery and, later, for small domestic applications (e.g., enhancing plant growth in greenhouses) [22]. Interestingly, the facility is still the largest commercial carbon capture facility worldwide, despite many improvements in technology and policy development in the CCUS sector [23]. As of 2010, 114 active carbon capture plants for enhanced oil recovery were operational across the United States [24]. However, it is expected that this number will decrease in the future, as social acceptance wanes for the use of carbon dioxide in enhanced oil recovery.

Research-wise, the first known consolidated academic study of carbon capture for decarbonization was carried out at Massachusetts Institute of Technology (MIT). The Carbon Capture and Sequestration Technologies Program, started in 1989, focused on the capture, use, and storage of carbon dioxide from large point sources. The research had multiple objectives and spanned all niche areas (i.e., technical, economic, social, and political aspects). The Intergovernmental Panel on Climate Change (IPCC) was instituted in 1988 upon the resolution moved by Malta in the UN General Assembly; however, no proper decarbonization protocol was in action during this research period [22].

Only in 1991 did governments start taxing activities related to carbon dioxide/greenhouse gas emissions. The Norwegian [25] and Swedish [26] governments levied taxes that motivated oil-producing companies to develop carbon capture and sequestration projects and cut down on CO_2 emissions. Statoil in Norway was the first oil company in 1996 to capture the CO_2 emitted from Sleipner, their oil and gas processing facility in the North Sea, and follow decarbonization protocols. Carbon dioxide gas was captured and injected into an underground aquifer [25]. One million tons of carbon dioxide per annum were still being separated by the early 2000s. This carbon dioxide capture and sequestration project is considered the first large-scale facility and now serves as an internationally important test case for monitoring geological CO_2 storage. In Sweden, the carbon tax rate has gradually increased from €24/ton of CO_2 in 1991 to €115/ton in 2019, apart from the subsidies for the adaptation of low carbon technologies [26]. This has motivated many businesses and industries to

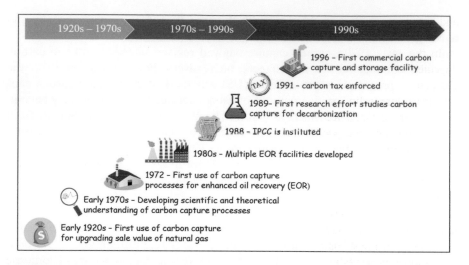

FIGURE 1.3   Time line of incremental events in use of carbon capture from natural gas upgrading (1920s) to decarbonization (late 1990s).

modify their practices to save on taxes incurred. This 'carrot and stick' model has been copied by many nations and is seemingly a successful model for implementing decarbonization protocols. Figure 1.3 shows a time line of various milestones in the development of carbon capture from a technology for natural gas treatment to a technology for decarbonization and climate change mitigation.

Since the Sleipner project, carbon capture has found many applications in different scenarios, sometimes purely motivated by carbon taxation and otherwise for technological requirements like enhanced oil recovery. However, the historic applications have developed a reflexive social resistance against future carbon capture applications. The public now views carbon capture as a technology that promotes the use of fossil fuels – a complete misunderstanding of the scientific excellence and application potential of carbon capture.

1.3   CURRENT STATE OF THE ART IN CARBON CAPTURE

Emissions of carbon dioxide are mostly associated with the combustion of fossil fuels and industrial processes that produce carbon dioxide via chemical reactions like lime/cement or steam-methane reforming (SMR). Still, the majority of the carbon dioxide emissions originate in combustion. Even 40% of the emissions from the lime sector and SMR units are a result of direct fuel combustion required for endothermic reactions to occur. Hence, with an exception or two, carbon dioxide capture technologies are broadly classified on the point where the CO_2 is extracted from the rest of the mixture. This can be either pre-combustion or post-combustion [6]. As the name implies, the CO_2 is either separated prior to combustion of the fuel or captured from the flue gas after combustion. For example, pre-combustion finds active application in natural gas sweetening and is equally good for biogas upgrading, hydrogen purification, syngas cleaning and many others. Post-combustion is usually associated with flue gas treatment from power generation, but the technology is similarly applicable to industrial CO_2 capture in the cement, iron/steel, paper, and pharmaceutical sectors. The choice of technologies depends on the type of process, carbon dioxide concentration, purity of the separated CO_2, and the degree of carbon capture required. Still, the large parasitic energy load of carbon capture technologies remains the main obstruction in their widespread adoption in post-combustion. Since the CO_2 content in the flue gas from coal-fired (7%–14% CO_2) and natural gas–fired (<4% CO_2) plants is comparably small to the overall volume of flue gas, a large equipment size is required for all carbon capture technologies. Another issue with post-combustion is the low pressure of the flue gas, which restricts the use of high pressure like adsorption and membranes, which are more energy efficient than chemical absorption [8]. This is a dilemma for the process design, as compression of the flue gas is energy intensive and may cancel out any benefits of energy-efficient adsorption. With developments in adsorbents, these drawbacks are claimed to be resolved [14], yet commercial systems are limited. Contrarily, the pre-combustion carbon capture enjoys inherent flexibility due to variance in feed gas conditions. Hence, all technology options can be applied. The choice of technology is modulated by the concentration of the carbon dioxide in the fuel, the degree of purification required, and the gas pressure. Absorption remains a technology of choice for low carbon containing natural gas (<15% CO_2). For a carbon dioxide concentration between 15% and 50%, adsorption, membrane, and absorption compete for application, defined by the end-user specification (usually downstream CO_2 content and gas pressure) [27].

Another recently developed method is to capture carbon dioxide directly from air, known as direct air carbon capture [13]. The air, containing minuscule quantities of air (currently ~415 ppmv and growing), is passed through a layer of adsorbents and/ or membranes, where the CO_2 in air binds to specially developed materials that have a high CO_2 affinity. The process is at the proof-of-concept stage, with some technologies reaching pilot scale [28].

The oxyfuel combustion and chemical looping processes are alternative strategies to pre- and post-combustion. In these processes, the nitrogen in air is separated, leaving almost pure oxygen. In the oxyfuel process, oxygen separation is achieved by cryogenics and/or adsorbents (not related to carbon capture), while the chemical looping process uses a metal oxide as an oxygen carrier to the combustion chamber. The pure (or nearly pure) oxygen is used for the fuel combustion. Hence, the primary products of such combustion are carbon dioxide and water. Since the water can be easily separated, the purification of carbon dioxide is not a technical constraint, bypassing all other technology options used for carbon capture. However, the technology is itself marred by high capital costs. Most of these capital costs are attributed to the use of high-temperature metals for containing the heat of reaction (as there is no inert nitrogen to lower the adiabatic flame temperature). The process was once considered as a technology option for conventional carbon capture. However, with massive improvements in other technology options and limited research in these fields, these processes are losing scientific and commercial importance. Yet, a couple of pilot-scale studies are underway in the Far East [29]. Figure 1.4 shows the process overview of the current state-of-the art technologies in carbon capture. The figure explains different routes for capturing carbon dioxide from the flue/fuel gases,

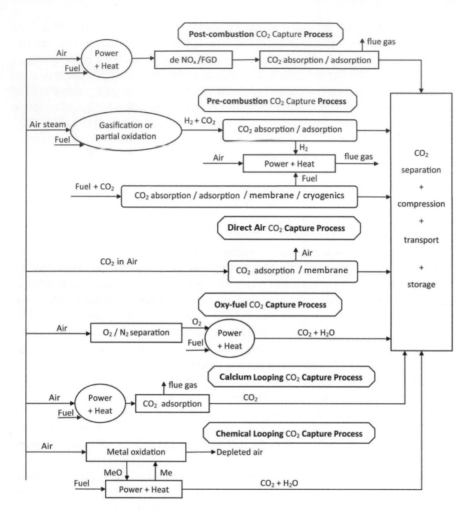

FIGURE 1.4 Technology overview of the current state-of-the-art technologies in carbon capture.

Source: **Adapted from [27].**

process industry, and air, namely, the post-combustion, pre-combustion, direct air carbon capture, oxyfuel, calcium looping, and chemical looping processes.

1.4 RECENT DEVELOPMENTS IN PILOT-SCALE AND COMMERCIAL CARBON CAPTURE SYSTEMS

Research and development in the field of carbon capture has increased since its potential in mitigating climate change was acknowledged. Research has reached medium to high technology readiness in the recent years. Many pilot and commercial CCS facilities have been installed, with mixed results. Research works are extensive and broad and are summarized here in tables that present the recent large-scale and pilot-scale developments in the field of carbon capture. Table 1.1 presents a summary of large-scale and/or commercial projects worldwide.

Apart from the large-scale projects, several pilot-scale studies are underway across the globe. Moreover, a few test centers have also been developed to support CCUS systems. Table 1.2 presents a list of recent pilot-scale projects and test centers worldwide.

Tables 1.1 and 1.2 show an interesting demographic distribution. The carbon capture is most developed in the United States, followed by the Europe. Many commercial CCS facilities are operational in United States, whereas the pilot studies are focused on the development of CCUS hubs (infrastructure development) rather than the technology itself. Europe seems to be catching up, as many pilot-scale facilities are underway with a few commercial projects at an advanced stage of development. The rest of the world is still in the nascent phase, with many pilot studies but negligible commercial projects. As expected, most of the CCS plants are situated in the developed countries, showing a paradigm shift towards low carbon technologies, while the developing and underdeveloped countries slowly follow in their footsteps.

Another area of recent development is direct air carbon capture. Recently, two companies have managed to install demonstration or pilot-scale projects. Climeworks in Switzerland has developed an amine functionalized filter to capture carbon dioxide directly from air with an installed capacity of 900 tons per annum. Similarly, Global Thermostat uses an amine-based monolith to capture around 1000 tons of carbon dioxide/year. Other companies like Carbon Engineering and Infinitree are using a mix of technologies to capture atmospheric carbon dioxide; however, these are at the proof-of concept stage [28]

After so many setbacks, carbon capture projects are now being approached with a new sense of boldness. There is a sound consensus among the scientific community and political policy that the carbon capture is an essential technology for the reduction of CO_2 emissions in the fossil fuel-based power generation and industrial processes that cannot be decarbonized in other ways.

1.5   THE CASE FOR SUSTAINABILITY

Power generation emits almost a quarter of global CO_2 emissions. Although developed economies have shifted from coal-based power generation to much cleaner natural gas-based electricity, emerging economies like India and China are adopting coal for power generation by choice. Assuming a generic power plant lifetime of 50 years, these new plants will survive well beyond the 2050 net-zero deadline. Overall CO_2 emissions may increase or, at best, remain at a standstill. For now, carbon capture is the only functional CO_2 emission reduction technology for existing power plants. Retrofitting of existing plants is necessary but expensive and time-consuming. It makes sense to install carbon capture units as early as possible to reduce capital costs. Indeed, being carbon capture–ready is much more socially acceptable than being ignorant. Therefore, the policymakers in such emerging economies are in favor of developing CCUS for reducing new CO_2 emissions. The high cost of carbon capture may offset implementation of a potential CCUS strategy, irrespective of the

TABLE 1.1

A Summary of Recent Large-Scale and Near-Commercial Carbon Capture Projects Worldwide [28–29]

Country	Project	Status	Activity / Objective
United Kingdom	A scalable full-chain industrial CCS project (ACORN)	Initial Stage	Upscaling of the ACT ACORN project
	Caledonia Clean Energy	Initial Stage	A CO_2 capture and storage facility that will store the captured gas in the North Sea near Scotland
	Drax BECCS Project	Initial Stage	Biomass-powered station with carbon capture and storage facility
	Hydrogen to Humber Saltend (H2H)	Initial Stage	CO_2 capture facility from hydrogen production (blue hydrogen)
	HyNet North West	Initial Stage	CO_2 capture system from hydrogen production; CO_2 transportation is also a part of the project
	Net Zero Teesside	Initial Stage	CCS installation to capture and store the CO_2 produced in industries in the Tees Valley, North East UK
	Northern Gas Network H21 (North of England)	Advanced Stage	Project to replace UK's methane-based gas grid with carbon-neutral hydrogen
	The Clean Gas Project	Initial Stage	Natural gas-based power generation with CO_2 capture, transportation, and storage under the southern North Sea
	TCE CCU plant	Initial Stage	Industrial facility to capture and convert CO_2 to marketable solid carbonate
	CO_2 MultiStore	Operational	CO_2 storage facility fed by multiple carbon capture units, run by Scottish Carbon Capture and Storage
	Joint Industrial Project (JIP)		
Norway	Langskip CCS–Fortum Oslo Varme	Advanced Stage	Waste to energy project with CCS facility
	Norway Full Chain CCS	Advanced Stage	Involves a number of studies to explore various opportunities in industrial-based CCS around Norway
	Sleipner CO_2 Storage	Operational	CO_2 storage facility accompanied with carbon capture from industrial sources
	CO_2 FieldLab Project	Operational	CO_2 capture and storage facility to determine the subsurface effects of carbon sequestration
	Snøhvit CO_2 Storage	Operational	CO_2 storage facility along with carbon capture from LNG

Country	Project	Stage	Description
The Netherlands	Hydrogen 2 Magnum (H2M)	Initial Stage	Hydrogen production for power generation accompanied with a CO_2 capture and storage system
	Port of Rotterdam CCUS Backbone Initiative (Porthos)	Advanced Stage	A carbon capture, transportation, utilization, and storage concept for managing large quantities of CO_2 (2–5 million tons per year)
Ireland	CATO Programme	Operational	Full CCS chain to manage its implementation in The Netherlands
	Ervia Cork CCS	Initial Stage	Natural gas–based power generation with CO_2 capture, transportation, and storage under the Kinsale gas field
France	DMX Demonstration in Dunkirk Project – part 2 (see part 1 in Table 1.2)	Initial Stage	Capturing CO_2 from the Arcelor-Mittal's steelworks plant in Dunkirk; part 2 of the project reaches a commercial level of 125 metric tons of CO_2 per hour
United States	Plant Daniel Carbon Capture	Advanced Stage	A carbon capture plant retrofit on a natural gas–fired powered plant
	Air Products SMR	Operational	Steam methane reformers for hydrogen production plant were fitted with carbon capture unit
	Cal Capture	Advanced Stage	Natural gas–based power generation is being installed with a CO_2 capture and will have options for enhanced oil recovery, or storage
	Clean Energy Systems Carbon Negative Energies – Central Valley	Initial Stage	Biomass-fed co-hydrogen and power generation units will be fitted with a CCS unit
	Core Energy CO_2-EOR	Operational	A carbon capture facility for the natural gas processing unit; the captured CO_2 is used for enhanced oil recovery
	PCS Nitrogen	Operational	A fertilizer plant with a carbon capture unit
	San Juan Power Gen	Advanced Stage	Natural gas–based power generation is being installed with a CO_2 capture
	Illinois CCS Facility	Operational	Ethanol production plant equipped with a CCS unit
	Lake Charles Methanol	Advanced Stage	Methanol production plant equipped with a CCS unit
	Project Interseqt	Initial Stage	Ethanol production plant equipped with a CCS unit
	OXY and Carbon Engineering DAC Unit	Initial Stage	Direct air carbon capture plant (1 million per ton capacity)
	Velocys-Bayou Fuels Negative Emission Project	Initial Stage	CCS from the biomass-to-fuels facility resulting in carbon-negative emissions
	Bonanza Bioenergy CCUS	Operational	Ethanol production plant equipped with a CCS unit

(Continued)

TABLE 1.1 (Continued)

A Summary of Recent Large Scale and Near-Commercial Carbon Capture Projects, Worldwide [28–29]

Country	Project	Status	Activity / Objective
	Project ECO2S	Initial Stage	Developing CCS networks in Kemper County by connecting a multitude of carbon dioxide emitters to usage and storage options
	Redtrail BECCS Project	Initial Stage	Ethanol-based power generation plant equipped with a CCS unit
	Dryforks Integrated CCS	Initial Stage	Natural gas–based power generation is being integrated with a CO_2 capture unit
	Century Plant	Operational	Natural gas processing facility with CCS infrastructure
	Coffeyville Gasification Plant	Operational	A retrofitted CCS unit for the gasification facility
	Great Plains Synfuels CCS	Operational	Synthetics natural gas production facility with CCS infrastructure
	Project Tundra	Advanced Stage	Coal-fired power generation with CCS facility
	Gerald Gentleman CCS	Advanced Stage	Natural gas–based power generation is being integrated with a CO_2 capture unit
	CarbonSAFE Illinois CCS Hubs	Advanced Stage	Developing CCS networks in Illinois by connecting a multitude of carbon dioxide emitters to usage and storage options
Australia	CarbonNet	Advanced Stage	Developing CCS networks in Australia by connecting a multitude of carbon dioxide emitters to usage and storage options
	Gorgon Carbon Dioxide Injection	Operational	A complete CCS facility to capture and inject carbon dioxide in deep subsurface wells
	Santos Cooper Basin CCS project	Advanced Stage	Natural gas processing facility with CCS infrastructure
Brazil	Petrobras Santos CCS Unit	Operational	Natural gas processing facility with CCS infrastructure
Canada	QUEST	Operational	Hydrogen-based power generation plant equipped with a CCS unit
	Alberta Carbon Trunk Line (ATCL)	Operational	Oil refining and fertilizer production facility with CCS infrastructure
	Boundary Dam CCS	Operational	Coal-fired power generation retrofitted with CCS facility
Qatar	Qatar LNG CCS	Operational	LNG processing facility with CCS infrastructure
New Zealand	Project Pouakai Hydrogen with CCS	Initial Stage	Hydrogen-based power generation plant equipped with a CCS unit
Indonesia	Gundih CCS Plant	Advanced Stage	Natural gas processing facility with CCS infrastructure

United Arab Emirates	Abu Dhabi CCS – Phase 1	Operational	A carbon capture facility for an iron and steel production facility; the captured carbon dioxide is used for enhanced oil recovery
	Abu Dhabi CCS – Phase 2	Advanced Stage	A CCS facility for an iron and steel production facility; the captured carbon dioxide will be stored underground
South Korea	Korea CCS 1 and 2	Initial Stage	An integrated gasification combined cycled power plant with many options for carbon capture; post-combustion/pre-combustion or oxyfuel combustion
China	Sinopec Qilu Petrochemical Plant	Advanced Stage	CCS unit for coke/coal gasification plant
	Sinopec Zhongyuan CCUS	Operational	CCS unit for a petrochemical plant
	CNPC Jilin Oil Filed CCS	Operational	Natural gas processing facility with CCS infrastructure
	Yanchang Integrated CCS project	Advanced Stage	CCS unit for a petrochemical plant
Saudi Arabia	Uthmaniya CCS facility	Operational	Natural gas processing facility with CCS infrastructure

TABLE 1.2

List of Recent Pilot-Scale Projects and Test Centers for Carbon Capture Worldwide [29–32]

Country	Pilot-Scale Study	Status
United Kingdom	Aberthaw Pilot Carbon Capture Facility	Completed
	ACT ACORN	Active
	Drax bioenergy carbon capture pilot plant	Active
	Ferrybridge Carbon Capture Pilot	Completed
	Pilot-Scale Advanced Capture Technology (PACT)	Test Center
	Renfrew Oxyfuel Project	Completed
Spain	ELCOGAS Pre-Combustion Carbon Capture Pilot Project at Puertollano	Completed
	CO_2 Capture and Transport Technology Development Plant (CIUDEN)	Completed
	La Pereda Calcium Looping Pilot Plant	Completed
Italy	Brindisi CO_2 Capture Pilot Plant	Completed
The Netherlands	Buggenum Carbon Capture (CO_2 Catch-Up) Pilot Project	Completed
	ConsenCUS pilot project	Active
	K12-B CO_2 Injection Project	Completed
France	C2A2 Field Pilot – Le Havre	Completed
	DMX Demonstration in Dunkirk – Part 1 (see part 2 in Table 1.1)	Active
	Lacq CCS Pilot Project	Completed
Iceland	CarbFix Project	Active
Denmark	CASTOR	Completed
	CESAR	Completed
	Hashøj Biogas CCS Plant	Completed
	ARC pilot and demonstration of zero energy CCS	Active
	$BioCO_2$ mobile pilot for advanced biogas upgrading	Active
Norway	CO_2 Capture Test Facility at Norcem Brevik	Completed
	Technology Centre Møngstad	Test Center
	CEMCAP	Completed
Croatia	Geothermal Plant With CO_2 Re-Injection	Completed
Sweden	Karlshamn Field Pilot	Completed
	STEPWISE (SEWGS Technology at Swerea/Mefos)	Active
Belgium	Low Emissions Intensity Lime and Cement Project (LEILAC)	Active
Germany	Ketzin Pilot Project	Completed
	Schwarze Pumpe Oxyfuel Pilot Plant	Completed
	Wilhelmshaven CO_2 Capture Pilot Plant	Completed
Canada	Pembina Cardum CO_2 Monitoring Plant	Completed
	CMC Research Institutes	Active
	CO_2 Solutions Valleyfield CCS Demonstration Project	Completed
	Svante Husky Energy VelxoTherm Capture Process Test	Active
	Shand Carbon Capture Test Facility	Test Center

Country	Pilot-Scale Study	Status
United States	Marshall County ECBM Project	Completed
	Oxy Combustion of Heavy Liquid Fuels Test Pilot	Completed
	NET Power Clean Energy Pilot Plant	Active
	Mountaineer Validation Facility	Completed
	CO_2 Sequestration Field Test: Deep unmineable lignite coal seam	Completed
	Bell Creek – Incidental CO_2 Storage Project	Active
	Plant Barry and Citronelle Integrated Project	Completed
	Cranfield Project	Active
	Frio Brine Pilot	Completed
	Illinois Basin Decatur Project	Completed
	MGSC Validation Phase for Sugar Creek Oilfield	Completed
	Michigan Basin CO_2 Sequestration Field Test	Completed
	National Carbon Capture Center	Test Center
	Kevin Dome CCS Development Project	Completed
	Wyoming Integrated Test Center	Test Center
	Fuel Cell Carbon Capture Pilot Plant	Active
Brazil	Miranga CO_2 Capture and Injection Project	Completed
China	Sinopec Shengli Oilfield CCUS Project	Active
	Haifneg Carbon Capture Test Platform	Active
	Chinese European Emission Reduction Solutions	Active
	Huazhong University's Oxyfuel Test Project	Active
	Petrochina Changqing Oilfield CCUS	Active
	Guohua Jinjie CCS Full Chain Demonstration	Active
	Daqing Oilfield Demonstration Project	Active
	Huaneng Green Gen IGCC Demonstration	Active
	ITRI Calcium Looping Pilot	Active
	Australia-China Post Combustion Carbon Capture Feasibility Study Project	Active
India	Solvay Vishnu Carbon Capture Plant	Completed
Japan	CO_2 Ultimate Reduction in Steelmaking	Active
	Tomakomai CCS Demonstration Project	Active
	Nagaoka CCS Project	Completed
	Osaki CoolGen Project	Active
	Mikawa Post-Combustion Carbon Capture Demonstration Plant	Active
Australia	Hydrogen Energy Supply Chain Project	Active
	CTSCo Surat Basin CCS Project	Active
	Hazelwood Carbon Capture and Mineral Sequestration Pilot Plant	Completed
	National Geo-Sequestration Laboratory	Test Center
	CO_2CRC Otway	Active
	Post-Combustion capture @ CSIRO	Test Center
	Wallumbilla Renewable Methane Demonstration Project	Active
South Korea	Boryeong – KoSol Process for Carbon Capture	Completed
	Hadong Dry Sorbent CO_2 Capture System Test	Completed
South Africa	Zululan Basin CCS Pilot Project	Active

time frame, which must be brought down. With the current state of the art, only the environmental and social benefits (for long-term use) offer some promise of sustainability [33].

Many studies have been performed that point to the unsustainable nature of carbon capture in economic terms. Governments should provide subsidies to support carbon capture (in the same way that electric cars are subsidized) or taxation (a carbon tax for gasoline and diesel cars). Another optimistic possibility is a technology breakthrough that can reduce the cost below the economic value of CO_2. The desirable European price target for carbon capture (coupled with storage or utilization) is below €30/ton (DMX Demonstration in Dunkirk Project) [34]. In a futuristic scenario of rapid emissions reduction or fully net-zero scale, the prices of carbon dioxide will rise [35]. This could make carbon capture attractive from the financial perspective. Until then, there is a little incentive for any business to deploy carbon capture. It is expected that with the use of fossil fuels reducing to zero and the rise in the price of carbon dioxide, carbon capture will become sustainable economically.

The sources of carbon dioxide emissions are not equal. Hence, a carbon capture system cannot be copied from one application to another. For example, CO_2 capture from coal-powered stations is not economically viable, but the same technology becomes affordable for the sweetening of natural gas. So, the cost of carbon capture cannot be estimated on the basis of cost per ton of CO_2 but is regulated by many other factors, such as the presence of other harmful impurities, safety issues, human factors, and sometimes even by water vapor (e.g., in cryogenic operations). A full life-cycle assessment is therefore necessary [36].

From a macro-economic perspective, a committed drive to achieve major emissions reduction in greenhouse gases is much better than not having one. But the global financial infrastructure does not determine or reward companies/nations that act for everybody's good. Additionally, cutthroat competition forces companies and business institutions to reap maximum profits. Although governments regulate the market forces in most countries, a common international law regarding fair or green practices is absent. Nations must reach a consensus to develop strategies that put a penalty for releasing new CO_2 emissions into the atmosphere [36].

Alternatively, society can add substantial social value to carbon dioxide. The current inaction of the society towards CO_2 emissions is an active contributor to the increase in atmospheric CO_2 levels. Further passivity will make the climate emergency more acute. Small behavioral actions can bring a large change to climate efforts, e.g. avoiding single-use plastics, or buying from green/sustainable businesses.

1.6 CURRENT SUSTAINABILITY LEVEL OF CARBON CAPTURE IN TERMS OF LCA

Life cycle assessment (LCA) is a tool that has been developed to assess the environmental impacts of a product or technology over its whole life cycle, that is, from the extraction of resources for production of the materials and hardware over its use to its decommissioning. The life cycle comprises all the processes that are needed for the technology to deliver its function, and an LCA maps all emissions from these processes and quantifies their contribution to all relevant environmental impacts, from local noise problems over regional acidification and photochemical ozone formation to global climate change. The systems perspective and the broad coverage of environmental impacts allow LCA to identify any problem shifting from one part of the life cycle to another or trade-offs between environmental impacts [37]. This makes it the tool of choice for assessment of the environmental sustainability of the different approaches from carbon capture storage (CCS) or utilization (CCU). The economy is an important parameter to consider, but if the overall emissions grow by using a certain new CO_2 capture technology, the activity is quite irrelevant if the purpose is to support a net removal of CO_2 from the atmosphere. By analyzing the whole life cycle of the system, LCA tells whether the net CO_2 account is negative or positive. An example that is often discussed is the case of bioenergy with carbon capture and storage (BECCS). If the biomass applied in BECCS is not sustainably produced, then from an LCA standpoint, that type of BECCS should not be used. If trees are cut and replanted, but it takes 25 years before the trees start accumulating CO_2 again, then trees should not be used in BECCS. Such BEECS would not reduce the CO_2 level in the atmosphere. The same can be said for a potential CO_2 capture technology. If the medium used for CO_2 capture emits more CO_2 while producing, using, or disposing of it, it should not be considered for CO_2 capture.

With its broad coverage of environmental impacts, LCA also reveals whether a CCS or CCU technique that offers a net benefit for climate change does so at the expense of other negative environmental impacts, for example, by releasing chemicals that are toxic to the environment or by using land in a way that leads to loss of biodiversity.

The cradle-to-grave perspective offered by LCA is obligatory. Benchmarking technologies in terms of their emissions and environmental impacts should consider the whole life cycle from production, to use, and eventual decommissioning.

1.7 ROLE OF CARBON CAPTURE IN MEETING THE GOALS OF THE PARIS AGREEMENT

The IPCC was established from an initiative of Malta, supported by the United Nations Environment Programme (UNEP) and the World Meteorological Organization (WMO) during the UN General Assembly in December 1988. The IPCC determines the state of knowledge on climate change. It identifies where there is agreement in the scientific community on topics related to climate change and where further research is needed. IPCC creates reports that are drafted and reviewed in several stages, thus guaranteeing objectivity and transparency. The IPCC does not conduct its own research. IPCC reports are neutral and policy relevant.

Since 1988, the IPCC has had five assessment cycles and delivered five assessment reports; a sixth report is ready for 2022. It has also produced a range of methodology reports, special reports, and technical papers. These reports often set the agenda for public debate and have been used as guiding factors for many governments to leverage climate policies. The United Nations Framework Convention on Climate Change

(UNFCCC) is, among other things, responsible for intergovernmental negotiation in relation to climate change and keeps track of greenhouse gas emissions from world states in a public database [38]. The UNFCCC obligations are to follow the states' development to reach the climate goal set by the Paris Agreement. UNFCCC is responsible for organization of the Conference of the Parties (COP) events, which are very visible in the media and attended by highly influential stakeholders.

Of the many IPCC reports, two special reports stand out. The first was published in 2005 on the topic of CCS [39] and outlines the potential and cost of the full CCS chain in terms of emitter, capture technologies, transportation, storage, and utilization of CO_2. At that point in time (2005), fossil fuel CCS was the focus, and very little emphasis was given to the need for negative emissions and carbon recycling. During 2017 and 2018, the COP conferences inspired the politicians to focus on reaching a maximum of 2°C temperature rise, with a focus of trying to make it 1.5°C. Currently, the temperature increase is in the order of 1°C (2021); this means the politicians have accepted that the climate will become significantly much worse in the coming decades.

A second IPCC report [40] was published end of 2018 that clearly highlights the need for carbon dioxide removal (CDR). During the 13 years from 2005 to 2018, there has been a noticeable switch from fossil emphasis into a need for reaching a zero-emission strategy by 2050 and a negative emission need after 2050. Direct air capture and BECCS, in terms of CCS from biomass combustion, are the only tools for reaching a sound climate strategy in this century. The IPCC envisages in this second study that depending on the political speed of changing the society, one out four scenarios (P1 to P4) will likely occur. P1 has the potential for almost no CCS, and P4 requiring a staggering 22 gigatons (GT; 22,000 million tons CO_2) CCS per year for many years. Already now, we see that P1 will most likely not be reached due to the lack of climate strategies by large world stakeholders.

To what extent CCS will be implemented is not certain, but it is clear that the level is anything from 1–5 GT up to 22 GT or more. CCS is a tool that cannot stand alone to save the climate; CCS will be the last resort. When society has removed the need for fossil fuels, there are still sectors (e.g., farming and heavy industry) that cannot change to a fully fossil-free production. This is where CCS will help us – the last step – in securing a negative emission strategy according to the IPCC predictions.

1.8 ROLE OF CARBON CAPTURE IN DECARBONIZING THE FUELS OF THE FUTURE

Since 2000s, CCS has been misrepresented as a technology that sustains the use of fossil fuels. This could have been true before the decarbonization protocols were ratified (i.e., 1990s), but not in the world that is slowly but substantially cutting down on the use of fossil fuels or at least choosing cleaner ones (natural gas over coal). It is widely assumed that carbon capture will be eliminated once fossil fuels are replaced by other energy alternatives. However, this is not true for the foreseeable future. This section presents a discussion on how carbon capture will still be required for decarbonizing the fuels of the future, even in a net-zero scenario.

1.8.1 CARBON CAPTURE FOR THE FUELS IN THE ENERGY SECTOR

Figure 1.5 forecasts primary energy consumption for year 2050, assuming three different scenarios, and compares it to the real value for the year 2018. The first scenario (business as usual), assumes no reduction in carbon dioxide emissions. The second scenario (rapid) assumes rapid reduction in carbon dioxide emissions but not achieving a net zero solution. The third scenario (net-zero) assumes zero net carbon dioxide emissions by the year 2050. The figure shows the use of coal below 10 exajoules (EJ) in terms of energy by the year 2050, and perhaps down to almost zero by the year 2100 [35].

However, to achieve this, large economies heavily based on coal today must either convert to other sources of energy or resort to CCS. Even with the most modern coal-based power plants with energy efficiencies above 45%, post-combustion CO_2 capture will reduce energy efficiency considerably. As an example, a recent NETL report [41] shows a 150 MWe coal-based power plant to have an efficiency of 40.3% without CO_2 capture and 29.9% with 83.6% CO_2 capture. IEAGHG, in 2014, published a report studying the cost of oxy-combustion, post-combustion, and IGCC with pre-combustion capture [42]. All cases showed energy efficiencies of about 35% (LHV basis) and an efficiency loss of about 9 percentage points compared to no CO_2 capture. Recently, IEA published a report discussing the present status of CCUS in power production [43]. The report forecasts 310 commercial CCUS facilities by year 2030 and a huge anticipated increase up to 1320 commercial CCUS project by the year 2040, while present-day plants are scarce (three commercial units).

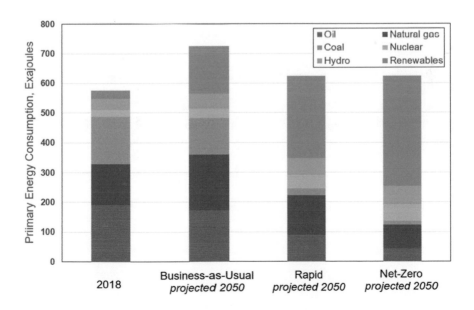

FIGURE 1.5 Forecasted primary energy consumption by source, year 2050.

Source: **Data from [35].**

However, much experience has been gathered. A feasibility study by the International CCS Knowledge Center (based on the Boundary Dam project, started in 2014, and the Petra Nova project, started in 2017) shows that second-generation facilities could be built at 67% lower capital cost, with a CO_2 recovery cost of $45 per ton and with a capture rate of 95%. As a base number for viable CO_2 capture, a removal rate of 90% has been used. This number is based on economic optimization and is not a technological limit. According to IEAGHG studies, increasing the removal rate from 90 to 99% can be realized at reasonable additional costs (e.g., 4% for coal-based exhaust and 10% for natural gas-based exhaust [42]).

Over recent decades, Europe has seen a conversion from coal-based power to natural gas-based power. Energy efficiency is higher, and the generation of CO_2 per kWh produced is about half that of coal-based power. With a sulfur-free natural gas and low-NOx burners, the exhaust gas lends itself to relatively problem-free CO_2 capture, even though the CO_2 content in the gas is much lower than in coal exhaust. The most optimistic case estimates a use of approximately 130 EJ of fossil fuel. Of this, about 100 EJ will be based on natural gas. Assuming that all the natural gas is used for power generation using a combined cycle (60% efficiency), 10 billion tons of CO_2 will still be produced. This is in line with the forecasted CO_2 emission shown in Figure 1.6 for the rapid scenario. This shows that there should be room for CCS even in 2050, both regarding CO_2 price and removal need [35].

Biomass can be used for energy production and CO_2 captured from the exhaust gases. The exhaust will normally contain CO_2 at concentrations of 10–15 vol% and resembles coal-based exhaust. The Drax power plant, fueled by 100% biomass, has established a PCCC pilot plant capturing about 1.5 tons of CO_2 per day. Pending

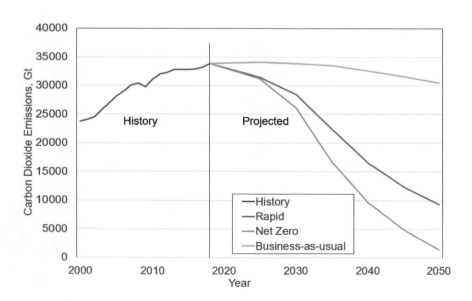

FIGURE 1.6   Projected CO_2 emissions from energy use up to year 2050.

Source: Data from [35].

successful operation, Drax plans a full-scale plant and to begin construction in 2024. It would then be the world's first bio-negative power plant [44].

Under certain conditions, this process removes carbon from the atmosphere, but if biomass for energy is produced instead of food crops, this may have very negative effects. Using agricultural wastes may seem to be a better option but may be a competitor to using the waste for improvement of and carbon storage in the soil. Boreal regions are the only regions that naturally store carbon in the ground. There, the oxidation rate of organic material is so low that it assembles and produces moss and later peat. Very large quantities of peat are stored in the Northern Hemisphere, and it is important not to extract more for purposes that will oxidize the carbon to CO_2 (e.g., for gardening and constructional work). A frightening scenario is the thawing of the Russian tundra, which will release captured methane in addition to CO_2 [45].

Direct air capture has been studied over several decades. From thermodynamics, we know that it is energetically more expensive to remove CO_2 from air than exhaust gases because of its low concentration (0.04% versus 4%–15%). Estimates show that removing 1 GT/year would require about 10% of the world's total energy consumption. In addition, this energy must be 'clean,' causing no new CO_2 emissions. Many companies have researched this option over recent decades, and absorption with nonvolatile absorbents is used already to maintain CO_2 levels in submarines below about 1000 ppmv. If inexpensive solar heat could be used for regeneration, this may be viable.

1.8.2 Carbon Capture for Alternative Automotive Fuels

The transportation sector is a major contributor to CO_2 emissions, and unfortunately, carbon capture from mobile sources is not a small source of emissions. The current trend is to convert, with or without carbon capture, fossil fuels to other energy carriers like hydrogen or electricity. A revolution has taken place regarding battery-driven cars for personal transportation, and in Sweden construction is underway introducing electrical rails in the road for larger vehicles. By 2025, all new cars in Norway should be electrically driven [46]. Ferries run on both electricity and hydrogen, and even planes are electrically driven. Production of electricity is already mentioned, but hydrogen production may be even more interesting seen from a carbon capture point of view [47].

Hydrogen can be produced by the electrolysis of water (green hydrogen), from natural gas (gray hydrogen), and from coal (brown hydrogen). Water electrolysis is currently about 80% effective (PEM cells), and a little more than half the cost of production is caused by the electricity used. The cost will thus largely depend on the electricity price, which tends to fluctuate widely in the current market, with continuously more renewable power (PV and wind). During periods of low prices, hydrogen production may be a good option. From fossil sources, hydrogen is produced by steam reforming and shift to obtain CO_2 and hydrogen and small quantities of other gases. The synthesis gas is produced at high pressure, and CO_2 can be captured by both membranes and absorption at reasonable cost (blue hydrogen), which will be the better solution economically.

One interesting aspect of hydrogen production, if this comes about at very large scale, is the possibility to produce liquid fuels for transportation. In a circular process, hydrogen comes from electrolysis and CO_2 must be captured at a similar scale. This will create a possibly permanent need for carbon capture.

1.9   SOCIETAL CHALLENGES OF CARBON CAPTURE

The world and society have for more than 100 years depended on the need for fossil energy consumption. The requirement for new products, heating, and transportation has increased considerably over the last 50 years. Figure 1.7 illustrates the increase in energy consumption during the past century, and there is no evidence of any change. But there is light in the dark. We see a beginning (albeit slow) transition towards substitution of fossil energy by more green sustainable alternatives.

There is still a long way to a complete change from fossil fuel into a sustainable society. We especially see a growing need for fossil energy in China and India, facing large challenges in the coming decades to reach a zero-emissions society (see Figure 1.8). Part of the change will be to remove CO_2 using CCS. But society should be careful. Using CCS to prolong the dependence of fossil fuel is not the right track. Sustainable electrification and circular-resource reuse are the only possible strategies, but this will come with an expense that the third world may not be able to support and may need help from other wealthier states.

The wealthy part of the world must accept that the majority of the world cannot simply pay 2–10 times more for the same goods. Cement prices for CO_2-free CCS cement will likely be twice as high. Electrofuels under current electricity prices will be 3–10 times higher. In any case, it will be very difficult for any sustainable product

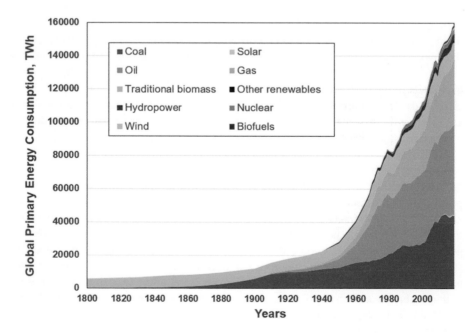

FIGURE 1.7   Trends of global primary energy consumption by source from 1800 to 2019.

to compete with fossil fuel implementation without a supporting global CO_2 tax or similar support function for green technology. Are we prepared for this economic burden?

A help towards a sustainable climate is the corporate social responsibility (CSR) of many fossil fuel–producing companies. It is common to see one company followed by the next in haste for changing names, giving the impression of a societal change towards acceptable production. But is this a real change, and can we expect a better climate from these initiatives?

When CCS is implemented, it is mandatory that global stakeholders understand that the infrastructure cannot just be pushed into the society without a prior clarification to which extent it will not influence the environment, well-being of people, and possibly induce a risk on CO_2 leakage. There is a natural fear towards the use of new technology, and especially a technology that is very visible in the landscape. Pipelines must be installed with right materials insight and right safety measures that can prove to people living around them that they are safe. Installation must be carried out with the acceptance and understanding of society, which does not always act rationally. The same is the question for storage of CO_2. The uninformed and concerned public asks, 'Will the CO_2 ever come up?' – a typical and understandable question. It is a binding task for CCS-informed states and the research community to inform and surveil CO_2 storage activities sufficiently for them to be accepted.

The CO_2 community has had a natural focus on technology development. There is also a need to support the understanding CCS in terms of social readiness level (SRL), all the way from stakeholder identification to full societal adaptation. The last

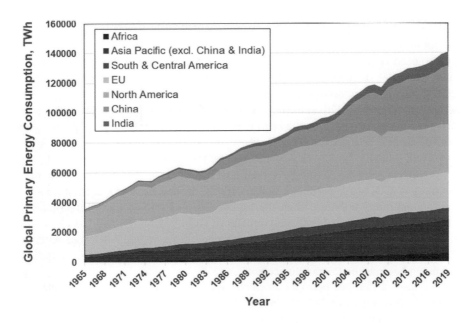

FIGURE 1.8 Trends of global primary energy consumption by region from 1965 to 2019.

part is still immature from a CCS perspective, and there is a bright future to make sure people understand what CCS is.

1.10 INFRASTRUCTURE FOR CARBON CAPTURE

In many places around the world, CCS has become a mandatory technology to reach a set emissions target by 2050. Companies are actively building pilots and planning full-scale plants and larger initiatives for storing CO_2. In Europe, Norway, the UK, and The Netherlands are especially active [29]. Recently, Iceland and Denmark have also shown initiative for storage sites. In the future, many other countries will follow: France, Spain, Germany, and Italy may soon become active.

The rest of the world needs to act but is probably hesitant because the real need for change will not happen before the period 2030 to 2050. China, the United States, and Australia have an urgent need to perform a relevant maturation of the underground CO_2 storage systems.

Some countries are in a situation where storage is not an easy solution due to the mere lack of the proper underground for CO_2 storage. In any case, there will be a need for cross-country and therefore cross-border transfer of CO_2. This is a problem in many ways. To what extent is the transfer of CO_2, legal, and how is CO_2 storage counted towards the statistics of the UNFCCC? There must be significant political development within a short period of time to clarify these mandatory questions and to make sure the basis for any infrastructure is in place.

A challenge for CCS is the traditional game of 'who dares to start and thereby take the larger risk.' Let's take an example. Imagine a company willing to co-invest in a CO_2 capture facility capable of removing 1 million tons per year (mpta) CO_2 from a gas. Who will take over this CO_2? Or from the opposite side, imagine a company willing to store or utilize 1 mpta CO_2. Who will deliver it? There is an obligation for the countries involved to supply the right resources just in time.

The need is quite similar to the electricity infrastructure or gas infrastructure in many countries. If there is a necessity for the resource, the state typically steps in, supporting the effort, taking on a significant risk on investment, and making sure the activity is performed. This same need will most likely be required from a CO_2 infrastructure. States will need, in the beginning, to support any effort of either capturing, storing, or converting CO_2. This is where infrastructure becomes important. The states will most likely also need to manage the infrastructure.

CO_2 infrastructures should be versatile to flexibly transfer CO_2 from one hub to the next. A CO_2 stock exchange is most likely needed on a global scale, possibly similar and more ambitious than the Europeans emissions trading scheme (ETS). This will allow for global trade and transfer of CO_2.

The CO_2 infrastructure must be able to cope with a versatile setup of both very small and very large customers – a bit like the electricity grid. How this is handled is unknown but must be taken into account.

The main difference between an electricity infrastructure and a CO_2 infrastructure is the restrictions on battery limits. This means the quality and the purity of the CO_2 from when it is captured until it is conditioned, transported, stored, or converted. Some impurities are quite irrelevant, but few can harm the complete infrastructure.

An example is the content of water. Any excess water can corrode the pipeline and holding containers, giving rise to expensive renovations. Sulfur is another important parameter. If the CO_2 is intended for conversion in a utilization facility (CCU) or power-2-X (P2X) plant, then any sulfur could potentially damage the catalysis most likely in place at such units.

Boundary conditions for CO_2 will become a competitive process parameter also setting the border for cost. It is not irrelevant if the CO_2 needs to be cleaned to ppb levels of an impurity or if it actually can go with several percent of specific inert impurities. For this reason, there is possibly a need for a separate low- and high-quality CO_2 infrastructure in the future.

The economic aspects are quite important where hubs of CO_2 storage facilities are far offshore. It is clear that the price goes up by several orders of magnitude in this process. If the CO_2 storage can be brought more near-shore, or even with time onshore (if social acceptance allows it), the price can come down significantly, making CCS more attractive and sustainable.

If the CO_2 must be transported far distances, it is also quite important, from a social acceptance point of view, what emission is attached to the transport. Most likely, it doesn't matter to the climate if the CO_2 transport contributes significantly to the overall emission. It is the mere storytelling around the CO_2 transport concept which is important. It will be quite important what fuel or emission is created during the transport.

Reuse or energy will most likely also be related to a CO_2 hub. In the conditioning and conversion, significant energy streams could be needed or produced, which must be sustainably distributed.

Collection of CO_2 hubs into clusters where companies and states agree on a set of boundary conditions will not be an easy task to settle. These political and technical challenges can take time to agree on and could in the end become a showstopper for large CO_2 infrastructures that are needed in the future.

REFERENCES

1. NOAA Global Monitoring Laboratory US Department of Commerce, "Global monitoring laboratory – carbon cycle greenhouse gases," [Online]. Available: https://gml.noaa.gov/ccgg/trends/ (accessed Jul. 3, 2021).
2. H. L. van Soest, M. G. J. den Elzen and D. P. van Vuuren, "Net-zero emission targets for major emitting countries consistent with the Paris agreement," *Nature Communications*, vol. 12, no. 1, pp. 1–9, Dec. 2021, doi:10.1038/s41467-021-22294-x.
3. S. Fawzy, A. I. Osman, J. Doran and D. W. Rooney, "Strategies for mitigation of climate change: A review," *Environmental Chemistry Letters*, vol. 18, no. 6. Springer Science and Business Media Deutschland GmbH, pp. 2069–2094, Nov. 1, 2020, doi:10.1007/s10311-020-01059-w.
4. R. Heede and N. Oreskes, "Potential emissions of CO_2 and methane from proved reserves of fossil fuels: An alternative analysis," *Global Environmental Change*, vol. 36, pp. 12–20, Jan. 2016, doi:10.1016/j.gloenvcha.2015.10.005.
5. International Energy Agency, "About CCUS – analysis – IEA," 2021. [Online]. Available: www.iea.org/reports/about-ccus (accessed Jul. 14, 2021).
6. A. R. Kohl and R. Nielsen, *Gas Purification*, 5th ed. Gulf Professional Publishing, 1997.

7. S. Y. Lee and S. J. Park, "A review on solid adsorbents for carbon dioxide capture," *Journal of Industrial and Engineering Chemistry*, vol. 23, pp. 1–11, Mar. 2015, doi:10.1016/J.JIEC.2014.09.001.

8. D. Aaron and C. Tsouris, "Separation of CO_2 from flue gas: A review," *Separation Science and Technology*, vol. 40, no. 1–3, pp. 321–348, 2005, doi:10.1081/SS-200042244.

9. N. S. Sifat and Y. Haseli, "A critical review of CO_2 capture technologies and prospects for clean power generation," *Energies*, vol. 12, no. 21. MDPI AG, p. 4143, Oct. 30, 2019, doi:10.3390/en12214143.

10. J. Zheng, Z. R. Chong, M. F. Qureshi and P. Linga, "Carbon dioxide sequestration via gas hydrates: A potential pathway toward decarbonization," *Energy and Fuels*, vol. 34, no. 9. American Chemical Society, pp. 10529–10546, Sept. 17, 2020, doi:10.1021/acs.energyfuels.0c02309.

11. F. Hussin and M. K. Aroua, "Recent trends in the development of adsorption technologies for carbon dioxide capture: A brief literature and patent reviews (2014–2018)," *Journal of Cleaner Production*, vol. 253. Elsevier Ltd, p. 119707, Apr. 20, 2020, doi:10.1016/j.jclepro.2019.119707.

12. M. Priestnall, "Method and system of activation of mineral silicate minerals – Google Patents," WO2015154887A1, 2015.

13. E. S. Sanz-Pérez, C. R. Murdock, S. A. Didas and C. W. Jones, "Direct capture of CO_2 from Ambient Air," *Chemical Reviews*, vol. 116, no. 19. American Chemical Society, pp. 11840–11876, Oct. 12, 2016, doi:10.1021/acs.chemrev.6b00173.

14. M. Ding, R. W. Flaig, H. L. Jiang and O. M. Yaghi, "Carbon capture and conversion using metal-organic frameworks and MOF-based materials," *Chemical Society Reviews*, vol. 48, no. 10. Royal Society of Chemistry, pp. 2783–2828, May 21, 2019, doi:10.1039/c8cs00829a.

15. R. Khalilpour, K. Mumford, H. Zhai, A. Abbas, G. Stevens and E. S. Rubin, "Membrane-based carbon capture from flue gas: A review," *Journal of Cleaner Production*, vol. 103. Elsevier Ltd, pp. 286–300, Sep. 15, 2015, doi:10.1016/j.jclepro.2014.10.050.

16. R. Nasir, H. Mukhtar, Z. Man and D. F. Mohshim, "Material advancements in fabrication of mixed-matrix membranes," *Chemical Engineering and Technology*, vol. 36, no. 5. John Wiley & Sons, Ltd, pp. 717–727, May 1, 2013, doi:10.1002/ceat.201200734.

17. C. C. Dean, J. Blamey, N. H. Florin, M. J. Al-Jeboori and P. S. Fennell, "The calcium looping cycle for CO_2 capture from power generation, cement manufacture and hydrogen production," *Chemical Engineering Research and Design*, vol. 89, no. 6, pp. 836–855, June 2011, doi:10.1016/J.CHERD.2010.10.013.

18. J. I. Lee, F. D. Otto and A. E. Mather, "Equilibrium between carbon dioxide and aqueous monoethanolamine solutions," *The Journal of Chemical Technology & Biotechnology*, vol. 26, no. 10, pp. 541–549, Jan. 1976, doi:10.1002/jctb.5020260177.

19. J. I. Lee, F. D. Otto and A. E. Mather, "The solubility of H_2S and CO_2 in aqueous monoethanolamine solutions," *The Canadian Journal of Chemical Engineering*, vol. 52, no. 6, pp. 803–805, Dec. 1974, doi:10.1002/cjce.5450520617.

20. J. F. Perez and O. C. Sandall, "Carbon dioxide solubility in aqueous carbopol solutions at 24°, 30°, and 35°," *The Journal of Chemical & Engineering Data*, vol. 19, no. 1, pp. 51–53, Jan. 1974, doi:10.1021/je60060a011.

21. J. D. Lawson and A. W. Garst, "Gas sweetening data: Equilibrium solubility of hydrogen sulfide and carbon dioxide in aqueous monoethanolamine and aqueous diethanolamine solutions," *The Journal of Chemical & Engineering Data*, vol. 21, no. 1, pp. 20–30, Jan. 1976, doi:10.1021/je60068a010.

22. Center for Climate and Energy Solutions, "Carbon capture," [Online]. Available: www.c2es.org/content/carbon-capture/ (accessed Jul. 3, 2021).

23. W. Carton, A. Asiyanbi, S. Beck, H. J. Buck and J. F. Lund, "Negative emissions and the long history of carbon removal," *Wiley Interdisciplinary Reviews: Climate Change*, vol. 11, no. 6. Wiley-Blackwell, p. e671, Nov. 1, 2020, doi:10.1002/wcc.671.

24. Department of Energy, "Enhanced oil recovery," [Online]. Available: www.energy.gov/fe/science-innovation/oil-gas-research/enhanced-oil-recovery (accessed Jul. 3, 2021).

25. E. Heiskanen, "Work package 2-historical and recent attitude of stakeholders case 24: Snohvit CO_2 capture & storage project," 2006. [Online]. Available: https://www.esteem-tool.eu/fileadmin/esteem-tool/docs/CASE_24_def.pdf.

26. P. Criqui, M. Jaccard and T. Sterner, "Carbon taxation: A tale of three countries," *Sustainability*, vol. 11, no. 22, p. 6280, Nov. 2019, doi:10.3390/su11226280.

27. D. Y. C. Leung, G. Caramanna and M. M. Maroto-Valer, "An overview of current status of carbon dioxide capture and storage technologies," *Renewable and Sustainable Energy Reviews*, vol. 39. Elsevier Ltd, pp. 426–443, Nov. 1, 2014, doi:10.1016/j.rser.2014.07.093.

28. M. J. Regufe, A. Pereira, A. F. P. Ferreira, A. M. Ribeiro and A. E. Rodrigues, "Current developments of carbon capture storage and/or utilization – looking for net-zero emissions defined in the Paris agreement," *Energies*, vol. 14, no. 9, p. 2406, May 2021, doi:10.3390/en14092406.

29. Global CCS Institute, "Carbon capture and storage facilities," [Online]. Available: https://co2re.co/FacilityData (accessed Jul. 3, 2021).

30. T. V. Jensen, "Large EU project to develop carbon capture technologies," 2021. [Online]. Available: www.dtu.dk/english/news/2021/04/stort-eu-projekt-udvikler-teknologier-til-co2-fangst??id=8d9189f3-5ffb-4a77-8918-24f1b9b0a43c (accessed Jul. 9, 2021).

31. C. O. Carlsson, "Net zero carbon capture på ARC," 2021. [Online]. Available: www.cere.dtu.dk/research-and-projects/framework-research-projects/net-zero-carbon-capture-paa-arc (accessed Jul. 9, 2021).

32. C. O. Carlsson, "Biogas upgrading for high-purity CO_2 and natural gas distribution," 2020. [Online]. Available: www.cere.dtu.dk/research-and-projects/framework-research-projects/bioco2 (accessed Jul. 9, 2021).

33. H. Lund and B. V. Mathiesen, "The role of carbon capture and storage in a future sustainable energy system," *Energy*, vol. 44, no. 1, pp. 469–476, Aug. 2012, doi:10.1016/j.energy.2012.06.002.

34. M. Cozier, "Recent developments in carbon capture utilisation and storage," *Greenhouse Gases: Science and Technology*, vol. 9, no. 4, pp. 613–616, Aug. 2019, doi:10.1002/ghg.1909.

35. British Petroleum, "Energy Outlook 2020 Edition," 2020. [Online]. Available: www.bp.com/content/dam/bp/business-sites/en/global/corporate/pdfs/energy-economics/energy-outlook/bp-energy-outlook-2020.pdf (accessed Jul. 9, 2021).

36. P. E. Hardisty, M. Sivapalan and P. Brooks, "The environmental and economic sustainability of carbon capture and storage," *The International Journal of Environmental Research and Public Health*, vol. 8, no. 5, pp. 1460–1477, May 2011, doi:10.3390/ijerph8051460.

37. M. Z. Hauschild, R. K. Rosenbaum and S. I. Olsen, *Life Cycle Assessment: Theory and Practice*. Springer International Publishing, 2017.

38. European Environment Agency, "Open database – inventories for United Nations framework convention on climate change," 2021. [Online]. Available: https://cdr.eionet.europa.eu/dk/Air_Emission_Inventories/Submission_UNFCCC/ (accessed Jul. 9, 2021).

39. B. Metz, O. Davidson, H. de Coninck, M. Loos, and L. Meyer, "IPCC special report on carbon dioxide capture and storage," 2005. [Online]. Available: www.ipcc.ch/site/assets/uploads/2018/03/srccs_wholereport-1.pdf (accessed Jul. 3, 2021).

40. Intergovernmental Panel on Climate Change, "IPCC special report on global warming of 1.5 °C," 2018. [Online]. Available: www.ipcc.ch/site/assets/uploads/sites/2/2019/05/SR15_Chapter1_Low_Res.pdf (accessed Jul. 3, 2021).

41. E. Power, J. D. Miller, and T. J. Held, "Pre-FEED-conceptual design report a low carbon supercritical CO_2 power cycle/pulverized coal power plant integrated with energy storage: Compact, efficient and flexible coal power recipient organization," 2020. [Online]. Available: https://netl.doe.gov/sites/default/files/2020-03/89243319CFE000022-PreFEED Design Basis Report_200228.pdf (accessed Jul. 9, 2021).

42. F. Wheeler, L. Mancuso, and N. Ferrari, "CO_2 capture at coal-based power and hydrogen plants," 2014. [Online]. Available: https://ieaghg.org/publications/technical-reports/reports-list/9-technical-reports/990-2014-03-co2-capture-at-coal-based-power-and-hydrogen-plants (accessed Jul. 9, 2021).
43. R. Malischek, "CCUS in power – analysis – IEA," 2020. [Online]. Available: www.iea.org/reports/ccus-in-power (accessed Jul. 9, 2021).
44. S. George, "Drax unveils plans to host UK's largest carbon capture project by 2027," 2021. [Online]. Available: www.edie.net/news/12/Drax-unveils-plans-to-host-UK-s-largest-carbon-capture-project-by-2027/ (accessed Jul. 9, 2021).
45. J. Mulligan, G. Ellison, K. Levin, K. Lebling and A. Rudee, "6 ways to remove carbon pollution from the sky," 2020. [Online]. Available: www.wri.org/insights/6-ways-remove-carbon-pollution-sky (accessed Jul. 9, 2021).
46. "Norwegian EV policy | Norsk elbilforening," [Online]. Available: https://elbil.no/english/norwegian-ev-policy/ (accessed Jul. 14, 2021).
47. Equinor, "There will be no net zero without carbon capture and hydrogen – FT and Equinor UK energy – equinor.com," 2021. [Online]. Available: www.equinor.com/en/magazine/uk-energy.html (accessed Jul. 14, 2021).

2 Reactive Chemical Absorption of CO₂ by Organic Molecules

Graeme Puxty, Marcel Maeder, and Robert Bennett

CONTENTS

2.1 INTRODUCTION

The absorption of carbon dioxide (CO_2) by reactive chemical absorption and its release by thermal swing is the most technically mature of the CO_2 separation technologies available. Aqueous alkanolamines are the ubiquitous absorbent for this application that has been in use since the 1930s [1]. Initial applications of this technology were in an industry where CO_2 removal from a process stream was necessary (e.g., natural gas sweetening and ammonia production). Reactive chemical absorption is particularly suited to applications where low residual CO_2 in the treated gas stream is required, as the chemical driving force facilitates removal even as the physical driving force is depleted.

DOI: 10.1201/9781003162780-2

Recent decades have seen a renewed focus on the development of reactive chemical absorbents born from their use to mitigate CO_2 emissions to the environment from large point sources such as coal- and gas-fired electricity generation and heavy industry such as cement manufacturing and steel smelting. This now also includes direct air capture for the mitigation of historical emissions, which has the additional challenge of being an extremely dilute source of CO_2, which is well suited to reactive chemical absorbents. When combined with geo-sequestration, this technology provides a way to redirect CO_2 emissions from the atmosphere to geological storage. Carbon capture and storage has been recognized as an important technology to meet the 1.5°C global warming targets set as part of the Paris Agreement international treaty [2]. It also provides a source of CO_2 for utilization applications such as conversion to liquid fuels and building materials. It is this use to control emissions of CO_2 to the atmosphere at a massive scale that has been the catalyst for better understanding of the chemical and physical properties of reactive chemical absorbents and the drive to improve these properties to minimize both capital and operating costs, maximize absorbent longevity, and minimize the risk of fugitive emissions.

This chapter introduces the chemistry of the CO_2 molecule and a review of the more recent developments in liquid-based reactive chemical absorbents based on organic molecules. Investigations of amines represent the bulk of research in this area. This includes their use in aqueous solution and includes amino acids as well as water-lean and biphasic systems. Finally, new chemistries are introduced based on the reaction of CO_2 with alkoxides and frustrated Lewis pairs. Discussion of ionic liquids is avoided, as this topic is covered in another chapter of this book.

2.2 THE CHEMISTRY OF CARBON DIOXIDE

To comprehend the chemistry of CO_2, we need to understand the molecule, its geometry, and in particular its electronic structure.

CO_2 is a linear, highly symmetric molecule, as depicted in Figure 2.1. The central carbon atom is bonded to two oxygen atoms with strong double bonds. The central carbon atom has a formal charge of +4 while the two oxygen atoms each have a formal charge of –2, leaving a neutral CO_2 molecule. As stated, these full charges are only formal, and they are identical (for symmetry reasons) partial negative charges $\delta-$ on the oxygen atoms, which leaves a $2\delta+$ partial positive charge on the central carbon atom, resulting in a neutral molecule.

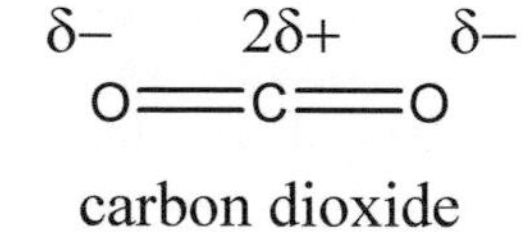

$\delta-$ $2\delta+$ $\delta-$

O $=\!\!=$ C $=\!\!=$ O

carbon dioxide

FIGURE 2.1 The structure and partial charges of CO_2.

Negatively charged reactants will tend to attack the positive charge of the central carbon atom, while positively charged reactants tend to interact with the partial negative charge on the oxygen atoms.

Next, we must distinguish two different types of reactions that can occur on the CO_2 molecule. The first type is the so-called redox reaction. In a redox reaction, the formal oxidation state of the carbon atom is reduced from +4 to a lower oxidation state: for example, +3 in formic acid, HOCO; +2 in CO; 0 in carbon, C; −2 in methanol; and even −4, which is the formal oxidation state of the carbon in methane, CH_4. All these redox reactions are important for the usage of CO_2 but not for CO_2 capture applications. They tend to be slow, and they are not easily reversible.

The easily reversible reactions relevant for CO_2 absorption by chemical reaction are substitution reactions, which leave the formal oxidation state of C at +4. All these reactions involve substitution at the central carbon atom, and to be efficient, the reactants must carry a partial or full negative charge that will react with the positive charge of the carbon. Molecules with a partial negative charge that are attracted by a positive charge are called nucleophiles and/or Lewis bases. The positively charged (full or partial) reaction partners, which would react with the partial negative charge on the oxygen atoms, are called electrophiles and/or Lewis acids. The proton, H^+, is of course a Lewis acid as well as a Brønsted acid.

2.2.1 Reaction of CO_2 with Lewis Bases

A generalized view of the reaction of a Lewis base R_1R_2HX with CO_2 via a substitution reaction is shown in Figure 2.2. Lewis acids all have one or more free electron pairs, indicated by the balloon with two dots (electrons) on the nucleophilic center X. Such a free electron pair can carry a full or partial negative charge, which will attack at the partial positive charge of the carbon atom of CO_2. The brackets indicate the reaction intermediate before the X–C bond formation and C–O and X–H bond breakage. Note that in the case of hydroxide and alkoxides, no proton is involved.

The most common Lewis bases reacting with the Lewis acid CO_2 are the water molecule, H_2O, and the hydroxide ion, OH^-. As a result of the negative charge, OH^- is expected to be a much stronger reactant than water, which only has a partial negative charge on the oxygen atom. The second-order rate constant for the reaction of CO_2 with OH^- to form the bicarbonate ion at 25 °C is 1.2×10^4 $M^{-1}s^{-1}$, while the equivalent second-order rate constant for the reaction of water with CO_2 is 6.6×10^{-4} $M^{-1}s^{-1}$ [3]. Because the concentration of water molecules in aqueous solution is much larger than the concentration of the hydroxide ions, CO_2 reacts faster with water at pH lower than approximately 8.5, while at higher pH values the hydroxide path is faster, even if there is still much more water in the solution compared to hydroxide. Table 2.1 is a collection of types of Lewis bases, which may react with CO_2 according to the mechanism of Figure 2.2.

The overall reaction scheme is summarized in Eq. 2.1. The second-order forward reaction rate constant is indicated by k_+, the reverse reaction rate constant k_-, and

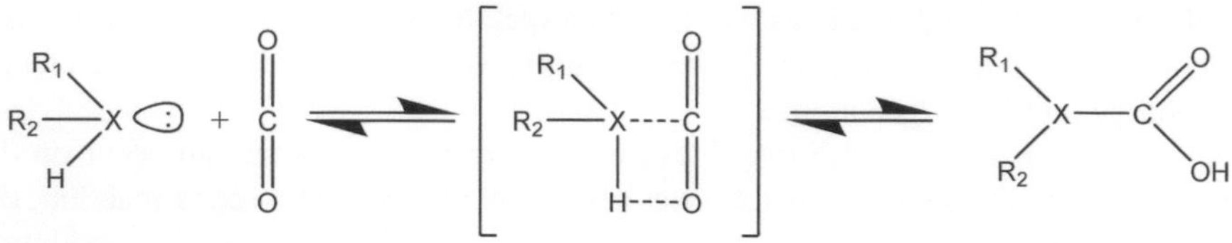

FIGURE 2.2 Generalized reaction mechanism between Lewis bases, R_1R_2HX, and CO_2 forming the corresponding acids.

TABLE 2.1

Types of Lewis Bases and Their Products of Reaction with CO_2

Lewis base	X	R_1	R_2	Product	
Hydroxide	OH⁻	O	—	—	HO–CO–O⁻	carbonate
Water	H_2O	O	H	—	HO–CO–OH	carbonic acid
Alkoxides	RO⁻	O	R	—	RO–CO–O⁻	ester
Ammonia	NH_3	N	H	H	H_2N–CO–OH	carbamic acid
Primary amines	RNH_2	N	R	H	R–NH–CO–OH	carbamic acid
Secondary amines	R_1R_2NH	N	R_1	R_2	R_1R_2–NH–CO–OH	carbamic acid

Note: X = nucleophilic center, and R_1 and R_2 = substituents, according to the reaction mechanism of Figure 2.2.

---

overall stability constant K. The forward rate constant, k_+, is crucially important, as it defines the rate at which the CO_2 molecule reacts with the different reaction partners. The faster this reaction, the faster CO_2 is absorbed (diffusion limitations due to viscosity aside). This rate constant is strongly influenced by the Lewis basicity – the 'size' or 'strength' of the electron pair. Promising new absorbents must react fast to be able to compete with existing absorbents, thus they must have a strong electron pair. Similarly, they must also form a stable product under absorption conditions, which is also a function of the strength of the electron pair and defined by the overall stability constant K. Finally, it must be possible to destabilize the product by changing the conditions (typically increasing the temperature), reversing the absorption reaction and releasing CO_2.

$$R_1R_2XH + CO_2 \underset{k_-}{\overset{k_+, K}{\rightleftharpoons}} R_1R_2XCOOH, \quad K = \frac{k_+}{k_-} \qquad Eq.\ 2.1$$

The relationship between the rate constant, k_+, stability constant K, and the Lewis basicity of a series of amines, as defined by the protonation constant, has been investigated by Conway *et al.* [4–5]. One example is shown in Figure 2.3, which relates the rate and equilibrium constants of a series of amines with their protonation constants. There are several groups of amines: a series of primary amines with different patterns of substitutions

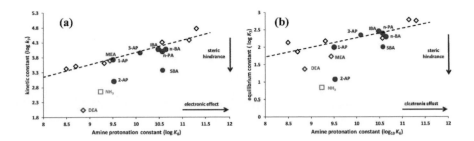

FIGURE 2.3 The linear relationship between the Lewis basicity, as defined by the proton-ation constant logK_6 and the reaction rate constant logk_7 (a) and the overall carbamic acid stability constant logK_7 (b) for sterically unhindered amines is visualized by the dashed line. Sterically hindered amines react slower and form a less stable reaction product with CO$_2$. Note that k_7 and K_7 are equivalent to k_+ and K, respectively, from Eq. 2.1.

Source: Reprinted with permission from W. Conway, X. G. Wang, D. Fernandes, R. Burns, G. Lawrance, G. Puxty and M. Maeder. "Toward the understanding of chemical absorption processes for post-combustion capture of carbon dioxide: Electronic and steric considerations from the kinetics of reactions of CO$_{2(aq)}$ with sterically hindered amines." *Environmental Science and Technology*, vol. 47, no. 2, pp. 1163–1169, 2013.

marked by red circles; a series of cyclic, sterically unhindered secondary amines marked by black diamonds; and a few commercially relevant amines, monoethanolamine (MEA), diethanolamine (DEA), and ammonia (NH$_3$). There is a distinct linear relationship between the protonation constants and both k_1 and K for most amines; this relationship is indicated by the dashed black lines. Within the collection of amines represented in this graph, there are very few outliers; notably, they all feature reduced reactivities relative to the others with similar protonation constants. This is straightforwardly explained by steric hindrance. The proton, which defines the protonation constant, is much smaller than CO$_2$. Sec. butylamine (SBA) and 2-amino-1-propanol (2-AP) can be seen as MEA analogues with sterically bulky methyl groups next to the amine nitrogen, thus they are sterically hindered. DEA is a secondary amine, which is not restrained. The remaining ammonia, NH$_3$, is not sterically hindered, but its interactions with the solvent, H$_2$O, are substantially different from the other amines; for this reason, it does not follow the general trend.

2.3 AMINES

The ubiquitous organic molecule for the reactive chemical absorption of CO$_2$ is an amine, the reason being that generally, amines contain nitrogen atoms with a free lone pair of electrons (Lewis base) that in solution can react directly with CO$_2$. This substitution reaction results in the loss of a proton and the formation of a carbamic acid (the replacement of an N–H bond by an N–C bond). Depending upon the reaction conditions and presence of a Brønsted base, the lost proton will end up associated with another molecule or one of the oxygens of the covalently bound CO$_2$. In aqueous solutions, the proton will generally be associated with the nitrogen of a second amine group, as they are weak bases and carbamic acids are weak acids. Thus, the final products are carbamate and protonated amine. In nonaqueous solutions, particularly in aprotic solvents, the proton may remain on the carbamic acid.

Regarding the reaction mechanism given in Figure 2.2, X = N, and for primary amines R_1 is a proton and R_2 is a carbon chain possibly containing other functionality. For secondary amines, both R_1 and R_2 are carbon-based chains. As mentioned, the proton lost from the amino group will generally be associated with a second amine molecule (or other Brønsted bases if present) in protic solvents such as water or result in carbamic acid formation in aprotic solvents where protons do not readily exchange between molecules. The two options are shown in Figure 2.4, with the fate of the proton illustrated in red. Where the proton resides is ultimately dependent on the relative basicity of the carbamate, amine, and other Brønsted bases present and the ability of protons to move through the solution.

Tertiary amines (i.e., those with three N–C bonds associated with the amino group) are unable to directly react with the CO_2, the reason being there is no proton available to act as a leaving group, and the reaction of replacing an N–C bond with another N–C bond would be extremely slow. They are, however, still bases and do have the ability to accept protons released from reactions between CO_2 and other molecules (e.g. primary and secondary amines, H_2O or OH⁻).

In addition to their reactivity with CO_2 and the associated absorption capacity that yields, several other factors are important for an amine molecule to function as an effective CO_2 absorbent. A nonexhaustive list includes the following:

> *Rate of reaction with CO_2*: Fast reaction with CO_2 is desirable, as it facilitates the rapid mass transfer and reduces the surface area of gas-liquid contacting equipment.
>
> *Viscosity*: Large viscosity reduces rates of mass transfer via reduced diffusivity of CO_2 and makes the effective wetting of gas-liquid contactors more difficult.
>
> *Vapor-liquid equilibria*: The ability to cyclically bind with and release CO_2 is a critical feature of an absorbent. The cyclic capacity (i.e., the amount of CO_2 that can be reversibly absorbed and released) is governed by how the CO_2 absorption capacity changes as a function of conditions such as

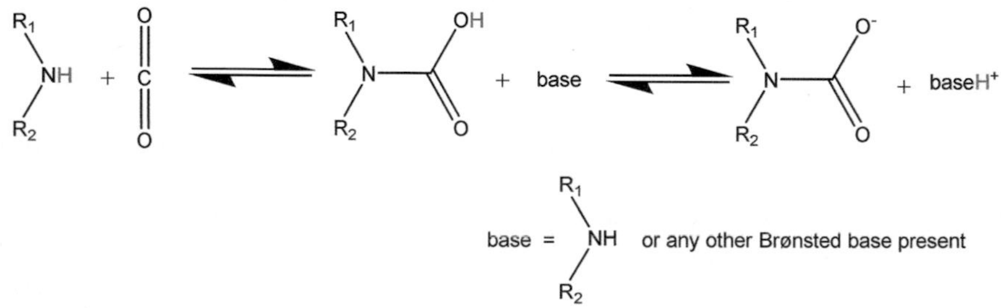

FIGURE 2.4 Illustration of where proton may reside following reaction of an amine with CO_2 to form carbamic acid.

the gas-phase CO_2 partial pressure, temperature, pH, solvent polarity, and applied electrical potential. The goal is a large change in absorption capacity with minimal energy input.

Chemical stability: The gas environments from which CO_2 is absorbed often contain other components such as O_2, SO_x, NO_x, H_2S, and metal-containing particulates to which the absorbent is exposed. For thermal swing-based CO_2 capture processes, the absorbent is also exposed to high temperatures. Long-term stability in these operating environments is critical.

Toxicity and biodegradability: Any deployment at scale will result in many thousands of tons of amines being in use. Given that some exposure of workers to the amines is likely, and some emissions to the environment will undoubtedly occur, avoiding the use of compounds that are highly toxic to humans, animals, or the environment is desirable, as is the use of compounds that biodegrade rather than bioaccumulate.

2.3.1 AQUEOUS AMINES

The most common amine-based solution for CO_2 absorption is an aqueous solution of single or multiple amines, including but not limited to alkanolamines (an amine that also contains one or more alcohol groups). The classic reactive chemical CO_2 absorbent is MEA. It was proposed in the original patent from 1930 [1] for an amine process to separate acid gases and has found widespread use in industrial gas treating [6]. As the carbamic acid formed from most amines in an aqueous solution is a weak acid, the overall reaction of an aqueous primary or secondary amine to form a carbamate can be summarized by Eq. 2.2.

$$CO_2 + R_1R_2NH \overset{K_1}{\rightleftharpoons} R_1R_2NCOO^- + H^+ \qquad \text{Eq. 2.2}$$

The proton lost via carbamate formation is then accepted by a Brønsted base present in the solution.

$$base + H^+ \overset{K_2}{\rightleftharpoons} baseH^+ \qquad \text{Eq. 2.3}$$

For aqueous MEA this is a second amine molecule, base = MEA. In the case of amine mixtures, the reaction of Eq. 2.3 will proceed via the more basic of the amines present (considering concentration effects). This may also be a tertiary amine, which can undergo acid-base reactions but not direct reaction with CO_2. In the case of poly-amines, each amino group may be able to form a carbamate and/or act as a Brønsted base. In the case of amines and traditional thermal swing CO_2 absorption processes, the increasing reversibility of these reactions with temperature is what drives the ability to absorb CO_2 at ~40°C and release it at ~100–120°C. The carbamate stability and the basicity at low temperature define the absorption capacity and the thermodynamics of these reactions (especially the reaction enthalpy) defines the cyclic capacity. In an aqueous solution, the enthalpy of protonation of the base is of particular importance, with this reaction accounting for around 60% of the total enthalpy of CO_2 absorption [7].

As already noted, in an aqueous solution CO_2 also reacts with the Lewis bases water and hydroxide ions. The reaction with water shown in Eq. 2.4 is very slow and not particularly favorable in terms of its equilibrium. As such, it contributes little to absorption. The reaction with hydroxide of Eq. 2.5 is rapid (comparable in rate to the fastest reacting amines) but the hydroxide ion concentration is generally low in the pH range of absorbents. At high pH, this reaction can contribute but it is not readily reversible as the bicarbonate product is very stable as a function of temperature (small reaction enthalpy). It is worth noting that water and hydroxide can also act as Brønsted bases as per Eq. 2.3, but this is generally a negligible contribution in the presence of concentrated amines.

$$CO_2 + H_2O \overset{K_3}{\rightleftharpoons} HCO_3^- + H^+ \qquad\qquad \text{Eq. 2.4}$$

$$CO_2 + OH^- \overset{K_4}{\rightleftharpoons} HCO_3^- \qquad\qquad \text{Eq. 2.5}$$

The role of the alcohol group in aqueous alkanolamine solutions is to improve the amine solubility and reduce its vapor pressure through the ability of the -OH group to hydrogen bond with water. The alcohol group itself is unreactive to CO_2 under these conditions, except possibly to a very small extent [8]. It may however modify the physical properties of a molecule and the electronic environment of the amino group (generally by withdrawing electron density from the amino group and reducing its basicity) which will affect reactivity with CO_2. Under appropriate conditions, alcohol groups exist as alkoxides, which can be directly reactive with CO_2. This will be discussed in the next section on alternative Lewis bases to amines.

Given the extensive literature spanning decades regarding the use of aqueous amines such as MEA, 2-amino-2-methyl-propanol (AMP), N-methyldiethanolamine (MDEA), piperazine (PZ), and DEA for CO_2 absorption [9], the focus will be on promising new developments that have occurred during the last 10 years (i.e., the period 2010–2020). These developments can be roughly divided into three main areas: new amines and their formulations, and in particular the use of polyamines; formulations with low/no water content (water lean) by the addition of an organic diluent; and formulations that undergo an aqueous/organic phase separation as a function of CO_2 content or temperature. Amine functionalized ionic liquids will not form part of this discussion, as they are covered in detail in a separate chapter.

The structures of the most common amines used in the formulation of aqueous amine CO_2 absorbents over the preceding decades are shown in Figure 2.5 [9]. The primary and secondary amines MEA, PZ, and DEA have been used as single amine solutions as the sterically free environment around the amino group can react rapidly with CO_2 to form a carbamate. Alternatively, the steric crowding of the amino group of AMP limits its ability to form carbamate and MDEA being a tertiary amine cannot directly interact with CO_2. Thus, AMP and MDEA only undergo acid-base reaction at the amino group leading to very slow CO_2 absorption in aqueous solution via bicarbonate formation. They are typically formulated with a fast-reacting primary or secondary amine (e.g., PZ-AMP and MEA-MDEA) to act as a proton-accepting base in the presence of a carbamate-forming amine. These compounds have structural

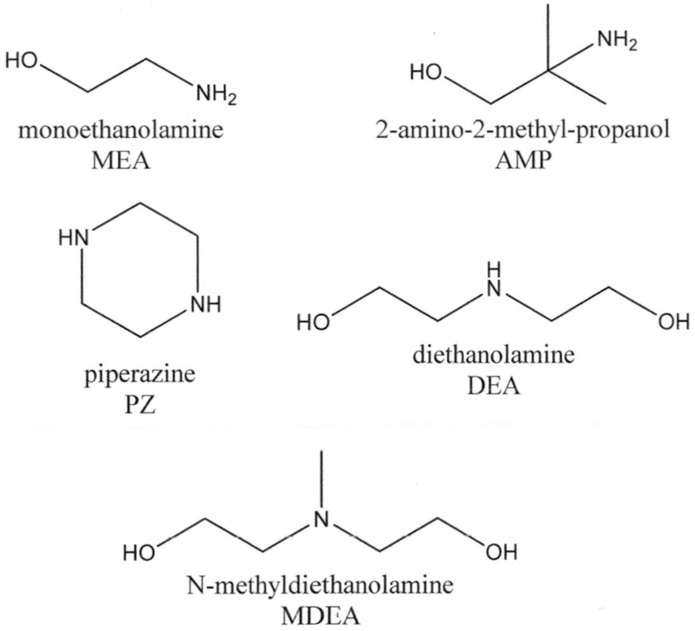

FIGURE 2.5 Common amines used as aqueous amine-based CO_2 absorbents.

features in common: they all contain alcohol groups and a single amino group (other than PZ), and the spacing between the alcohol and/or amino groups is two carbons. PZ is a special case in that it is a diamine and has no alcohol group. The presence of a second amino group helps play the role of an alcohol group in terms of improved solubility and reduced vapor pressure while also providing a second reactive amino group.

The published literature in the period 2010–2020 has seen aqueous solutions of more than 70 different amine structures proposed for CO_2 absorption applications. The discussion is limited to new compounds that have appeared in at least five publications within that period (excluding those of Figure 2.5), which reduces the number to 23. These 23 compounds can be broadly divided into three distinct categories: derivatives of the alkanolamine MEA with additional branching and/or heteroatoms; acyclic aliphatic polyamines; and cyclic amines.

2.3.1.1 Derivatives of the Alkanolamine MEA

Amines that represent derivatives of MEA are shown in Figure 2.6. The amine 2-(methylamino)ethanol (MAE) [10–17] is more basic than MEA and reacts more rapidly with CO_2 as a consequence. It demonstrates a larger CO_2 absorption capacity at low temperature and similar capacity to MEA at elevated temperature, resulting in a larger cyclic capacity. It has been characterized as a single amine and in a formulation with 2-(2-aminoethylamino)ethanol (AEEA). Being a secondary amine, it may

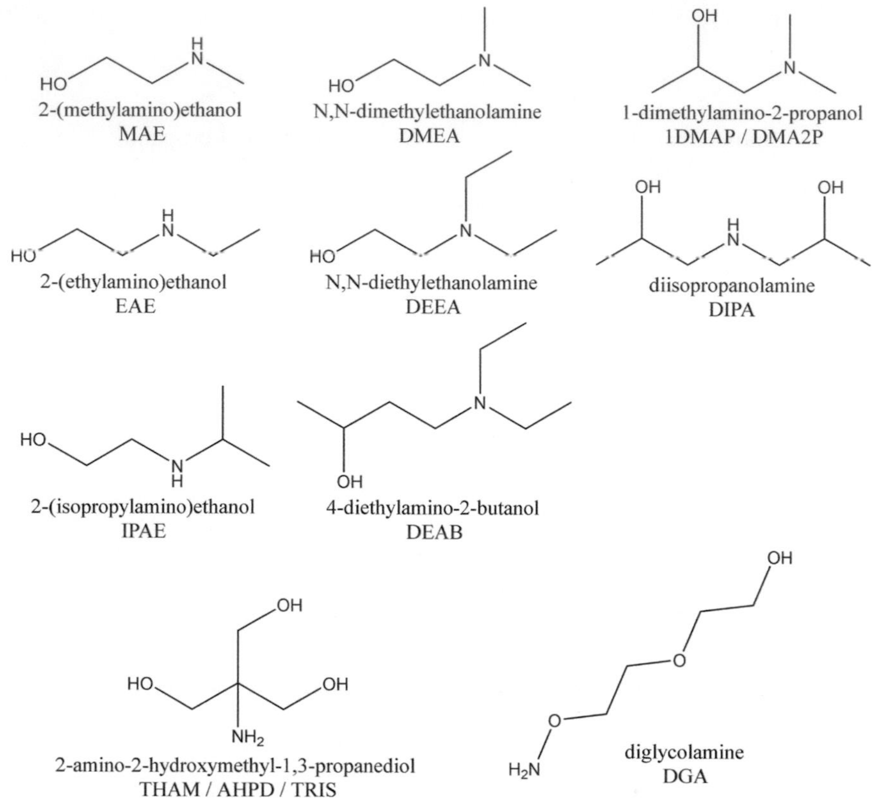

FIGURE 2.6   Amines that represent derivatives of MEA that have appeared in five or more publications in the period 2010–2020.

have a propensity to form nitrosamines in the presence of NO_x that may be present in some flue gases, and little is known about its long-term stability and degradation. The amine 2-(ethylamino)ethanol (EAE) [18–25] is structurally very similar to MAE and shares similar properties. Its main point of difference is that it is known to react very rapidly with CO_2, and it has been used as an absorbent rate promoter in formulations with tertiary amines 3-(dimethylamino)propan-1-ol and diglycolamine. A drawback is that it is also a secondary amine and would likely form nitrosamines, and there is little information about its stability or degradation. The amine 2-(isopropylamino) ethanol (IPAE) [20, 26–27] is a sterically hindered derivative of EAE. The steric hindrance enhances its CO_2 absorption and cyclic capacity but decreases its rate of reaction with CO_2 relative to EAE. It has seen small-scale pilot-plant operation with some success, and the thermal stability has been assessed and found to be somewhat better than MEA. It is also a secondary amine, but the presence of steric hindrance may reduce its affinity for nitrosamine formation, although this has not been investigated.

MAE, EAE, and IPAE are all relatively low viscosity (similar to MEA) and are completely miscible with water and do not form precipitates.

Diglycolamine (DGA) has been used for CO_2 capture applications for some time [6] and has continued to see sustained interest over the period 2010–2020 [28–36]. Its chemical, physical, and mass transfer properties have been well characterized, and as a consequence, it is present as a component in a number of the commonly used process modeling packages such as ProTreat® (Optimised Gas Treating Inc.), ProMax® (Bryan Research and Engineering, LLC), and Aspen PLUS® (AspenTech). Its stability and degradation characteristics are also reasonably well studied. DGA is more stable than MEA but has slower rates of mass transfer and smaller absorption capacity at low CO_2 partial pressure. Previously, DGA has been applied to natural gas sweetening, in particular low pressure but high CO_2 fraction applications as a single aqueous amine. The more recent publications focus on its use in biogas upgrading applications, which have a similar low pressure but high CO_2 fraction composition. Similarly, diisopropanolamine (DIPA) has been used for high-pressure CO_2 capture applications [6], it is well characterized and is incorporated into process modeling packages. More recent work involving DIPA has focused on characterizing its formulation with other amines, typically to improve mass transfer performance, which is slow, but still with high-pressure applications in mind [37–55].

The amine 2-amino-2-hydroxymethyl-1,3-propanediol (THAM, AHPD or TRIS) is sterically hindered and is similar in structure to AMP, with the addition of an alcohol group to each methyl group attached to the alpha carbon [56–67]. Its prominence is unusual, as the additional alcohol groups reduce the basicity of the amino group due to their electron-withdrawing effect, limiting capacity at low CO_2 partial pressure and making it more suited to high-pressure CO_2 removal applications. Surprisingly, the additional alcohol groups do not seem to have a particularly detrimental impact on viscosity (at least in the absence of CO_2), and the vapor pressure in an aqueous solution is low due to enhanced hydrogen bonding capacity. For most applications, formulation with a faster reacting primary or secondary amine that is not sterically hindered will be necessary to achieve useful mass transfer rates. Its thermodynamics are typical of non-carbamate-forming sterically hindered amines, but little is known about its stability.

Even though they are tertiary amines, and therefore do not undergo direct reaction with CO_2, N,N-dimethylethanolamine (DMEA) [25, 68–76] and N,N-diethylethanolamine (DEEA) [68–69, 71, 77–108] have seen considerable activity in terms of publications over the past decade. Tertiary amines appear primarily in formulations with other primary or secondary amines to achieve useful rates of mass transfer. Even though they are structurally very similar, DEEA is more basic and lends itself to more rapid mass transfer of CO_2. Consequently, it has been studied in more detail and appeared in publications about three times as often as DMEA. Its primary appeal is that it has a smaller CO_2 absorption enthalpy and viscosity and is more basic than both AMP and MDEA (the sterically hindered and tertiary amine workhorses), which yields mass transfer and absorption capacity benefits. However, relative to AMP, it exhibits a reduced thermally driven CO_2 cyclic capacity and is not as resistant to degradation. DEEA certainly provides an alternative to AMP as a reasonably robust and low-cost amine for formulation with primary and secondary amines, but it would have alternative optimum operating conditions due to its different thermodynamic properties.

Another tertiary amine that has been the focus of considerable research recently is 1-dimethylamino-2-propanol (1DMAP or DMA2P) [69, 73, 109–126]. It is structurally similar to DMEA but with the replacement of hydrogen by a methyl group at the 2-position. This methyl group is unlikely to interact with the amino group but is close to the alcohol group of the molecule. This will likely affect the solvation environment, particularly in an aqueous solution where the alcohol group will hydrogen bond with water. It has similar basicity to the structurally similar tertiary amines already discussed. However, there is some discrepancy in the literature concerning whether it has an unusually large or unusually small enthalpy of CO_2 absorption. The focus so far has been on the determination of its properties, and it has been considered only primarily in formulations with MEA, EAE, and PZ. It certainly warrants further investigation to better understand its thermodynamic properties and the role of the additional methyl group and its stability. DEAB (4-diethylamino-2-butanol) is similar in structure to 1DMAP but with an additional carbon in the chain separating the amino and alcohol functionality and ethyl substituents on the amino group [110, 127–135]. The net effect of the greater distance to the electron-withdrawing alcohol is greater basicity compared to DEAB and the other amines already discussed (pK_a of conjugate acid >10.2 at 25°C, compared to 9.55 for 1DMAP). This imparts it with greater absorption capacity and a more rapid mass transfer behavior compared to other tertiary amines (due to more basic conditions in solution). However, formulation with other primary or secondary amines is still necessary to achieve practically useful mass transfer. Little information is currently available about its thermodynamics and therefore its cyclic capacity or stability.

2.3.1.2 Acyclic Aliphatic Polyamines

Acyclic aliphatic di- and triamines have also appeared frequently in the literature during the period 2010–2020. This section only discusses those used in traditional aqueous solution, however, they also feature prominently in water-lean and biphasic absorbents discussed later. Those that have appeared in five or more publications are listed in Figure 2.7. The advantage of polyamines is that the CO_2-reactive amino functional groups are a larger fraction of each molecule. In principle, this can allow the same molecule to form a carbamate and accept a proton or form multiple carbamate functional groups or accept multiple protons. The extent to which this occurs in practice is dependent upon the properties of each amino group and their proximity to each other. If nearby, carbamate formation or protonation of one amino group will affect the reactivity of another. The amino groups may also interact via intramolecular interactions, such as intramolecular ring formation stabilizing a carbamate or allowing more rapid proton transfer. A drawback of acyclic aliphatic polyamines is that the presence of multiple reactive amino groups also leads to greater oxidative and thermal degradation.

Ethylenediamine (EDA) is by far the most mature of the acyclic polyamines in terms of its application [31, 136–149]. Structurally, it is MEA with the alcohol group replaced by a second amino group. It has been used in pilot-plant trials but also continues to be the focus of fundamental studies of the characteristics of linear

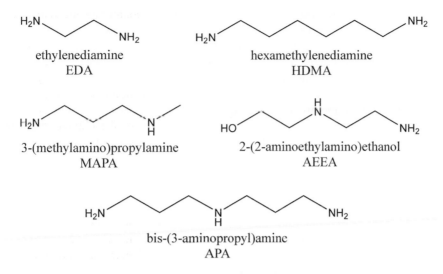

FIGURE 2.7 Acyclic aliphatic polyamines that have appeared in five or more publications in the period 2010–2020.

diamines. The amino groups have similar properties to that of MEA, although EDA is more basic and does react more rapidly with CO_2. The proximity of the amino groups means that once one is protonated, protonation of the second amino group occurs at a pH value lower than is relevant for CO_2 capture. However, it can form a di-carbamate. EDA is more volatile and is less stable than MEA, which limits its utility. There has been an interest in the use of hexamethylenediamine (HDMA) [141, 150–158]. It is similar to EDA but has six instead of two carbons in the chain separating the amino groups. This greater separation allows the amino groups to function almost independently, with the protonation constant of both amino groups remaining in a useful range. The useful range for protonation constants as defined by Eq. 2.3 is qualitatively $8<\log_{10}K_2<10.5$ as below 8 appreciable formation of $CO_{2(g)}$ occurs at pH values where protons will be accepted, and above 10.5 stripping of CO_2 is hindered by excess basicity. HDMA is marginally more basic than EDA and has a larger absorption capacity and a faster reaction rate with CO_2, making it a potential rate promoter in conjunction with tertiary or sterically hindered amines. More work is needed to determine its cyclic capacity and its stability.

The amine 3-(Methylamino)propylamine (MAPA) has been the focus of quite a bit of research activity [89, 159–175]. Compared to EDA, it has one additional carbon atom separating the amino groups and is of similar basicity. In addition, one of the groups is methylated, resulting in MAPA containing a primary and a secondary amino group. It is more basic than EDA, and while the second protonation constant is significantly smaller than the first, it remains in a useful range for CO_2 absorption. MAPA has exhibited fast reaction kinetics with CO_2 and good cyclic capacity. It has been considered generally for use in formulations with other tertiary and sterically hindered amines and sarcosine, as high viscosity limits its use alone at high concentration. Its thermodynamics have been well characterized, and its cyclic capacity is good. A downside appears to be its stability, as it degrades rapidly in the presence of oxygen. The presence of a secondary amino group also limits its utility in the presence of NO_x due to the potential for nitrosamine formation. AEEA, or 2-(2-Aminoethylamino) ethanol, shares characteristics with both EDA and MAPA [36, 141, 176–179]. The spacing between amino groups is the same as EDA, and it contains a primary and secondary amino group like MAPA. The point of difference is the addition of an ethyl alcohol substituent. As with EDA, the proximity of the amino groups means only the first protonation constant is of relevance for CO_2 capture. The second occurs at too low a pH. Also, the addition of the alcohol group reduces the basicity relative to both EDA and MAPA. AEEA reacts rapidly with CO_2 although not as rapidly as HDMA, most likely due to its reduced basicity. Most studies have focused on vapor-liquid equilibria and CO_2 reaction kinetics. Little is known about its stability or its thermodynamic and cyclic capacity properties across a broad temperature range.

Bis-(3-aminopropyl) amine (APA) is the only triamine identified among the group of linear polyamines. It has been studied less than the other polyamines already discussed, and mostly by the same authors [180–184]. Its structure is symmetrical with a central secondary amine and two propylamine substituents attached. It certainly shows promise, as it has low viscosity and a fast reaction with CO_2, and the separation of the amino groups mean they all have the potential to be active proton acceptors and carbamate formers. The data collected to date is focused on physical properties, equilibrium studies of capacity, and a small amount on absorption rate. More data needs to be collected, especially as a function of temperature, to fully characterize its performance as an absorbent. Investigation of the chemical speciation would also provide valuable information. No stability data has been collected, and given the number of reactive amino groups, it may be expected to degrade quite rapidly in the presence of oxygen and high temperatures and form nitrosamines at the secondary amine site in the presence of NO_x.

2.3.1.3 Cyclic Amines

The final class of amine to be discussed in this section are cyclic amines. The expression 'cyclic amine' is defined here to mean an amine compound that contains a cycle (all six-atom cycles in this case). The amine itself may or may not be incorporated into the cycle. These include both mono- and diamines and all but one is aliphatic, with a single aromatic example. The structures are given in Figure 2.8. From a chemistry perspective, the presence of a cycle within a molecule generally reduces the degrees of freedom for conformations that can be adopted and enhances stability. The primary reason molecules adopt cyclic structures is that they are a lower energy configuration than their acyclic equivalent. It can be expected that cyclic molecules would be less susceptible to degradation due to ring-opening being more difficult than cleaving or substitution of acyclic molecules.

PZ is by far the most studied cyclic amine of the previous decade, exhibiting stability and fast reaction with CO_2. Although it contains two secondary amino groups that can form nitrosamines in the presence of NO_x, its stability allows CO_2 stripping at higher temperatures that also destroys any nitrosamines formed [185]. The first

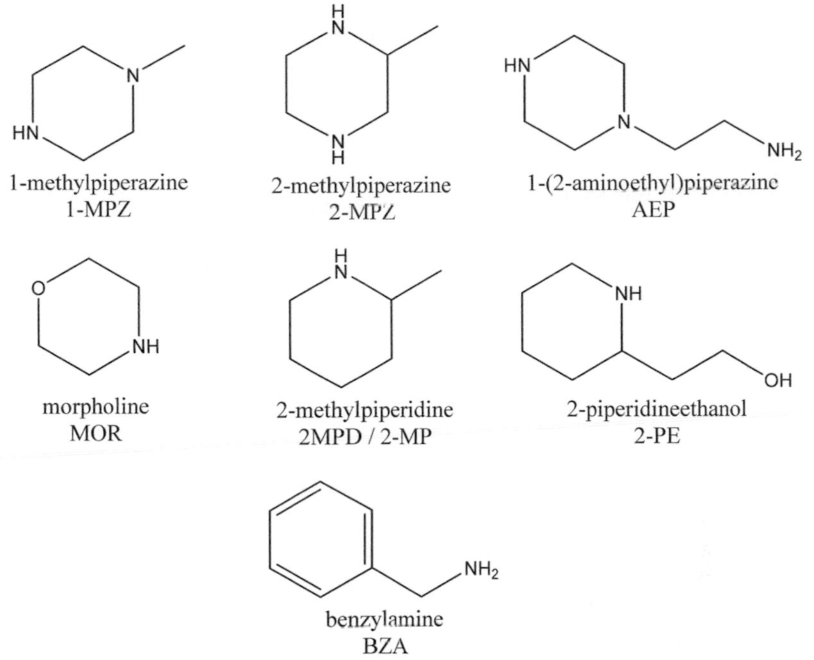

FIGURE 2.8 Cyclic amines that have appeared in five or more publications in the period 2010–2020.

row of compounds shown in Figure 2.8 are derivatives of PZ. The first, 1-methyl-piperazine (1-MPZ), has seen limited study and has been used generally to better understand the properties of PZ [186–191]. It has secondary and tertiary amino functionality and is less basic than PZ [192] but retains the fast reaction with CO_2. It is less stable than PZ but still is one of the more stable amines studied to date. It does seem to have a reduced CO_2 absorption capacity likely due to the inability to form a dicarbamate. The next, 2-methylpiperazine (2-MPZ), is very similar, but the location of the methyl group means it retains two secondary amino groups, but one is steri-cally hindered. There are more studies of 2-MPZ than 1-MPZ [151, 187, 189, 193–199]; it is more stable than 1-MPZ but still less stable than PZ. CO_2 mass transfer also has been found to be slower for 2-MPZ, even though its basicity sits between that of 1-MPZ and PZ [192]. It is unclear what factors are responsible for the differences in rate, as both 1-MPZ and 2-MPZ have a sterically free secondary amine able to react with CO_2, and 2-MPZ can form a dicarbamate at high CO_2 loadings. This points to why there is particular interest in 2-MPZ, as it has better CO_2 cyclic capacity than 1-MPZ and PZ while still retaining good rates of reaction with CO_2 and stability bet-ter than most other amines. There has also been some interest in 1-(2-aminoethyl) piperazine (AEP) [180, 187, 200–204], particularly as a rate promoter. It is a triamine consisting of a primary, secondary, and tertiary amino group but does not appear to show any advantage over the smaller 1-MPZ and 2-MPZ molecules. The mass transfer benefits of the other PZ derivatives are lost, and speciation studies indicate carbamate formation occurs principally at the primary amino group. It is unclear why this is the case, given the secondary amines of PZ are known to react very rapidly with CO_2. A possibility is that the dangling primary amino group can interfere with the secondary amino group within the ring. Although stability has not been investigated extensively, it is likely AEP would also be less stable than the other PZ derivatives due to the additional primary ethyl amino substituent, which would be more susceptible to degradation.

The second row of Figure 2.8 contains three different derivatives of piperidine, which is similar to PZ but with only a single secondary amino group within the ring structure. Piperidine itself is not suited to CO_2 capture applications, as its basicity is too great and carbamate too stable (with a small negative reaction enthalpy) to allow adequate reversibility of the absorption reactions under moderate thermal swing [205–206]. In essence, the CO_2 absorption capacity remains high at elevated temperature. Morpholine (MOR) results from the insertion of an oxygen atom within the ring structure of piperidine [5, 189, 204–205, 207–212]. This leads to a significant reduction in basicity compared to piperidine (amine protonation constant reduced from 11.1 to 8.5 at 25 °C [206]) to a value more useful for reversible CO_2 absorption. MOR is more stable than PZ, likely due to the stability benefits of the ring structure and fewer reactive amine groups, but suffers from a lower rate of reaction with CO_2 consistent with its smaller basicity. Still, it provides reasonable rates of mass transfer and CO_2 absorption and cyclic capacity to warrant ongoing study. An alternative piperidine derivative is 2-methylpiperdine (2MPD or 2-MP), with a methyl group in position 2 of the ring [213–218]. The proximity of the methyl group on a carbon alpha to the amino group results in steric hindrance, which suppresses the formation of the carbamate (carbamate forms to a small extent). The methyl group reduces the basicity of the amino group only marginally compared to piperidine, which will similarly limit its effective cyclic capacity [219]. The derivative 2-piperdineethanol (2-PE) has an ethanol substituent in position 2 rather than a methyl substituent [5, 205, 220–222]. This results in steric hindrance and minimal carbamate formation, but the alcohol functionality also reduces the basicity to a more useful value. More work is needed to better characterize its cyclic capacity and to determine if the pendant arm reduces the stability compared to MOR.

Benzylamine (BZA) is the last of the amines containing a cyclic structure found to have appeared in journal literature five or more times between 2010 and 2020 [223–231]. It is different from those discussed so far in that it contains an aromatic benzene ring with a pendant methyl amino group. Unlike the other cyclic structures, the amino group is primary, which limits the possibility of nitrosamine formation. In addition, the combination of cyclic structure and aromaticity means the ring is extremely robust, and any degradation will occur via the pendant arm. Having a single carbon spacing between the ring and the amino group is the minimum required for reactivity with CO_2, as direct attachment of the amino group, such as in aniline, results in the lone pair of the nitrogen atom becoming delocalized within the ring and unreactive towards CO_2. BZA has been found to react rapidly with CO_2 and have a good absorption capacity and excellent cyclic capacity. Its enthalpy of protonation and subsequently enthalpy of absorption are large and negative, yielding good cyclic capacity but shifting the optimal operating conditions in a process compared to the more typical thermodynamic values of other amines. It also forms a precipitate at high concentration upon reaction with CO_2, which can be problematic. Also, while being apparently stable, more experimental study of its stability is needed, particularly in comparison to the other cyclic amines discussed.

2.3.2 AMINO ACIDS

Amino acids, which are amines that contain carboxylic acid functionality, have been investigated as CO_2 absorbents. They have favorable characteristics in terms of vapor pressure, as they are ionic in their neutralized form used for CO_2 absorption, and many are resistant to oxidative degradation [232]. Amino acids undergo the same chemistry with CO_2 as amines, with the amino group being the reactive center. A challenge to their use is they are often of limited solubility and may form precipitates. To date, pilot-plant trials of single-phase, amino acid–based systems have not produced favorable results [233–234]. Therefore amino acid–based liquid-solid processes are also being developed [235], which represent a form of biphasic absorbent. In this case, the precipitate formed can be either the products containing neutral amino acids or CO_2, depending upon the amino acid used.

Little has changed in terms of the structure and therefore the chemistry of amino acids used for CO_2 capture over the period 2010–2020. This is not surprising, as the pool of naturally occurring amino acids is limited. The focus during this period has been more on the development of processes that are amenable in their operation to the use of amino acids (e.g. the use of membrane contactors). The reader is referred to a recent review for more information on developments in amino acid-based technology [236].

2.3.3 WATER-LEAN ABSORBENTS

A comprehensive review of water-lean absorbents by Wanderley et al. [237] was recently published. The reader is referred to this work for a more complete discussion of this area, including a historical perspective. This section will only provide a summary of more recent work.

Water-lean absorbents generally consist of solutions containing many of the amines already discussed. To be water-lean, some (if not all) of the water is replaced by an organic solvent. The motivation for doing this is to reduce the heat capacity and therefore the latent heat requirements of CO_2 stripping, minimize water vaporization during CO_2 stripping, and increase the physical solubility of CO_2. Ionic liquids would also fall into this class of absorbents, but they will not be discussed here as they appear in more detail in another chapter of this book. Also, molecules that react with CO_2 via alternative chemistries to amines are discussed later in this chapter. A general drawback of water-lean absorbents is they tend to have a high viscosity (>100 cP), which increases upon reaction with CO_2, hindering both mass and heat transfer. Notably, most aqueous amines used in traditional CO_2 capture processes have viscosities in the 2–30 cP range. Above 30 cP heat and mass transfer, pumping, flow characteristics and wetting of packing materials can become problematic.

Maintaining the water balance in processes where the feed gas contains moisture may also be challenging. On the flip side, the lack of water can require the use of a stripping gas, which can be problematic for applications requiring high-purity CO_2. They also introduce other factors such as changes in the solvation environment and polarity that can both positively and negatively influence amine-CO_2 reaction kinetics and equilibria and the solubility of ionic species. The vapor-liquid-equilibria behavior of water-lean absorbents differs from aqueous absorbents in that carbamic acid/carbamate formation is somewhat destabilized in a lower polarity solvent environment, leading to reduced absorption capacity due to chemical reaction at low CO_2 partial pressure. Alternatively, at high CO_2 partial pressure, the capacity is enhanced due to the increased physical solubility of CO_2. Technically, deep eutectic absorbents, in which an amine is added to a deep eutectic solvent such as choline chloride [238–240], fit the definition of a water-lean absorbent. A deep eutectic solvent is a mixture in which the melting point of the mixture is lower than that of the individual components. However, they are currently far from being practically useful due to very high viscosity and poor absorption at low to moderate CO_2 partial pressure. As such, they are mentioned for completeness but will not be discussed.

Alcohols feature prominently as diluents for water-lean absorbents [241–259]. Methanol and ethanol are the most common, as they provide low viscosity, good amine and reaction product solubility, and generally enhanced physical CO_2 solubility relative to water only. Glycols also feature, such as di-, tri-, and polyethylene glycol. Examples of their structures are given in Figure 2.9. Alcohols are appealing because they are generally miscible with water and amines. Being protic solvents, the reactions between CO_2 and amines proceed via essentially the same mechanism as in water, with protons free to migrate between molecules. A drawback is that the vapor pressure of smaller alcohols can be high, leading to losses that need to be managed. A trade-off is that alcohols of larger molecular weight have reduced vapor pressure but also tend to have larger viscosity.

A broad range of other organic compounds have been tested as diluents during 2010–2020, but one that stands out in terms of prominence is N-methyl-2-pyrrolidone [255, 260–271] (NMP), which is shown in Figure 2.10. NMP features a large physical solubility for CO_2, is known to be stable, is of low viscosity, and is moderately polar. It has been used as a physical absorbent for high-pressure CO_2 applications. Mass transfer in mixtures with amines such as MEA remains good. Studies of stability under conditions that represent long-term operation or even long-term operation

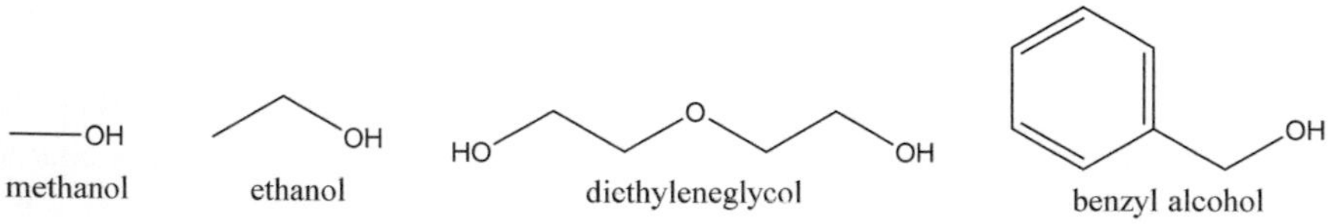

methanol   ethanol   diethyleneglycol   benzyl alcohol

FIGURE 2.9 Examples of common alcohol diluents used between 2010 and 2020 in water-lean formulations.

itself are currently lacking for water-lean absorbents. It is unclear what the impact of water-lean compositions will be on the degradation of amines. The diluent itself may undergo degradation or reaction with amine degradation products. Another interesting hydrophobic and low-vapor-pressure example of a water-lean system not based on typical alcoholic diluents is that consisting of equimolar fluorinated amine and fluorinated alcohol [272], shown in Figure 2.10. CO_2 appears to react with the amine via the typical mechanism already described, with the fluorinated alcohol acting as a diluent only. The viscosities under CO_2-free and CO_2-loaded conditions are acceptable (<30 cP at 40°C). Little is known about the system in terms of mass transfer or stability, but it has an interesting feature of the large variability of the absorption enthalpy with temperature. It also demonstrates a low CO_2 affinity at elevated temperature and may be able to release CO_2 at high pressure.

A recent water-lean example from Zheng *et al.* [273] tested a system using the single component N-(2-ethoxyethyl)-3-morpholinopropan-1-amine (2-EEMPA), shown in Figure 2.11. It stands out and warrants special mention due to its unusual single component composition. Water loading is small, at only around 1.5 wt%. It is a derivative of MOR and contains both a secondary and tertiary amino group. Based on vapor–liquid equilibria results, it has been inferred that 2-EEMPA reacts to form separate carbamate and protonated amine products in a manner equivalent to aqueous amines. This is surprising, given the extremely water-lean conditions and the fact that the tertiary amino group should have useful basicity to act as a proton acceptor. Additional investigation of the speciation upon reaction with CO_2 would help in

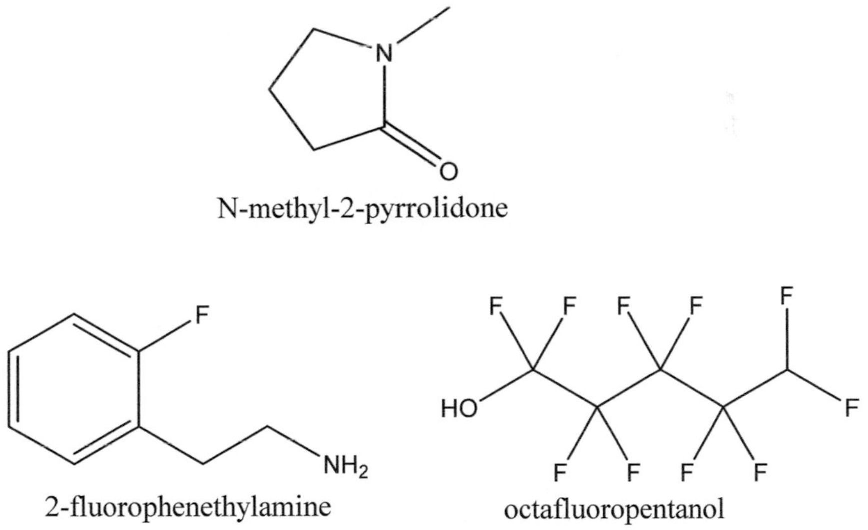

N-methyl-2-pyrrolidone

2-fluorophenethylamine

octafluoropentanol

FIGURE 2.10 The structure of N-methyl-2-pyrrolidine, which featured prominently as a diluent for use in water-lean absorbents between 2010 and 2020, and a fluorinated amine and fluorinated alcohol pair used to make a unique water-lean absorbent.

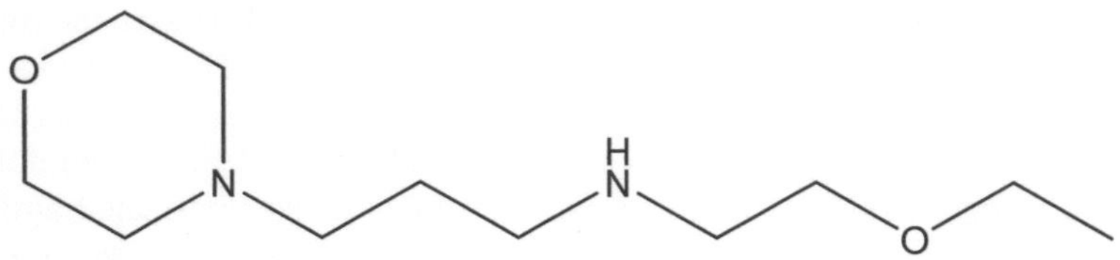

N-(2-ethoxyethyl)-3-morpholinopropan-1-amine
2-EEMPA

FIGURE 2.11　Single component absorbent system consisting of only this molecule.

understanding the chemical behavior. This aside, techno-economic analysis was performed with a simple stripper configuration and compared with MEA and the proprietary absorbent from Cansolv. Reboiler duty was estimated to be 2.27 GJ per tonne, 30% lower than 30 wt% aqueous MEA and 10% lower than the Cansolv absorbent. Viscosity is relatively high (~100 cP at 30 °C and high CO_2 loading), so capital cost is higher, giving a total estimated 5% benefit relative to the Cansolv technology. The wettability of 2-EEMPA means that it is possible to use plastic packing in the absorber for further cost reduction. Investigations of the potential of nitrosamine formation via the secondary amino group and absorbent degradation still need to be undertaken.

2.3.4　BIPHASIC ABSORBENTS

Biphasic absorbents are similar to water-lean absorbents in that they have a reduced or no water content and have a high concentration of a compound that has nonpolar/organic character [274]. This compound may be an amine or other organic diluent. The typical scenario for biphasic absorbents is that upon absorption of CO_2, phase separation occurs into an organic and an aqueous/polar phase, with ionic species bound to CO_2 residing in the aqueous/polar phase. The CO_2-containing phase is then processed for CO_2 stripping. Because the CO_2 is concentrated in a smaller amount of absorbent, the stripping energy requirement is less than if processing both phases. Although there are alternatives to this, such as separate phases initially that form a single phase upon reaction, phase separation occurring as a function of temperature, and formation of a solid CO_2-containing phase. These are still quite a new form of absorbent, so little information is available regarding stability or long-term operation.

　　Aqueous amine biphasic absorbents generally consist of aqueous solutions of amines with more organic character or lower polarity. Upon reaction with CO_2, the resulting ionic species of protonated amine, carbamate and bicarbonate, are insoluble in the neutral amines, and phase separation occurs with the ions favoring the aqueous phase. For example, aqueous binary mixtures of 3-(methylamino) propylamine (MAPA, Figure 2.7) or dibutylamine (DBA) with N,N-diethlylethanolamine (DEEA, Figure 2.6) form two phases upon reaction with CO_2 [275]. It may be that in an aqueous solution, the neutral amines tend to form clusters solvated by water, but as the

di-N-propylamine   N,N-dimethylcyclohexylamine   N-methylcyclohexylamine

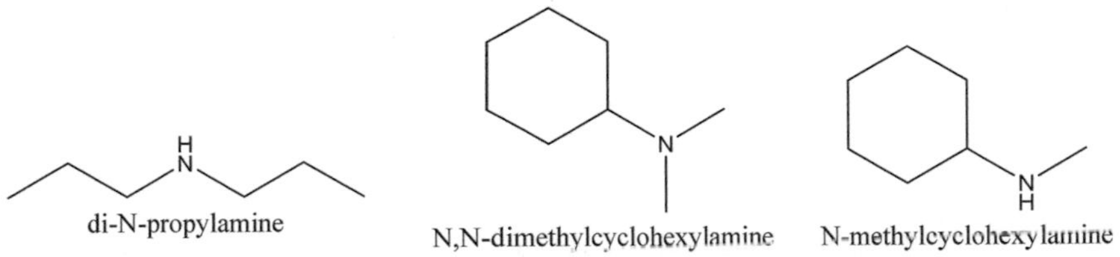

FIGURE 2.12 Examples of amines used in thermomorphic biphasic absorbents.

ionic strength increases following reaction with CO_2, these clusters break down and phase separation occurs. This form of biphasic absorbent is the most studied in the period 2010–2020 [94, 274–294].

Nonaqueous biphasic absorbents are quite similar in composition to aqueous ones but replace water with alcohol or another organic compound [275]. The compounds used are like those used for water-lean absorbents, although long-chain alcohols are favored, such as 1-heptanol or 1-octanol, to facilitate phase separation into two liquid phases upon reaction with CO_2. Smaller alcohols tend to result in a solid–liquid phase transition. For the longer-chain alcohols, the CO_2 products form a separate ionic-liquid-like phase with the alcohol in the second phase. Both single amines and mixtures of those commonly used in aqueous solution have been used. N-methyl-2-pyrrolidone (NMP) has also been used with AMP to form a solid-liquid biphasic absorbent [263].

Thermomorphic biphasic absorbents are a unique form of biphasic absorbent in which phase separation occurs as a function of CO_2 absorption and temperature [295–298]. These absorbents consist of aqueous solutions of relatively nonpolar amines (not alkanolamines) that only have limited molecular interactions with water at room temperature. Some examples are shown in Figure 2.12. The interactions are strong enough to result in a homogeneous solution, although the aqueous solubility is lower than more polar amines overall. Upon heating, these molecular interactions break down and phase separation occurs. When CO_2 is introduced, the resulting ions favor the aqueous phase upon phase separation. These absorbents and the thermomorphic phase change behavior are very sensitive to the composition, which would make process operation a challenge and somewhat more complex than a typical process.

2.4 ALTERNATIVE LEWIS BASES

2.4.1 ALKOXIDE

The Lewis bases given in Table 2.1 suggest alkoxide chemistry can be an alternative to amine chemistry for reaction with CO_2. A hurdle for the applicability of alkoxides is the stability of oxygen anions in solution. Ionic liquids address this and are discussed in another chapter of this book. One way of making the oxygen atom, a good nucleophile for reaction with CO_2 in standard solutions, is demonstrated in the CO_2BOL class of absorbents.

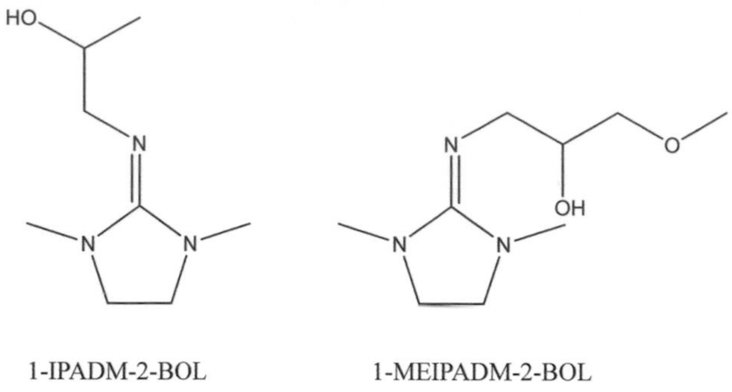

1-IPADM-2-BOL   1-MEIPADM-2-BOL

FIGURE 2.13   Example CO_2BOL molecules combining the guanidine and alcohol functionality in single molecules.

A CO_2BOL is a CO_2-binding organic liquid. It is distinct from an amine-based aqueous system because the amino group is not the site for reaction with CO_2 and is distinct from an ionic liquid because no ions are present until CO_2 is absorbed. They are based around a guanidine core structure with an alkanolamidine group attached, the guanidine group providing a highly basic nitrogen lone pair. In the liquid phase, the base will deprotonate the alcohol moiety, giving alkoxide, which may then react with CO_2. The concept originated from combining guanidine with aliphatic alcohols to promote reaction with CO_2 [299]. In the molecules in Figure 2.13, both functions are combined in one [300]. The ring structure surrounding the guanidine lends stability against hydrolysis to the guanidinium ion generated in CO_2 binding.

As already noted, high viscosity is the common hurdle for water-lean absorbents, which also applies to CO_2BOLs. To avoid this, Malhotra et al. [300] used chemical synthesis and computer modeling to study how hydrogen bonding, degrees of freedom, and cation/anion charge solvation affected absorbent viscosity. They found that intramolecular hydrogen bonding leads to lower viscosity under CO_2-loaded conditions and studied which components of a molecule led to which kind of hydrogen bonding. Using this approach, they developed the molecule 1-MEIPADM-2-BOL (Figure 2.13), which has 60% lower viscosity than the original 1-PADM-2-BOL at 25% CO_2 loading, reduced from 171 cP to 75 cP. CO_2 uptake for the molecules was 9.0wt% and 7.3 wt% (44 mol% and 35.5 mol%, respectively). The absorption capacity of CO_2BOLs is similar to amine-based absorbents, but high regeneration temperatures are required for conventional thermal regeneration (~159 °C).

A feature of CO_2BOLs is that they can operate in thermal swing–assisted or polarity swing–assisted mode [301]. Adding an anti-solvent like hexadecane can destabilize the ions and thereby remove CO_2 from the absorbent. This can reduce the required temperature for regeneration from 159 °C to 86 °C with a theoretical minimum of 65 °C. Simulations were carried out in ASPEN Plus to compare CO_2BOLs with polarity assist to the US Department of Energy (US

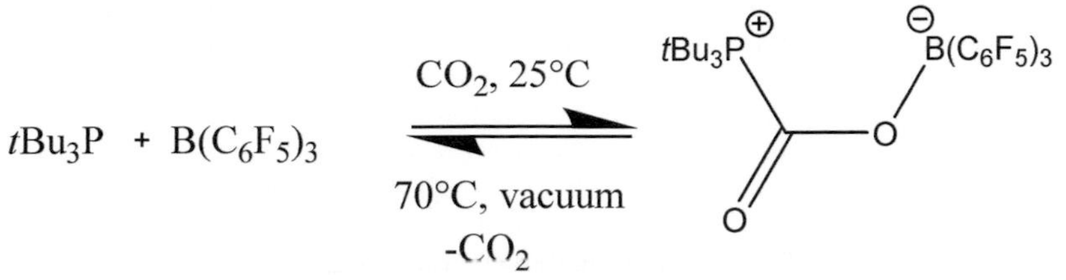

FIGURE 2.14 An example of frustrated Lewis pair concept gives alternative Lewis base (P) for CO_2 capture [302].

DOE) reference case 10 based on 30 wt% aqueous MEA. Relative to the reference case, the use of a CO_2BOL was estimated to reduce the parasitic energy demand on a power station by 16.7% for thermal desorption only and 43.3% when combined with polarity swing. The lower desorption operating temperature when using polarity swing could also reduce solvent attrition due to thermal degradation, although thermal and oxidative stability of CO_2BOLs is not well investigated.

2.4.2 FRUSTRATED LEWIS PAIRS

Another alternative chemistry is based around the concept of frustrated Lewis pairs [302–305]. They consist of a Lewis base and acid pair where the steric environment surrounding the reaction center prevents neutralization of the Lewis base by the Lewis acid. This leaves the Lewis base free to react with CO_2, and the resulting carbamate anion is stabilized by the Lewis acid. For carbon capture applications, the phosphine-borane pair have been investigated primarily with an example solution consisting of $B(C_6F_5)_3$ and $PtBu_3$ dissolved in C_6H_5Br, as shown in Figure 2.14. Exposing this solution to CO_2 causes a white solid to precipitate. The interaction of the phosphorous with the carbon of CO_2, and of the borane with a CO_2-based O, was confirmed by X-ray crystallography. This solid could be dissolved in bromobenzene and heated to 80 °C to return half of the captured CO_2. The use of frustrated Lewis pairs for CO_2 absorption is still at the proof of principle stage, but it offers quite a novel chemistry concept for reactive chemical absorption.

REFERENCES

1. R. R. Bottoms, *Process for Separating Acid Gases*, U. S. P. Office, 1930.
2. J. Rogelj *et al.*, "Mitigation pathways compatible with 1.5 °C in the context of sustainable development," in *Global Warming of 1.5 °C: An IPCC Special Report on the Impacts of Global Warming of 1.5 °C Above Pre-Industrial Levels and Related Global Greenhouse Gas Emission Pathways, in the Context of Strengthening the Global Response to the Threat of Climate Change, Sustainable Development, and Efforts to Eradicate Poverty*, V. Masson-Delmotte *et al.*, Eds. The Intergovernmental Panel on Climate Change, 2018, In Press.

3. X. G. Wang, W. Conway, R. Burns, N. McCann and M. Maeder, "Comprehensive study of the hydration and dehydration reactions of carbon dioxide in aqueous solution" (in English), *Journal of Physical Chemistry A*, vol. 114, no. 4, pp. 1734–1740, Feb. 4, 2010.

4. W. Conway *et al.*, "Toward the understanding of chemical absorption processes for post-combustion capture of carbon dioxide: Electronic and steric considerations from the kinetics of reactions of CO_2(aq) with sterically hindered amines" (in English), *Environmental Science & Technology*, vol. 47, no. 2, pp. 1163–1169, Jan. 15, 2013.

5. W. Conway *et al.*, "Toward rational design of amine solutions for PCC applications: The kinetics of the reaction of CO_2(aq) with cyclic and secondary amines in aqueous solution," *Environmental Science & Technology*, vol. 46, no. 13, pp. 7422–7429, July 2012.

6. A. Kohl and R. Nielsen, *Gas Purification*, 5th ed. Gulf Publishing Company, 1997.

7. N. McCann, M. Maeder and M. Attalla, "Simulation of enthalpy and capacity of CO_2 absorption by aqueous amine systems" (in English), *Industrial & Engineering Chemistry Research*, vol. 47, no. 6, pp. 2002–2009, Mar. 19, 2008.

8. E. Kessler *et al.*, "Speciation in CO_2-loaded aqueous solutions of sixteen triaceto-neamine-derivates (EvAs) and elucidation of structure-property relationships" (in English), *Chemical Engineering Science*, vol. 229, Jan. 16, 2021.

9. K. A. Mumford, Y. Wu, K. H. Smith and G. W. Stevens, "Review of solvent based carbon-dioxide capture technologies" (in English), *Frontiers of Chemical Science and Engineering*, vol. 9, no. 2, pp. 125–141, June 2015.

10. I. Folgueira, I. Teijido, A. Garcia-Abuin, D. Gomez-Diaz and A. Rumbo, "2-(Methylamino) ethanol for CO_2 absorption in a bubble reactor," *Energy & Fuels*, vol. 28, no. 7, pp. 4737–4745, July 2014.

11. H. A. M. Haider, R. Yusoff and M. K. Aroua, "Equilibrium solubility of carbon dioxide in 2(methylamino)ethanol," *Fluid Phase Equilibria*, vol. 303, no. 2, pp. 162–167, Apr. 2011.

12. Y. H. Jhon, J. G. Shim, J. H. Kim, J. H. Lee, K. R. Jang and J. Kim, "Nucleophilicity and accessibility calculations of alkanolamines: Applications to carbon dioxide absorption reactions," *Journal of Physical Chemistry A*, vol. 114, no. 49, pp. 12907–12913, Dec. 2010.

13. X. Luo *et al.*, "Density, viscosity, and N_2O solubility of aqueous 2-(Methylamino)ethanol solution," *Journal of Chemical and Engineering Data*, vol. 62, no. 1, pp. 129–140, Jan. 2017.

14. K. Maneeintr, R. Nimcharoen and T. Charinpanitkul, "Absorption process with aqueous solution of 2-(methylamino) ethanol for carbon dioxide removal from gas stream," in *International Conference on Sustainable Energy and Green Technology 2018*, vol. 268, C. W. Tong, W. ChinTsan and B. S. L. Huat, Eds. IOP Conference Series-Earth and Environmental Science, 2019.

15. R. Pacheco, A. Sanchez, M. D. La Rubia, A. B. Lopez, S. Sanchez and F. Camacho, "Thermal effects in the absorption of pure CO_2 into aqueous solutions of 2-methyl-amino-ethanol" (in English), *Industrial & Engineering Chemistry Research*, vol. 51, no. 13, pp. 4809–4818, Apr. 4, 2012.

16. D. Pandey and M. K. Mondal, "Experimental data and modeling for viscosity and refractive index of aqueous mixtures with 2-(Methylamino)ethanol (MAE) and amino-ethylethanolamine (AEEA)," *Journal of Chemical and Engineering Data*, vol. 64, no. 8, pp. 3346–3355, Aug. 2019.

17. D. Pandey and M. K. Mondal, "Equilibrium CO_2 solubility in the aqueous mixture of MAE and AEEA: Experimental study and development of modified thermodynamic model," *Fluid Phase Equilibria*, vol. 522, Nov. 2020, Art. no. 112766.

18. I. Folgueira, I. Teijido, A. Garcia-Abuin, D. Gomez-Diaz and A. Rumbo, "Carbon dioxide absorption behavior in 2-(ethylamino)ethanol aqueous solutions," *Fuel Processing Technology*, vol. 131, pp. 14–20, Mar. 2015.

19. K. Fu, P. Zhang and D. Fu, "Absorption capacity and CO_2 removal efficiency in tray tower by using 2-(ethylamino)ethanol activated 3-(dimethylamino)propan-1-ol aqueous solution," *Journal of Chemical Thermodynamics*, vol. 139, Dec. 2019, Art. no. 105862.

20. S. J. Hwang, J. Kim, H. Kim and K. S. Lee, "Solubility of carbon dioxide in aqueous solutions of three secondary amines: 2-(Butylamino)ethanol, 2-(Isopropylamino) ethanol, and 2-(Ethylamino)ethanol secondary alkanolamine solutions," *Journal of Chemical and Engineering Data*, vol. 62, no. 8, pp. 2428–2435, Aug. 2017.

21. P. N. Sutar, A. Jha, P. D. Vaidya and E. Y. Kenig, "Secondary amines for CO$_2$ capture: A kinetic investigation using N-ethylmonoethanolamine" (in English), *Chemical Engineering Journal*, vol. 207, pp. 718–724, Oct. 1, 2012.

22. M. Xiao, H. L. Liu, J. L. Wang, X. Luo, H. X. Gao and Z. W. Liang, "An experimental and modeling study of physical N$_2$O solubility in 2-(ethylamino)ethanol," *Journal of Chemical Thermodynamics*, vol. 138, pp. 34–42, Nov. 2019.

23. H. Yamada, F. A. Chowdhury, Y. Matsuzaki, K. Goto, T. Higashii and S. Kazama, "Effect of alcohol chain length on carbon dioxide absorption into aqueous solutions of alkanolamines," in *GHGT-11*, vol. 37, T. Dixon and K. Yamaji, Eds. Energy Procedia, 2013, pp. 499–504.

24. H. Yamada, Y. Matsuzaki and K. Goto, "Quantitative spectroscopic study of equilibrium in co2-loaded aqueous 2-(ethylamino)ethanol solutions," *Industrial & Engineering Chemistry Research*, vol. 53, no. 4, pp. 1617–1623, Jan. 2014.

25. C. Y. Zhu, X. J. Liu, T. T. Fu, X. Q. Gao and Y. G. Ma, "Density, viscosity and excess properties of N, N-dimethylethanolamine+2-(ethylamino) ethanol + H$_2$O at T = (293.15 to 333.15) K," *Journal of Molecular Liquids*, vol. 319, Dec. 2020, Art. no. 114095.

26. K. Goto, S. Kodama, T. Higashii and H. Kitamura, "Evaluation of amine-based solvent for post-combustion capture of carbon dioxide," *Journal of Chemical Engineering of Japan*, vol. 47, no. 8, pp. 663–665, Aug. 2014.

27. J. Kim, J. Lee, Y. Lee, H. Kim, E. Kim and K. S. Lee, "Evaluation of aqueous polyamines as CO$_2$ capture solvents," *Energy*, vol. 187, Nov. 2019, Art. no. 115908.

28. P. Biernacki, S. Steinigeweg, W. Paul and A. Brehm, "Eco-efficiency analysis of biomethane production," *Industrial & Engineering Chemistry Research*, vol. 53, no. 50, pp. 19594–19599, Dec. 2014.

29. X. Chen, F. Closmann and G. T. Rochelle, "Accurate screening of amines by the Wetted Wall Column," in *10th International Conference on Greenhouse Gas Control Technologies*, vol. 4, J. Gale, C. Hendriks and W. Turkenberg, Eds. Energy Procedia, 2011, pp. 101–108.

30. O. Dixit and N. Mollekopf, "Designing absorption processes with aqueous diglycolamine," *Chemical Engineering & Technology*, vol. 37, no. 9, pp. 1583–1592, Sept. 2014.

31. S. A. Freeman, J. Davis and G. T. Rochelle, "Degradation of aqueous piperazine in carbon dioxide capture," *International Journal of Greenhouse Gas Control*, vol. 4, no. 5, pp. 756–761, Sept. 2010.

32. A. Kazemi, A. K. Joujili, A. Mehrabani-Zeinabad, Z. Hajian and R. Salehi, "Influence of CO$_2$ residual of regenerated amine on the performance of natural gas sweetening processes using alkanolamine solutions," *Energy & Fuels*, vol. 30, no. 5, pp. 4263–4273, May 2016.

33. V. Kubacz and H. Fahlenkamp, "Improved procedure for the evaluation of the operating suitability of an amine based process for CO$_2$ removal of flue gases generated by coal-fired power plants," in *10th International Conference on Greenhouse Gas Control Technologies*, vol. 4, J. Gale, C. Hendriks, and W. Turkenberg, Eds. Energy Procedia, 2011, pp. 2564–2571.

34. B. Morero, E. S. Groppelli and E. A. Campanella, "Evaluation of biogas upgrading technologies using a response surface methodology for process simulation," *Journal of Cleaner Production*, vol. 141, pp. 978–988, Jan. 2017.

35. A. Nuchitprasittichai and S. Cremaschi, "Optimization of CO$_2$ capture process with aqueous amines – A comparison of two simulation-optimization approaches," *Industrial & Engineering Chemistry Research*, vol. 52, no. 30, pp. 10236–10243, July 2013.

36. A. T. Zoghi, F. Feyzi and S. Zarrinpashneh, "Experimental investigation on the effect of addition of amine activators to aqueous solutions of N-methyldiethanolamine on the rate of carbon dioxide absorption," *International Journal of Greenhouse Gas Control*, vol. 7, pp. 12–19, Mar. 2012.

37. C. Dell'Era, P. Uusi-Kyyny, J. P. Pokki, M. Pakkanen and V. Alopaeus, "Solubility of carbon dioxide in aqueous solutions of diisopropanolamine and methyldiethanolamine," *Fluid Phase Equilibria*, vol. 293, no. 1, pp. 101–109, June 2010.

38. A. Haghtalab and A. Afsharpour, "Solubility of CO$_2$ + H$_2$S gas mixture into different aqueous N-methyldiethanolamine solutions blended with 1-butyl-3-methylimidazolium acetate ionic liquid," *Fluid Phase Equilibria*, vol. 406, pp. 10–20, Nov. 2015.

39. A. Haghtalab, H. Eghbali and A. Shojaeian, "Experiment and modeling solubility of CO$_2$ in aqueous solutions of diisopropanolamine+2-amino-2-methyl-1-propanol + piperazine at high pressures," *Journal of Chemical Thermodynamics*, vol. 71, pp. 71–83, Apr. 2014.

40. A. Haghtalab and V. Gholami, "Carbon dioxide solubility in the aqueous mixtures of diisopropanolamine plus L-arginine and diethanolamine plus L-arginine at high pressures," *Journal of Molecular Liquids*, vol. 288, Aug. 2019, Art. no. 111064.

41. A. Haghtalab and A. Izadi, "Simultaneous measurement solubility of carbon dioxide plus hydrogen sulfide into aqueous blends of alkanolamines at high pressure," *Fluid Phase Equilibria*, vol. 375, pp. 181–190, Aug. 2014.

42. A. Haghtalab and A. Izadi, "Solubility and thermodynamic modeling of hydrogen sulfide in aqueous (diisopropanolamine+2-amino-2-methyl-1-propanol + piperazine) solution at high pressure," *Journal of Chemical Thermodynamics*, vol. 90, pp. 106–115, Nov. 2015.

43. A. Haghtalab and M. B. Z. Talavaki, "Measurement of carbon dioxide solubility in aqueous diisopropanolamine solutions blended by N-(2-aminoethyl) ethanolamine piperazine and density measurement of solutions," *Journal of Natural Gas Science and Engineering*, vol. 46, pp. 242–250, Oct. 2017.

44. J. Kim, J. Na and H. Y. Shin, "Measurement and correlation of density and excess volume for water + DIPA, DIPA + MDEA and water + DIPA + MDEA systems," *Korean Chemical Engineering Research*, vol. 57, no. 2, pp. 198–204, Apr. 2019.

45. A. B. Lopez, M. D. La Rubia, R. Pacheco, S. Sanchez, J. M. Navaza and D. Gomez-Diaz, "Carbon dioxide absorption by aqueous mixtures of diisopropanolamine and triethanolamine," *Chemical Engineering and Processing-Process Intensification*, vol. 110, pp. 73–79, Dec. 2016.

46. A. B. Lopez, R. Pacheco, M. D. La Rubia, A. Sanchez, S. Sanchez and F. Camacho, "Kinetics of the absorption of pure CO$_2$ by mixtures of diisopropanolamine and triethanolamine in aqueous solution," *International Journal of Chemical Kinetics*, vol. 49, no. 6, pp. 398–408, June 2017.

47. C. Mendez-Alvarez, V. Plesu, A. E. B. Ruiz, J. B. Ruiz, P. Iancu and J. Llorens, "Distillation energy assessment for solvent recovery from carbon dioxide absorption," in *26th European Symposium on Computer Aided Process Engineering*, vol. 38B, Z. Kravanja and M. Bogataj, Eds. Computer Aided Chemical Engineering, 2016, pp. 1917–1922.

48. J. Na, B. M. Min, J. H. Moon, J. S. Lee and H. Y. Shin, "Isothermal vapor-liquid equilibrium data for water plus diisopropanolamine, water plus N-methyldiethanolamine and diisopropanolamine plus N-methyldiethanolamine systems," *International Journal of Thermophysics*, vol. 39, no. 8, Aug. 2018, Art. no. 96.

49. M. S. Ojala, N. F. Serrano, P. Uusi-Kyyny and V. Alopaeus, "Comparative study: Absorption enthalpy of carbon dioxide into aqueous diisopropanolamine and monoethanolamine solutions and densities of the carbonated amine solutions," *Fluid Phase Equilibria*, vol. 376, pp. 85–95, Aug. 2014.

50. A. Padurean, C. C. Cormos, A. M. Cormos and P. S. Agachi, "Technical assessment of CO$_2$ capture using alkanolamines solutions," *Studia Universitatis Babes-Bolyai Chemia*, vol. 55, no. 1, pp. 55–63, 2010.

51. A. Penttila, C. Dell'Era, P. Uusi-Kyyny and V. Alopaeus, "The Henry's law constant of N$_2$O and CO$_2$ in aqueous binary and ternary amine solutions (MEA, DEA, DIPA, MDEA, and AMP)," *Fluid Phase Equilibria*, vol. 311, pp. 59–66, Dec. 2011.

52. A. Shojaeian and A. Haghtalab, "Solubility and density of carbon dioxide in different aqueous alkanolamine solutions blended with 1-butyl-3-methylimidazolium acetate ionic liquid at high pressure," *Journal of Molecular Liquids*, vol. 187, pp. 218–225, Nov. 2013.

53. M. Shokouhi, A. H. Jalili, F. Samani and M. Hosseini-Jenab, "Experimental investigation of the density and viscosity of CO$_2$-loaded aqueous alkanolamine solutions," *Fluid Phase Equilibria*, vol. 404, pp. 96–108, Oct. 2015.

54. V. D. Spasojevic, S. P. Šerbanovic, B. D. Djordjevic and M. L. Kijevčanin, "Densities, viscosities, and refractive indices of aqueous alkanolamine solutions as potential carbon dioxide removal reagents," *Journal of Chemical and Engineering Data*, vol. 58, no. 1, pp. 84–92, Jan. 2013.

55. L. Zong and C. C. Chen, "Thermodynamic modeling of CO$_2$ and H$_2$S solubilities in aqueous DIPA solution, aqueous sulfolane-DIPA solution, and aqueous sulfolane-MDEA solution with electrolyte NRTL model," *Fluid Phase Equilibria*, vol. 306, no. 2, pp. 190–203, July 2011.

56. F. Bougie and M. C. Iliuta, "CO$_2$ absorption into mixed aqueous solutions of 2-amino-2-hydroxymethyl-1,3-propanediol and piperazine," *Industrial & Engineering Chemistry Research*, vol. 49, no. 3, pp. 1150–1159, Feb. 2010.

57. F. Bougie and M. C. Iliuta, "Analysis of regeneration of sterically hindered alkanolamines aqueous solutions with and without activator," *Chemical Engineering Science*, vol. 65, no. 16, pp. 4746–4750, Aug. 2010.

58. E. S. Espiritu, A. N. Soriano and M. H. Li, "Thermophysical property characterization of ternary system containing (glycol (DEG/TEG/T(4)EG)+2-amino-2-hydroxymethyl-1,3-propanediol plus water)," *Journal of Chemical Thermodynamics*, vol. 59, pp. 121–126, Apr. 2013.

59. M. Ghulam, S. A. Mohd and B. M. Azmi, "Solubility of carbon dioxide in aqueous solution of 2-amino-2-hydroxymethyl-1, 3-propanediol at elevated pressures," *Research Journal of Chemistry and Environment*, vol. 17, no. 10, pp. 41–45, Oct. 2013.

60. R. Muraleedharan, A. Mondal and B. Mandal, "Absorption of carbon dioxide into aqueous blends of 2-amino-2-hydroxymethyl-1,3-propanediol and monoethanolamine," *Separation and Purification Technology*, vol. 94, pp. 92–96, June 2012.

61. G. Murshid, S. A. Mohd, K. L. Kok, B. M. Azmi and A. Faizan, "Thermophysical analysis of aqueous solutions of 2-amino-2-hydroxymethyl-1, 3-propanediol (Potential CO$_2$ removal solvent from gaseous streams)," *Research Journal of Chemistry and Environment*, vol. 15, no. 2, pp. 819–822, June 2011.

62. G. Murshid, A. M. Shariff, K. K. Lau, M. A. Bustam and F. Ahmad, "Physical properties and thermal decomposition of aqueous solutions of 2-amino-2-hydroxymethyl-1, 3-propanediol (AHPD)," *International Journal of Thermophysics*, vol. 32, no. 10, pp. 2040–2049, Oct. 2011.

63. G. Murshid, A. M. Shariff, K. K. Lau, M. A. Bustam and F. Ahmad, "Physical properties of piperazine (PZ) activated aqueous solutions of 2-amino-2-hydroxymethyl-1,3-propanediol (AHPD plus PZ)," *Journal of Chemical and Engineering Data*, vol. 57, no. 1, pp. 133–136, Jan. 2012.

64. R. Oktavian, M. Taha and M. J. Lee, "Experimental and computational study of CO$_2$ storage and sequestration with aqueous 2-amino-2-hydroxymethyl-1,3-propanediol (TRIS) solutions," *Journal of Physical Chemistry A*, vol. 118, no. 49, pp. 11572–11582, Dec. 2014.

65. L. Rodier, K. Ballerat-Busserolles and J. Y. Coxam, "Enthalpy of absorption and limit of solubility of CO_2 in aqueous solutions of 2-amino-2-hydroxymethyl-1,3-propanediol, 2–2-(dimethyl-amino)ethoxy ethanol, and 3-dimethyl-amino-1-propanol at T = (313.15 and 353.15) K and pressures up to 2 MPa," *Journal of Chemical Thermodynamics*, vol. 42, no. 6, pp. 773–780, June 2010.

66. C. S. Ume and E. Alper, "Reaction kinetics of carbon dioxide with 2-amino-2-hydroxy-methyl-1,3-propanediol in aqueous solution obtained from the stopped flow method," *Turkish Journal of Chemistry*, vol. 36, no. 3, pp. 427–435, 2012.

67. C. S. Ume, M. C. Ozturk and E. Alper, "Kinetics of CO_2 absorption by a blended aqueous amine solution," *Chemical Engineering & Technology*, vol. 35, no. 3, pp. 464–468, Mar. 2012.

68. W. Conway *et al.*, "CO_2 absorption into aqueous amine blended solutions containing monoethanolamine (MEA), N,N-dimethylethanolamine (DMEA), N,N-diethylethanolamine (DEEA) and 2-amino-2-methyl-1-propanol (AMP) for post-combustion capture processes," *Chemical Engineering Science*, vol. 126, pp. 446–454, Apr. 2015.

69. H. X. Gao, Z. Y. Wu, H. Liu, X. Luo and Z. W. Liang, "Experimental studies on the effect of tertiary amine promoters in aqueous monoethanolamine (MEA) solutions on the absorption/stripping performances in post-combustion CO_2 capture," *Energy & Fuels*, vol. 31, no. 12, pp. 13883–13891, Dec. 2017.

70. Z. Idris, J. Chen and D. A. Eimer, "Densities of aqueous 2-dimethylaminoethanol solutions at temperatures of (293.15 to 343.15) K," *Journal of Chemical and Engineering Data*, vol. 62, no. 3, pp. 1076–1082, Mar. 2017.

71. W. S. Jiang *et al.*, "A comparative kinetics study of CO_2 absorption into aqueous DEEA/MEA and DMEA/MEA blended solutions," *AIChE Journal*, vol. 64, no. 4, pp. 1350–1358, Apr. 2018.

72. H. Ling, H. X. Gao and Z. W. Liang, "Comprehensive solubility of N_2O and mass transfer studies on an effective reactive N,N-dimethylethanolamine (DMEA) solvent for post-combustion CO_2 capture," *Chemical Engineering Journal*, vol. 355, pp. 369–379, Jan. 2019.

73. H. Ling, S. Liu, H. X. Gao, H. Y. Zhang and Z. W. Liang, "Solubility of N_2O, equilibrium solubility, mass transfer study and modeling of CO_2 absorption into aqueous monoethanolamine (MEA)/1-dimethylamino-2-propanol (1DMA2P) solution for post-combustion CO_2 capture," *Separation and Purification Technology*, vol. 232, Feb. 2020, Art. no. 115957.

74. H. Ling, S. Liu, T. Y. Wang, H. X. Gao and Z. W. Liang, "Characterization and correlations of CO_2 absorption performance into aqueous amine blended solution of monoethanolamine (MEA) and N,N-dimethylethanolamine (DMEA) in a packed column," *Energy & Fuels*, vol. 33, no. 8, pp. 7614–7625, Aug. 2019.

75. D. D. D. Pinto, J. Monteiro, B. Johnsen, H. F. Svendsen and H. Knuutila, "Density measurements and modelling of loaded and unloaded aqueous solutions of MDEA (N-methyldiethanolamine), DMEA (N,N-dimethylethanolamine), DEEA (diethylethanolamine) and MAPA (N-methyl-1,3-diaminopropane)," *International Journal of Greenhouse Gas Control*, vol. 25, pp. 173–185, June 2014.

76. C. Tong, C. C. Perez, J. Chen, J. C. V. Marcos, T. Neveux and Y. Le Moullec, "Measurement and calculation for CO_2 solubility and kinetic rate in aqueous solutions of two tertiary amines," in *GHGT-11*, vol. 37, T. Dixon and K. Yamaji, Eds. Energy Procedia, 2013, pp. 2084–2093.

77. M. W. Arshad, P. L. Fosbøl, N. von Solms, H. F. Svendsen and K. Thomsen, "Equilibrium solubility of CO_2 in alkanolamines," in *7th Trondheim Conference on CO_2 Capture, Transport and Storage*, vol. 51, N. A. Rokke and H. Svendsen, Eds. Energy Procedia, 2014, pp. 217–223.

78. A. Baltar, D. Gomez-Diaz, J. M. Navaza and A. Rumbo, "Absorption and regeneration studies of chemical solvents based on dimethylethanolamine and diethylethanolamine for carbon dioxide capture," *AIChE Journal*, vol. 66, no. 1, Jan. 2020, Art. no. e16770.

79. I. M. Bernhardsen, I. R. T. Krokvik, C. Perinu, D. D. D. Pinto, K. J. Jens and H. K. Knuutila, "Influence of pKa on solvent performance of MAPA promoted tertiary amines," *International Journal of Greenhouse Gas Control*, vol. 68, pp. 68–76, Jan. 2018.

80. D. Fu, L. M. Wang and X. F. Tian, "Experiments and model for the surface tension of DEAE-PZ and DEAE-MEA aqueous solutions," *Journal of Chemical Thermodynamics*, vol. 105, pp. 71–75, Feb. 2017.

81. H. X. Gao, Z. W. Liang, H. Y. Liao and R. O. Idem, "Thermal degradation of aqueous DEEA solution at stripper conditions for post-combustion CO$_2$ capture," *Chemical Engineering Science*, vol. 135, pp. 330–342, Oct. 2015.

82. H. X. Gao, B. Xu, L. Han, X. Luo and Z. W. Liang, "Mass transfer performance and correlations for CO$_2$ absorption into aqueous blended of DEEA/MEA in a random packed column," *AIChE Journal*, vol. 63, no. 7, pp. 3048–3057, July 2017.

83. H. X. Gao, B. Xu, H. L. Liu and Z. W. Liang, "Effect of amine activators on aqueous N,N-diethylethanolamine solution for postcombustion CO$_2$ capture," *Energy & Fuels*, vol. 30, no. 9, pp. 7481–7488, Sept. 2016.

84. M. Garcia, H. K. Knuutila and S. Gu, "Thermodynamic modelling of unloaded and loaded N,N-diethylethanolamine solutions," *Green Energy & Environment*, vol. 1, no. 3, pp. 246–257, Oct. 2016.

85. H. Kierzkowska-Pawlak, "Kinetics of CO$_2$ absorption in aqueous N,N-diethylethanolamine and its blend with N-(2-aminoethyl)ethanolamine using a stirred cell reactor," *International Journal of Greenhouse Gas Control*, vol. 37, pp. 76–84, June 2015.

86. H. Kierzkowska-Pawlak and E. Kruszczak, "Revised kinetics of CO$_2$ absorption in aqueous N,N-diethylethanolamine (DEEA) and its blend with N-methyl-1,3-propane diamine (MAPA)," *International Journal of Greenhouse Gas Control*, vol. 57, pp. 134–142, Feb. 2017.

87. H. Kierzkowska-Pawlak and K. Sobala, "Heat of absorption of CO$_2$ in aqueous solutions of DEEA and DEEA plus MAPA blends-a new approach to measurement methodology," *International Journal of Greenhouse Gas Control*, vol. 100, Sept. 2020, Art. no. 103102.

88. I. Kim and H. F. Svendsen, "Comparative study of the heats of absorption of postcombustion CO$_2$ absorbents," *International Journal of Greenhouse Gas Control*, vol. 5, no. 3, pp. 390–395, May 2011.

89. H. K. Knuutila, R. Rennemo and A. F. Ciftja, "New solvent blends for post-combustion CO$_2$ capture," *Green Energy & Environment*, vol. 4, no. 4, pp. 439–452, Oct. 2019.

90. E. Kruszczak and H. Kierzkowska-Pawlak, "CO$_2$ capture by absorption in activated aqueous solutions of n,n-diethylethanoloamine," *Ecological Chemistry and Engineering S-Chemia I Inzynieria Ekologiczna S*, vol. 24, no. 2, pp. 239–248, June 2017.

91. E. Kruszczak and H. Kierzkowska-Pawlak, "CO$_2$ absorption into aqueous solutions of N-methyl-1,3-propane-diamine and its blends with N,N-diethylethanolamine-new kinetic data," *International Journal of Energy Research*, vol. 45, no. 3, pp. 4098–4111, Mar, 2021.

92. S. Kumar and M. K. Mondal, "Equilibrium solubility of CO$_2$ in aqueous blend of 2-(Diethylamine)ethanol and 2-(2-Aminoethylamine)ethanol," *Journal of Chemical and Engineering Data*, vol. 63, no. 5, pp. 1163–1169, May 2018.

93. S. Kumar and M. K. Mondal, "Equilibrium solubility of CO$_2$ in aqueous binary mixture of 2-(diethylamine)ethanol and 1, 6-hexamethyldiamine," *Korean Journal of Chemical Engineering*, vol. 35, no. 6, pp. 1335–1340, June 2018.

94. Y. Y. Li, C. J. Liu, R. Parnas, Y. Y. Liu, B. Liang and H. F. L. Lu, "The CO_2 absorption and desorption performance of the triethylenetetramine plus N,N-diethylethanolamine + H-2 O system," *Chinese Journal of Chemical Engineering*, vol. 26, no. 11, pp. 2351–2360, Nove 2018.

95. D. F. Ma, C. Y. Zhu, T. G. Fu, X. G. Yuan and Y. G. Ma, "An effective hybrid solvent of MEA/DEEA for CO_2 absorption and its mass transfer performance in microreactor," *Separation and Purification Technology*, vol. 242, July 2020, Art. no. 116795.

96. D. F. Ma, C. Y. Zhu, T. T. Fu, X. G. Yuan and Y. G. Ma, "Volumetric and viscometric properties for binary and ternary solutions of diethylenetriamine, N,N-diethylethanolamine, and water," *Journal of Chemical and Engineering Data*, vol. 65, no. 1, pp. 239–254, Jan. 2020.

97. J. Monteiro, H. Knuutila, N. Penders-van Elk, G. Versteeg and H. F. Svendsen, "Kinetics of CO_2 absorption by aqueous N,N-diethylethanolamine solutions: Literature review, experimental results and modelling," *Chemical Engineering Science*, vol. 127, pp. 1–12, May 2015.

98. J. Monteiro, H. Majeed, H. Knuutila and H. F. Svendsen, "Kinetics of CO_2 absorption in aqueous blends of N,N-diethylethanolamine (DEEA) and N-methyl-1,3-propanediamine (MAPA)," *Chemical Engineering Science*, vol. 129, pp. 145–155, June 2015.

99. J. Monteiro, D. D. D. Pinto, S. A. H. Zaidy, A. Hartono and H. F. Svendsen, "VLE data and modelling of aqueous N,N-diethylethanolamine (DEEA) solutions," *International Journal of Greenhouse Gas Control*, vol. 19, pp. 432–440, Nov. 2013.

100. S. Mouhoubi, L. Dubois, P. L. Fosbøl, G. De Weireld and D. Thomas, "Thermodynamic modeling of CO_2 absorption in aqueous solutions of N,N-diethylethanolamine (DEEA) and N-methyl-1,3-propanediamine (MAPA) and their mixtures for carbon capture process simulation," *Chemical Engineering Research & Design*, vol. 158, pp. 46–63, June 2020.

101. C. F. Patzschke, J. F. Zhang, P. S. Fennell and J. P. M. Trusler, "Density and viscosity of partially carbonated aqueous solutions containing a tertiary alkanolamine and piperazine at temperatures between 298.15 and 353.15 K," *Journal of Chemical and Engineering Data*, vol. 62, no. 7, pp. 2075–2083, July 2017.

102. K. Sobala and H. Kierzkowska-Pawlak, "Heat of absorption of CO_2 in aqueous N,N-diethylethanolamine plus N-methyl-1,3-propanediamine solutions at 313 K," *Chinese Journal of Chemical Engineering*, vol. 27, no. 3, pp. 628–633, Mar. 2019.

103. A. Verma, P. Kumar, A. Bindwal and S. Paul, "Polyamine-promoted aqueous DEEA for CO_2 capture: An experimental analysis," *Indian Journal of Chemical Technology*, vol. 26, no. 5, pp. 411–417, Sept. 2019.

104. T. Wang, F. Liu, K. Ge and M. X. Fang, "Reaction kinetics of carbon dioxide absorption in aqueous solutions of piperazine, N-(2-aminoethyl) ethanolamine and their blends," *Chemical Engineering Journal*, vol. 314, pp. 123–131, Apr. 2017.

105. B. Xu, H. X. Gao, M. L. Chen, Z. W. Liang and R. Idem, "Experimental study of regeneration performance of aqueous N,N-diethylethanolamine solution in a column packed with Dixon ring random packing," *Industrial & Engineering Chemistry Research*, vol. 55, no. 31, pp. 8519–8526, Aug. 2016.

106. Z. C. Xu, S. J. Wang and C. H. Chen, "Solubility of N_2O in density and viscosity of aqueous solutions of 1,4-butanediamine, 2-(Diethylamino)-ethanol, and their mixtures from (298.15 to 333.15) K," *Journal of Chemical and Engineering Data*, vol. 58, no. 6, pp. 1633–1640, June 2013.

107. Z. C. Xu, S. J. Wang and C. H. Chen, "Kinetics study on CO_2 absorption with aqueous solutions of 1,4-butanediamine, 2-(Diethylamino)-ethanol, and their mixtures," *Industrial & Engineering Chemistry Research*, vol. 52, no. 29, pp. 9790–9802, July 2013.

108. Z. C. Xu, S. J. Wang, G. J. Qi, A. A. Trollebo, H. F. Svendsen and C. H. Chen, "Vapor liquid equilibria and heat of absorption of CO_2 in aqueous 2-(diethylamino)-ethanol solutions," *International Journal of Greenhouse Gas Control*, vol. 29, pp. 92–103, Oct. 2014.

109. M. Afkhamipour and M. Mofarahi, "Rate-based modeling and sensitivity analysis of a packed column for post-combustion CO_2 capture into a novel reactive 1-dimethyl-amino-2-propanol (1DMA2P) solution," *International Journal of Greenhouse Gas Control*, vol. 65, pp. 137–148, Oct. 2017.

110. M. Afkhamipour and M. Mofarahi, "A modeling-optimization framework for assess-ment of CO_2 absorption capacity by novel amine solutions: 1DMA2P, 1DEA2P, DEEA, and DEAB," *Journal of Cleaner Production*, vol. 171, pp. 234–249, Jan. 2018.

111. M. Afkhamipour and M. Mofarahi, "Experimental measurement and modeling study on CO_2 equilibrium solubility, density and viscosity for 1-dimethylamino-2-propanol (1DMA2P) solution," *Fluid Phase Equilibria*, vol. 457, pp. 38–51, Feb. 2018.

112. A. Dey, S. K. Dash and B. Mandal, "Elucidating the performance of (N-(3-aminopropyl)-1, 3-propanediamine) activated (1-dimethylamino-2-propanol) as a novel amine formulation for post combustion carbon dioxide capture," *Fuel*, vol. 277, Oct. 2020, Art. no. 118209.

113. D. Fu, L. M. Wang and X. F. Tian, "Surface thermodynamics of DMA2P, DMA2P-MEA and DMA2P-PZ aqueous solutions," *Journal of Chemical Thermodynamics*, vol. 107, pp. 79–84, Apr. 2017.

114. H. X. Gao, N. Wang, J. G. Du, X. Luo and Z. W. Liang, "Comparative kinetics of car-bon dioxide (CO_2) absorption into EAE, 1DMA2P and their blends in aqueous solution using the stopped-flow technique," *International Journal of Greenhouse Gas Control*, vol. 94, Mar. 2020, Art. no. 102948.

115. Z. Idris, J. Chen and D. A. Eimer, "Densities of unloaded and CO_2-loaded 3-dimeth-ylamino-1-propanol at temperatures (293.15 to 343.15) K," *Journal of Chemical Thermodynamics*, vol. 97, pp. 289–296, June 2016.

116. Y. J. Liang, H. L. Liu, W. Rongwong, Z. W. Liang, R. Idem and P. Tontiwachiwuthikul, "Solubility, absorption heat and mass transfer studies of CO_2 absorption into aqueous solution of 1-dimethylamino-2-propanol," *Fuel*, vol. 144, pp. 121–129, Mar. 2015.

117. H. L. Liu, H. X. Gao, R. Idem, P. Tontiwachiwuthikul and Z. W. Liang, "Analysis of CO_2 solubility and absorption heat into 1-dimethylamino-2-propanol solution," *Chemical Engineering Science*, vol. 170, pp. 3–15, Oct. 2017.

118. H. L. Liu, R. Idem and P. Tontiwachiwuthikul, "Novel models for correlation of Solubility constant and diffusivity of N_2O in aqueous 1-dimethylamino-2-propanol," *Chemical Engineering Science*, vol. 203, pp. 86–103, Aug. 2019.

119. H. L. Liu, R. Idem, P. Tontiwachiwuthikul and Z. W. Liang, "Study of ion speciation of CO_2 absorption into aqueous 1-dimethylamino-2-propanol solution using the NMR technique," *Industrial & Engineering Chemistry Research*, vol. 56, no. 30, pp. 8697–8704, Aug. 2017.

120. H. L. Liu *et al.*, "Solubility, kinetics, absorption heat and mass transfer studies of CO_2 absorption into aqueous solution of 1-Dimethylamino-2-propanol," in *12th International Conference on Greenhouse Gas Control Technologies, GHGT-12*, vol. 63, T. Dixon, H. Herzog and S. Twinning, Eds. Energy Procedia, 2014, pp. 659–664.

121. H. L. Liu, X. Luo, Z. W. Liang and P. Tontiwachiwuthikul, "Determination of vapor-liquid equilibrium (VLE) plots of 1-dimethylamino-2-propanol solutions using the pH method," *Industrial & Engineering Chemistry Research*, vol. 54, no. 17, pp. 4709–4716, May 2015.

122. A. V. Rayer and A. Henni, "Heats of absorption of CO_2 in aqueous solutions of ter-tiary amines: N-methyldiethanolamine, 3-dimethylamino-1-propanol, and 1-dimeth-ylamino-2-propanol," *Industrial & Engineering Chemistry Research*, vol. 53, no. 12, pp. 4953–4965, Mar. 2014.

123. A. Sodiq, N. El Hadri, E. L. V. Goetheer and M. R. M. Abu-Zahra, "Chemical reaction kinet-ics measurements for single and blended amines for CO_2 postcombustion capture applica-tions," *International Journal of Chemical Kinetics*, vol. 50, no. 9, pp. 615–632, Sept. 2018.

124. L. M. Wang, X. F. Tian, D. Fu, X. Q. Du and J. H. Ye, "Experimental investigation on CO_2 absorption capacity and viscosity for high concentrated 1-dimethylamino-2-propanol-monoethanolamine aqueous blends," *Journal of Chemical Thermodynamics*, vol. 139, Dec. 2019, Art. no. 105865.

125. L. Wen *et al.*, "Comparison of overall gas-phase mass transfer coefficient for CO_2 absorption between tertiary amines in a randomly packed column," *Chemical Engineering & Technology*, vol. 38, no. 8, pp. 1435–1443, Aug. 2015.

126. P. Zhang, M. Y. Li, C. J. Lv, Y. J. Zhang, L. M. Wang and D. Fu, "Effect of partial pressure on CO_2 absorption performance in piperazine promoted 2-diethylaminoethanol and 1-dimethylamino-2-propanol aqueous solutions," *Journal of Chemical Thermodynamics*, vol. 150, Nov. 2020, Art. no. 106198.

127. H. L. Liu, M. Xiao, Z. W. Liang and P. Tontiwachiwuthikul, "The analysis of solubility, absorption kinetics of CO_2 absorption into aqueous 1-diethylamino-2-propanol solution," *AIChE Journal*, vol. 63, no. 7, pp. 2694–2704, July 2017.

128. S. M. Melnikov and M. Stein, "Molecular dynamics study of the solution structure, clustering, and diffusion of four aqueous alkanolamines," *Journal of Physical Chemistry B*, vol. 122, no. 10, pp. 2769–2778, Mar. 2018.

129. S. M. Melnikov and M. Stein, "Solvation and dynamics of CO_2 in aqueous alkanolamine solutions," *ACS Sustainable Chemistry & Engineering*, vol. 7, no. 1, pp. 1028–1037, Jan. 2019.

130. S. M. Melnikov and M. Stein, "The effect of CO_2 loading on alkanolamine absorbents in aqueous solutions," *Physical Chemistry Chemical Physics*, vol. 21, no. 33, pp. 18386–18392, Sept. 2019.

131. T. Sema, M. Edali, A. Naami, R. Idem and P. Tontiwachiwuthikul, "Solubility and diffusivity of N_2O in aqueous 4-(diethylamino)-2-butanol solutions for use in postcombustion CO_2 capture," *Industrial & Engineering Chemistry Research*, vol. 51, no. 2, pp. 925–930, Jan. 2012.

132. T. Sema *et al.*, "Comprehensive mass transfer and reaction kinetics studies of a novel reactive 4-diethylamino-2-butanol solvent for capturing CO_2," *Chemical Engineering Science*, vol. 100, pp. 183–194, Aug. 2013.

133. T. Sema *et al.*, "A novel reactive 4-diethylamino-2-butanol solvent for capturing CO_2 in the aspect of absorption capacity, cyclic capacity, mass transfer, and reaction kinetics," in *GHGT-11*, vol. 37, T. Dixon and K. Yamaji, Eds. Energy Procedia, 2013, pp. 477–484.

134. T. Sema, A. Naami, Z. W. Liang, R. Idem, H. Ibrahim and P. Tontiwachiwuthikul, "1D absorption kinetics modeling of CO_2-DEAB-H_2O system," *International Journal of Greenhouse Gas Control*, vol. 12, pp. 390–398, Jan. 2013.

135. T. Zarogiannis, A. I. Papadopoulos and P. Seferlis, "Systematic selection of amine mixtures as post-combustion CO_2 capture solvent candidates," *Journal of Cleaner Production*, vol. 136, pp. 159–175, Nov. 2016.

136. A. F. Ciftja, A. Hartono and H. F. Svendsen, "C-13 NMR as a method species determination in CO_2 absorbent systems," *International Journal of Greenhouse Gas Control*, vol. 16, pp. 224–232, Aug. 2013.

137. T. Davran-Candan, "DFT modeling of CO_2 interaction with various aqueous amine structures," *Journal of Physical Chemistry A*, vol. 118, no. 25, pp. 4582–4590, June 2014.

138. A. Hafizi, M. H. Mokari, R. Khalifeh, M. Farsi and M. R. Rahimpour, "Improving the CO_2 solubility in aqueous mixture of MDEA and different polyamine promoters: The effects of primary and secondary functional groups," *Journal of Molecular Liquids*, vol. 297, Jan. 2020, Art. no. 111803.

139. Y. E. Kim, J. H. Choi, S. C. Nam and Y. I. Yoon, "Carbon dioxide absorption into aqueous blends of potassium carbonate and amine," *Asian Journal of Chemistry*, vol. 24, no. 8, pp. 3386–3390, Aug. 2012.

140. J. N. Knudsen, J. Andersen, J. N. Jensen and O. Biede, "Evaluation of process upgrades and novel solvents for the post combustion CO_2 capture process in pilot-scale" (in English), *10th International Conference on Greenhouse Gas Control Technologies*, vol. 4, pp. 1558–1565, 2011.

141. S. Kumar and M. K. Mondal, "Selection of efficient absorbent for CO_2 capture from gases containing low CO_2," *Korean Journal of Chemical Engineering*, vol. 37, no. 2, pp. 231–239, Feb. 2020.

142. S. Kumar, R. Padhan and M. K. Mondal, "Equilibrium solubility measurement and modeling of CO_2 absorption in aqueous blend of 2-(diethyl amino) ethanol and ethylenediamine," *Journal of Chemical and Engineering Data*, vol. 65, no. 2, pp. 523–531, Feb. 2020.

143. H. P. Mangalapally and H. Hasse, "Pilot plant study of two new solvents for post combustion carbon dioxide capture by reactive absorption and comparison to monoethanolamine" (in English), *Chemical Engineering Science*, vol. 66, no. 22, pp. 5512–5522, Nov. 15, 2011.

144. P. Muchan, J. Narku-Tetteh, C. Saiwan, R. Idem and T. Supap, "Effect of number of amine groups in aqueous polyamine solution on carbon dioxide (CO_2) capture activities," *Separation and Purification Technology*, vol. 184, pp. 128–134, Aug. 2017.

145. M. Rabensteiner, G. Kinger, M. Koller, G. Gronald and C. Hochenauer, "Pilot plant study of ethylenediamine as a solvent for post combustion carbon dioxide capture and comparison to monoethanolamine," *International Journal of Greenhouse Gas Control*, vol. 27, pp. 1–14, Aug. 2014.

146. F. Sha, T. X. Zhao, B. Guo, X. X. Ju, L. H. Li and J. B. Zhang, "Density, viscosity and spectroscopic studies of the binary system 1,2-ethylenediamine+1,4-butanediol at T = (293.15 to 318.15) K," *Journal of Molecular Liquids*, vol. 208, pp. 373–379, Aug. 2015.

147. R. E. Tataru-Farmus, M. Dragan, S. Dragan and I. Siminiceanu, "Kinetics of carbon dioxide absorption into new amine solutions," *Studia Universitatis Babes-Bolyai Chemia*, vol. 58, no. 4, pp. 113–120, Dec. 2013.

148. B. Yoon and G. S. Hwang, "On the mechanism of predominant urea formation from thermal degradation of CO_2 loaded aqueous ethylenediamine," *Physical Chemistry Chemical Physics*, vol. 22, no. 30, pp. 17336–17343, Aug. 2020.

149. S. Zhou, X. Chen, T. Nguyen, A. K. Voice and G. T. Rochelle, "Aqueous ethylenediamine for CO_2 capture," *Chemsuschem*, vol. 3, no. 8, pp. 913–918, 2010.

150. S. Dinda, V. V. Goud, A. V. Patwardhan and N. C. Pradhan, "Kinetics of reactive absorption of carbon dioxide with solutions of 1,6-hexamethylenediamine in polar protic solvents," *Separation and Purification Technology*, vol. 75, no. 1, pp. 1–7, Sept. 2010.

151. Y. E. Kim *et al.*, "Comparison of the CO_2 absorption characteristics of aqueous solutions of diamines: Absorption capacity, specific heat capacity, and heat of absorption," *Energy & Fuels*, vol. 29, no. 4, pp. 2582–2590, Apr. 2015.

152. B. K. Mondal, S. S. Bandyopadhyay and A. N. Samanta, "Vapor-liquid equilibrium measurement and ENRTL modeling of CO_2 absorption in aqueous hexamethylenediamine," *Fluid Phase Equilibria*, vol. 402, pp. 102–112, Sept. 2015.

153. B. K. Mondal, S. S. Bandyopadhyay and A. N. Samanta, "Kinetics of CO_2 absorption in aqueous hexamethylenediamine," *International Journal of Greenhouse Gas Control*, vol. 56, pp. 116–125, Jan. 2017.

154. B. K. Mondal, S. S. Bandyopadhyay and A. N. Samanta, "Equilibrium solubility measurement and Kent-Eisenberg modeling of CO_2 absorption in aqueous mixture of N-methyldiethanolamine and hexamethylenediamine," *Greenhouse Gases-Science and Technology*, vol. 7, no. 1, pp. 202–214, Feb. 2017.

155. B. K. Mondal, S. S. Bandyopadhyay and A. N. Samanta, "Experimental measurement and Kent-Eisenberg modelling of CO_2 solubility in aqueous mixture of 2-amino-2-methyl-1-propanol and hexamethylenediamine," *Fluid Phase Equilibria*, vol. 437, pp. 118–126, Apr. 2017.

156. B. K. Mondal, S. S. Bandyopadhyay and A. N. Samanta, "Measurement of CO_2 absorption enthalpy and heat capacity of aqueous hexamethylenediamine and its aqueous mixture with N-methyldiethanolamine," *Journal of Chemical Thermodynamics*, vol. 113, pp. 276–290, Oct. 2017.

157. B. K. Mondal, S. S. Bandyopadhyay and A. N. Samanta, "Kinetics of CO_2 absorption in aqueous hexamethylenediamine blended N-methyldiethanolamine," *Industrial & Engineering Chemistry Research*, vol. 56, no. 50, pp. 14902–14913, Dec. 2017.

158. B. K. Mondal and A. N. Samanta, "Equilibrium solubility and kinetics of CO_2 absorption in hexamethylenediamine activated aqueous sodium glycinate solvent," *Chemical Engineering Journal*, vol. 386, Apr. 2020, Art. no. 121462.

159. U. E. Aronu, A. Hartono and H. F. Svendsen, "Kinetics of carbon dioxide absorption into aqueous amine amino acid salt: 3-(methylamino)propylamine/sarcosine solution," *Chemical Engineering Science*, vol. 66, no. 23, pp. 6109–6119, Dec. 2011.

160. U. E. Aronu, A. Hartono and H. F. Svendsen, "Density, viscosity, and N_2O solubility of aqueous amino acid salt and amine amino acid salt solutions," *Journal of Chemical Thermodynamics*, vol. 45, no. 1, pp. 90–99, Feb. 2012.

161. U. E. Aronu, K. A. Hoff and H. F. Svendsen, "Vapor-liquid equilibrium in aqueous amine amino acid salt solution: 3-(methylamino)propylamine/sarcosine," *Chemical Engineering Science*, vol. 66, no. 17, pp. 3859–3867, Sept. 2011.

162. M. W. Arshad, H. F. Svendsen, P. L. Fosbøl, N. von Solms and K. Thomsen, "Equilibrium total pressure and CO_2 solubility in binary and ternary aqueous solutions of 2-(Diethylamino)ethanol (DEEA) and 3-(Methylamino)propylamine (MAPA)," *Journal of Chemical and Engineering Data*, vol. 59, no. 3, pp. 764–774, Mar. 2014.

163. I. M. Bernhardsen, A. A. Trollebo, C. Perinu and H. K. Knuutila, "Vapour-liquid equilibrium study of tertiary amines, single and in blend with 3-(methylamino)propylamine, for post-combustion CO_2 capture," *Journal of Chemical Thermodynamics*, vol. 138, pp. 211–228, Nov. 2019.

164. S. Y. Choi, S. C. Nam, Y. I. Yoon, K. T. Park and S. J. Park, "Carbon dioxide absorption into aqueous blends of methyldiethanolamine (MDEA) and alkyl amines containing multiple amino groups," *Industrial & Engineering Chemistry Research*, vol. 53, no. 37, pp. 14451–14461, Sept. 2014.

165. H. K. Knuutila and A. Nannestad, "Effect of the concentration of MAPA on the heat of absorption of CO_2 and on the cyclic capacity in DEEA-MAPA blends," *International Journal of Greenhouse Gas Control*, vol. 61, pp. 94–103, June 2017.

166. C. P. Liao, R. B. Leron and M. H. Li, "Mutual diffusion coefficients, density, and viscosity of aqueous solutions of new polyamine CO_2 absorbents," *Fluid Phase Equilibria*, vol. 363, pp. 180–188, Feb. 2014.

167. S. Y. Lin, R. B. Leron and M. H. Li, "Molar heat capacities of diethylenetriamine and 3-(methylamino)propylamine, their aqueous binaries, and aqueous ternaries with piperazine," *Thermochimica Acta*, vol. 575, pp. 34–39, Jan. 2014.

168. J. Monteiro *et al.*, "Kinetics of CO_2 absorption by aqueous 3-(methylamino)propylamine solutions: Experimental results and modeling," *AIChE Journal*, vol. 60, no. 11, pp. 3792–3803, Nov. 2014.

169. J. Monteiro *et al.*, "Activity-based kinetics of the reaction of carbon dioxide with aqueous amine systems. Case studies: MAPA and MEA," in *GHGT-11*, vol. 37, T. Dixon and K. Yamaji, Eds. Energy Procedia, 2013, pp. 1888–1896.

170. C. Perinu, I. M. Bernhardsen, D. D. D. Pinto, H. K. Knuutila and K. J. Jens, "NMR speciation of aqueous MAPA, tertiary amines, and their blends in the presence of CO_2: Influence of pK(a) and reaction mechanisms," *Industrial & Engineering Chemistry Research*, vol. 57, no. 5, pp. 1337–1349, Feb. 2018.

171. C. Perinu, I. M. Bernhardsen, D. D. D. Pinto, H. K. Knuutila and K. J. Jens, "Aqueous MAPA, DEEA, and their blend as CO_2 absorbents: Interrelationship between NMR

speciation, pH, and heat of absorption data," *Industrial & Engineering Chemistry Research*, vol. 58, no. 23, pp. 9781–9794, June 2019.

172. C. Perinu, I. M. Bernhardsen, H. F. Svendsen and K. J. Jens, "CO_2 capture by aqueous 3-(Methylamino) propylamine in blend with tertiary amines: An NMR analysis," in *13th International Conference on Greenhouse Gas Control Technologies, GHGT-13*, vol. 114, T. Dixon, L. Laloui and S. Twinning, Eds. Energy Procedia, 2017, pp. 1949–1955.

173. A. Rahimi, A. T. Zoghi, F. Feyzi and A. H. Jalili, "Experimental study of density, viscosity and equilibrium carbon dioxide solubility in some aqueous alkanolamine solutions," *Journal of Solution Chemistry*, vol. 48, no. 4, pp. 489–501, Apr. 2019.

174. A. K. Voice, S. J. Vevelstad, X. Chen, T. Nguyen and G. T. Rochelle, "Aqueous 3-(methylamino)propylamine for CO_2 capture," *International Journal of Greenhouse Gas Control*, vol. 15, pp. 70–77, July 2013.

175. A. T. Zoghi, A. Rahimi, F. Feyzi and A. H. Jalili, "Measuring and modeling equilibrium solubility of carbon dioxide in aqueous solution of dimethylaminoethanol and 3-methylaminopropylamine," *Thermochimica Acta*, vol. 686, Apr. 2020, Art. no. 178565.

176. H. H. Cheng, J. F. Shen and C. S. Tan, "CO_2 capture from hot stove gas in steel making process," *International Journal of Greenhouse Gas Control*, vol. 4, no. 3, pp. 525–531, May 2010.

177. C. Guo, S. Y. Chen and Y. C. Zhang, "Solubility of carbon dioxide in aqueous 2-(2-aminoethylamine)ethanol (AEEA) solution and its mixtures with N-methyldiethanolamine/2-Amino-2-methyl-1-propanol," *Journal of Chemical and Engineering Data*, vol. 58, no. 2, pp. 460–466, Feb. 2013.

178. H. Kierzkowska-Pawlak, A. Chacuk and M. Siemieniec, "Reaction kinetics of CO_2 in aqueous 2-(2-aminoethylamino)ethanol solutions using a stirred cell reactor," *International Journal of Greenhouse Gas Control*, vol. 24, pp. 106–114, May 2014.

179. S. Y. Wu *et al.*, "Mass-transfer performance for CO_2 absorption by 2-(2-Aminoethylamino) ethanol solution in a rotating packed bed," *Energy & Fuels*, vol. 31, no. 12, pp. 14053–14059, Dec. 2017.

180. S. Balchandani, B. Mandal and S. Dharaskar, "Measurements and modeling of vapor liquid equilibrium of CO_2 in amine activated imidazolium ionic liquid solvents," *Fluid Phase Equilibria*, vol. 521, Oct. 2020, Art. no. 112643.

181. B. Das, B. Deogam, Y. Agrawal and B. Mandal, "Measurement and correlation of the physicochemical properties of novel aqueous bis(3-aminopropyl)amine and its blend with N-methyldiethanolamine for CO_2 capture," *Journal of Chemical and Engineering Data*, vol. 61, no. 7, pp. 2226–2235, July 2016.

182. B. Das, B. Deogam and B. Mandal, "Experimental and theoretical studies on efficient carbon dioxide capture using novel bis(3-aminopropyl)amine (APA)-activated aqueous 2-amino-2-methyl-1-propanol (AMP) solutions," *RSC Advances*, vol. 7, no. 35, pp. 21518–21530, 2017.

183. B. Das, B. Deogam and B. Mandal, "Absorption of CO_2 into novel aqueous bis(3-aminopropyl)amine and enhancement of CO_2 absorption into its blends with N-methyldiethanolamine," *International Journal of Greenhouse Gas Control*, vol. 60, pp. 172–185, May 2017.

184. B. K. Mondal, S. S. Bandyopadhyay and A. N. Samanta, "Equilibrium solubility and enthalpy of CO_2 absorption in aqueous bis (3-aminopropyl) amine and its mixture with MEA, MDEA, AMP and K_2CO_3," *Chemical Engineering Science*, vol. 170, pp. 58–67, Oct. 2017.

185. N. A. Fine, P. T. Nielsen and G. T. Rochelle, "Decomposition of nitrosamines in CO_2 capture by aqueous piperazine or monoethanolamine" (in English), *Environmental Science & Technology*, vol. 48, no. 10, pp. 5996–6002, May 20, 2014.

186. J. Chen, H. Li, Y. Le Moullec, J. H. Lu, J. C. V. Marcos and G. F. Chen, "Process simulation for CO_2 capture with the aqueous solution of 1-methylpiperazine and its mixture with piperazine," in *13th International Conference on Greenhouse Gas Control*

Technologies, GHGT-13, vol. 114, T. Dixon, L. Laloui and S. Twinning, Eds. Energy Procedia, 2017, pp. 1388–1393.

187. X. Chen and G. T. Rochelle, "Aqueous piperazine derivatives for CO_2 capture: Accurate screening by a wetted wall column," *Chemical Engineering Research & Design*, vol. 89, no. 9, pp. 1693–1710, Sept. 2011.

188. S. Delgado, B. Valentin, D. Bontemps and O. Authier, "Degradation of amine solvents in a CO_2 capture plant at lab-scale: Experiments and modeling," *Industrial & Engineering Chemistry Research*, vol. 57, no. 18, pp. 6057–6067, May 2018.

189. S. A. Freeman and G. T. Rochelle, "Thermal degradation of piperazine and its structural analogs," in *10th International Conference on Greenhouse Gas Control Technologies*, vol. 4, J. Gale, C. Hendriks and W. Turkenberg, Eds. Energy Procedia, 2011, pp. 43–50.

190. F. P. Gordesli and E. Alper, "The kinetics of carbon dioxide capture by solutions of piperazine and N-methyl piperazine," *International Journal of Global Warming*, vol. 3, no. 1–2, pp. 67–76, 2011.

191. H. Li *et al.*, "CO_2 solubility measurement and thermodynamic modeling for 1-methyl-piperazine/water/CO_2." *Fluid Phase Equilibria*, vol. 394, pp. 118–128, May 2015.

192. F. Khalili, A. Henni and A. L. L. East, "pK(a) values of some piperazines at (298, 303, 313, and 323) K" (in English), *Journal of Chemical and Engineering Data*, vol. 54, no. 10, pp. 2914–2917, Oct. 2009.

193. X. Chen and G. T. Rochelle, "Thermodynamics of CO_2/2-methylpiperazine/water," *Industrial & Engineering Chemistry Research*, vol. 52, no. 11, pp. 4229–4238, Mar. 2013.

194. X. Chen and G. T. Rochelle, "Modeling of CO_2 absorption kinetics in aqueous 2-methylpiperazine," *Industrial & Engineering Chemistry Research*, vol. 52, no. 11, pp. 4239–4248, Mar. 2013.

195. S. Gangarapu, G. J. Wierda, A. T. M. Marcelis and H. Zuilhof, "Quantum chemical studies on solvents for post-combustion carbon dioxide capture: Calculation of pK(a) and carbamate stability of disubstituted piperazines," *Chemphyschem*, vol. 15, no. 9, pp. 1880–1886, June 2014.

196. Y. E. Kim, J. H. Choi, S. H. Yun, S. C. Nam and Y. I. Yoon, "CO_2 capture using aqueous solutions of K_2CO_3 + 2-methylpiperazine and monoethanolamine: Specific heat capacity and heat of absorption," *Korean Journal of Chemical Engineering*, vol. 33, no. 12, pp. 3465–3472, Dec. 2016.

197. R. Ramezani, S. Mazinani, R. Di Felice and B. Van der Bruggen, "Experimental and correlation study of corrosion rate, absorption rate and CO_2 loading capacity in five blend solutions as new absorbents for CO_2 capture," *Journal of Natural Gas Science and Engineering*, vol. 45, pp. 599–608, Sept. 2017.

198. S. H. Yun, Y. E. Kim, J. H. Choi, S. C. Nam, J. Chang and Y. I. Yoon, "CO_2 absorption, density, viscosity and vapor pressure of aqueous potassium carbonate+2-methylpiperazine," *Korean Journal of Chemical Engineering*, vol. 33, no. 12, pp. 3473–3486, Dec. 2016.

199. R. Zhang, W. S. Jiang, Z. W. Liang, X. Luo and Q. Yang, "Study of equilibrium solubility, heat of absorption, and speciation of CO_2 absorption into aqueous 2-methylpiperazine (2MPZ) solution," *Industrial & Engineering Chemistry Research*, vol. 57, no. 51, pp. 17496–17503, Dec. 2018.

200. J. H. Choi, Y. E. Kim, S. C. Nam, S. H. Yun, Y. I. Yoon and J. H. Lee, "CO_2 absorption characteristics of a piperazine derivative with primary, secondary, and tertiary amino groups," *Korean Journal of Chemical Engineering*, vol. 33, no. 11, pp. 3222–3230, Nov. 2016.

201. A. Dey, S. K. Dash, S. C. Balchandani and B. Mandal, "Investigation on the inclusion of 1-(2-aminoethyl) piperazine as a promoter on the equilibrium CO_2 solubility of aqueous 2-amino-2methyl-1-propanol," *Journal of Molecular Liquids*, vol. 289, Sept. 2019, Art. no. 111036.

202. A. Dey, S. K. Dash and B. Mandal, "Equilibrium CO_2 solubility and thermophysical properties of aqueous blends of 1-(2-aminoethyl) piperazine and N-methyldiethanolamine," *Fluid Phase Equilibria*, vol. 463, pp. 91–105, May 2018.

203. G. Murshid, W. A. Butt and S. Garg, "Investigation of thermophysical properties for aqueous blends of sarcosine with 1-(2-aminoethyl) piperazine and diethylenetriamine as solvents for CO_2 absorption," *Journal of Molecular Liquids*, vol. 278, pp. 584–591, Mar. 2019.

204. Z. G. Tang, W. Y. Fei and Y. Oli, "CO_2 capture by improved hot potash process," in *10th International Conference on Greenhouse Gas Control Technologies*, vol. 4, J. Gale, C. Hendriks and W. Turkenberg, Eds. Energy Procedia, 2011, pp. 307–317.

205. D. Fernandes, W. Conway, R. Burns, G. Lawrance, M. Maeder and G. Puxty, "Investigations of primary and secondary amine carbamate stability by H-1 NMR spectroscopy for post combustion capture of carbon dioxide," *Journal of Chemical Thermodynamics*, vol. 54, pp. 183–191, Nov. 2012.

206. D. Fernandes *et al.*, "Protonation constants and thermodynamic properties of amines for post combustion capture of CO_2" (in English), *Journal of Chemical Thermodynamics*, vol. 51, pp. 97–102, Aug. 2012.

207. N. Dai, A. D. Shah, L. H. Hu, M. J. Plewa, B. McKague and W. A. Mitch, "Measurement of nitrosamine and nitramine formation from NO reactions with amines during amine-based carbon dioxide capture for postcombustion carbon sequestration," *Environmental Science & Technology*, vol. 46, no. 17, pp. 9793–9801, Sept. 2012.

208. B. Khosravi, F. Feyzi, M. R. Dehghani and S. Kaviani, "Experimental measurement and thermodynamic modeling of CO_2 solubility in aqueous solutions of morpholine," *Journal of Molecular Liquids*, vol. 214, pp. 411–417, Feb. 2016.

209. K. Liu, K. M. Jinka, J. E. Remias and K. L. Liu, "Absorption of carbon dioxide in aqueous morpholine solutions," *Industrial & Engineering Chemistry Research*, vol. 52, no. 45, pp. 15932–15938, Nov. 2013.

210. N. S. Matin, J. E. Remias, J. K. Neathery and K. Liu, "The equilibrium solubility of carbon dioxide in aqueous solutions of morpholine: experimental data and thermodynamic modeling," *Industrial & Engineering Chemistry Research*, vol. 52, no. 14, pp. 5221–5229, Apr. 2013.

211. S. A. Mazari *et al.*, "Thermal degradation kinetics of morpholine for carbon dioxide capture," *Journal of Environmental Chemical Engineering*, vol. 8, no. 3, June 2020, Art. no. 103814.

212. M. O. Ogidi, W. A. Thompson and M. M. Maroto-Valer, "Thermal degradation of morpholine in CO_2 post-combustion capture," in *13th International Conference on Greenhouse Gas Control Technologies, GHGT-13*, vol. 114, T. Dixon, L. Laloui and S. Twinning, Eds. Energy Procedia, 2017, pp. 1033–1037.

213. J. H. Choi, S. G. Oh, M. Jo, Y. I. Yoon, S. K. Jeong and S. C. Nam, "Absorption of carbon dioxide by the mixed aqueous absorbents using 2-methylpiperidine as a promoter," *Chemical Engineering Science*, vol. 72, pp. 87–93, Apr. 2012.

214. J. H. Choi, S. H. Yun, Y. E. Kim, Y. I. Yoon and S. C. Nam, "The effect of functional group position of the piperidine derivatives on the CO_2 absorption characteristics in the (H_2O-Piperidine-CO_2) system," *Korean Chemical Engineering Research*, vol. 53, no. 1, pp. 57–63, Feb. 2015.

215. Y. Coulier, A. Lowe, P. R. Tremaine, J. Y. Coxam and K. Ballerat-Busserolles, "Absorption of CO_2 in aqueous solutions of 2-methylpiperidine: Heats of solution and modeling," *International Journal of Greenhouse Gas Control*, vol. 47, pp. 322–329, Apr. 2016.

216. O. Fandino, S. Sasidharanpillai, D. V. Soldatov and P. R. Tremaine, "Carbamate formation in the system (2-methylpiperidine plus carbon dioxide) by Raman spectroscopy and x-ray diffraction," *Journal of Physical Chemistry B*, vol. 122, no. 48, pp. 10880–10893, Dec. 2018.

217. C. McGregor, M. S. Al-Abdul-Wahid, V. Robertson, J. S. Cox and P. R. Tremaine, "Formation constants and conformational analysis of carbamates in aqueous solutions of 2-methylpiperidine and CO_2 from 283 to 313 K by NMR spectroscopy," *Journal of Physical Chemistry B*, vol. 122, no. 39, pp. 9178–9190, Oct. 2018.

218. E. Moine, Y. Coulier, J. Y. Coxam, K. Ballerat-Busserolles and R. Privat, "Thermodynamic study of four (methylpiperidine plus water) systems: New experimental data and challenging modeling for the simultaneous representation of liquid-liquid equilibrium and energetic properties," *Journal of Chemical and Engineering Data*, vol. 64, no. 2, pp. 743–754, Feb. 2019.

219. K. Ballerat-Busserolles, M. R. Simond, Y. Coulier and J. Y. Coxam, "Protonation of alkanolamines and cyclic amines in water at temperatures from 293.15 to 373.15 K" (in English), *Pure and Applied Chemistry*, vol. 86, no. 2, pp. 233–243, Feb. 2014.

220. S. Garg, G. Murshid, F. S. Mjalli, A. Ali and W. Ahmad, "Experimental and correlation study of selected physical properties of aqueous blends of potassium sarcosinate and 2-piperidineethanol as a solvent for CO_2 capture," *Chemical Engineering Research & Design*, vol. 118, pp. 121–130, Feb. 2017.

221. A. Hartono *et al.*, "Characterization of 2-piperidineethanol and 1-(2-hydroxyethyl) pyrrolidine as strong bicarbonate forming solvents for CO_2 capture," *International Journal of Greenhouse Gas Control*, vol. 63, pp. 260–271, Aug. 2017.

222. H. Yamada, Y. Matsuzaki, H. Okabe, S. Shimizu and Y. Fujioka, "Quantum chemical analysis of carbon dioxide absorption into aqueous solutions of moderately hindered amines," in *10th International Conference on Greenhouse Gas Control Technologies*, vol. 4, J. Gale, C. Hendriks and W. Turkenberg, Eds. Energy Procedia, 2011, pp. 133–139.

223. W. Conway, Y. Beyad, G. Richner, G. Puxty and P. Feron, "Rapid CO_2 absorption into aqueous benzylamine (BZA) solutions and its formulations with monoethanolamine (MEA), and 2-amino-2-methyl-1-propanol (AMP) as components for post combustion capture processes," *Chemical Engineering Journal*, vol. 264, pp. 954–961, Mar. 2015.

224. J. Gao *et al.*, "Study on CO_2 absorption by aqueous benzylamine and its formulations with monoethanolamine as a component for post-combustion capture process," *China Petroleum Processing & Petrochemical Technology*, vol. 18, no. 3, pp. 7–14, Sept. 2016.

225. S. Mukherjee, S. S. Bandyopadhyay and A. N. Samanta, "Vapor-liquid equilibrium (VLE) of CO_2 in aqueous solutions of benzylamine: New data and modeling using ENRTL-equation," *International Journal of Greenhouse Gas Control*, vol. 56, pp. 12–21, Jan. 2017.

226. S. Mukherjee, S. S. Bandyopadhyay and A. N. Samanta, "Kinetic study of CO_2 absorption in aqueous benzylamine solvent using a stirred cell reaction calorimeter," *Energy & Fuels*, vol. 32, no. 3, pp. 3668–3680, Mar. 2018.

227. S. Mukherjee, S. S. Bandyopadhyay and A. N. Samanta, "Experimental measurements and modelling of CO_2 solubility in aqueous mixtures of benzylamine and N-(2-aminoethyl) ethanolamine," *Asia-Pacific Journal of Chemical Engineering*, vol. 13, no. 6, Nov.–Dec. 2018, Art. no. e2264.

228. S. Mukherjee and A. N. Samanta, "Heat of absorption of CO_2 and heat capacity measurements in aqueous solutions of benzylamine, N-(2-Aminoethyl)-ethanolamine, and their blends using a reaction calorimeter," *Journal of Chemical and Engineering Data*, vol. 64, no. 8, pp. 3392–3406, Aug. 2019.

229. G. Puxty *et al.*, "The evolution of a new class of CO_2 absorbents: Aromatic amines," *International Journal of Greenhouse Gas Control*, vol. 83, pp. 11–19, Apr. 2019.

230. G. Richner, "Promoting CO_2 absorption in aqueous amines with benzylamine," in *GHGT-11*, vol. 37, T. Dixon and K. Yamaji, Eds. Energy Procedia, 2013, pp. 423–430.

231. G. Richner, G. Puxty, A. Carnal, W. Conway, M. Maeder and P. Pearson, "Thermokinetic properties and performance evaluation of benzylamine-based solvents for CO_2 capture," *Chemical Engineering Journal*, vol. 264, pp. 230–240, Mar. 2015.

232. U. E. Aronu, H. F. Svendsen and K. A. Hoff, "Investigation of amine amino acid salts for carbon dioxide absorption" (in English), *International Journal of Greenhouse Gas Control*, vol. 4, no. 5, pp. 771–775, Sept. 2010.

233. M. Rabensteiner, G. Kinger, M. Koller, G. Gronald, S. Unterberger and C. Hochenauer, "Investigation of the suitability of aqueous sodium glycinate as a solvent for post combustion carbon dioxide capture on the basis of pilot plant studies and screening methods" (in English), *International Journal of Greenhouse Gas Control*, vol. 29, pp. 1–15, Oct. 2014.

234. M. Rabensteiner, G. Kinger, M. Koller and C. Hochenauer, "PCC pilot plant studies with aqueous potassium glycinate" (in English), *International Journal of Greenhouse Gas Control*, vol. 42, pp. 562–570, Nov. 2015.

235. E. S. Fernandez *et al.*, "Conceptual design of a novel CO_2 capture process based on precipitating amino acid solvents" (in English), *Industrial & Engineering Chemistry Research*, vol. 52, no. 34, pp. 12223–12235, Aug. 28, 2013.

236. V. S. Sefidi and P. Luis, "Advanced amino acid-based technologies for CO_2 capture: A Review" (in English), *Industrial & Engineering Chemistry Research*, vol. 58, no. 44, pp. 20181–20194, Nov. 6, 2019.

237. R. R. Wanderley, D. D. D. Pinto and H. K. Knuutila, "From hybrid solvents to water-lean solvents – a critical and historical review" (in English), *Separation and Purification Technology*, vol. 260, Apr. 1, 2021.

238. H. Wibowo *et al.*, "Study on the effect of operating parameters towards CO_2 absorption behavior of choline chloride-monoethanolamine deep eutectic solvent and its aqueous solutions," *Chemical Engineering and Processing-Process Intensification*, vol. 157, Nov. 2020, Art. no. 108142.

239. S. Sarmad, D. Nikjoo and J. P. Mikkola, "Amine functionalized deep eutectic solvent for CO_2 capture: Measurements and modeling," *Journal of Molecular Liquids*, vol. 309, July 2020, Art. no. 113159.

240. A. U. Maheswari and K. Palanivelu, "Carbon dioxide capture and utilization by alkanolamines in deep eutectic solvent medium," *Industrial & Engineering Chemistry Research*, vol. 54, no. 45, pp. 11383–11392, Nov. 2015.

241. S. M. Chen, S. Y. Chen, X. Y. Fei, Y. C. Zhang and L. Qin, "Solubility and characterization of CO_2 in 40 mass % N-ethylmonoethanolamine solutions: explorations for an efficient nonaqueous solution," *Industrial & Engineering Chemistry Research*, vol. 54, no. 29, pp. 7212–7218, July 2015.

242. Y. Dong, T. T. Ping, X. Q. Shi and S. F. Shen, "Density, viscosity and excess properties for binary mixtures of 2-(ethylamino)ethanol and 2-(butylamino)ethanol with 2-butoxyethanol at temperatures from (293.15 to 353.15) K," *Journal of Molecular Liquids*, vol. 312, Aug. 2020, Art. no. 113351.

243. F. P. Gordesli Duatepe, O. Y. Orhan and E. Alper, "Kinetics of carbon dioxide absorption by nonaqueous solutions of promoted sterically hindered amines," in *13th International Conference on Greenhouse Gas Control Technologies, GHGT-13*, vol. 114, T. Dixon, L. Laloui and S. Twinning, Eds. Energy Procedia, 2017, pp. 57–65.

244. C. Guo, S. Y. Chen, Y. C. Zhang and G. B. Wang, "Solubility of CO_2 in nonaqueous absorption system of 2-(2-Aminoethylamine)ethanol plus benzyl alcohol," *Journal of Chemical and Engineering Data*, vol. 59, no. 6, pp. 1796–1801, June 2014.

245. K. S. Hwang, S. W. Park, D. W. Park, K. J. Oh and S. S. Kim, "Absorption of carbon dioxide into diisopropanolamine solutions of polar organic solvents," *Journal of the Taiwan Institute of Chemical Engineers*, vol. 41, no. 1, pp. 16–21, Jan. 2010.

246. S. Kadiwala, A. V. Rayer and A. Henni, "Kinetics of carbon dioxide (CO_2) with ethylenediamine, 3-amino-1-propanol in methanol and ethanol, and with 1-dimethylamino-2-propanol and 3-dimethylamino-1-propanol in water using stopped-flow technique," *Chemical Engineering Journal*, vol. 179, pp. 262–271, Jan. 2012.

247. M. K. Kang, S. B. Jeon, J. H. Cho, J. S. Kim and K. J. Oh, "Characterization and comparison of the CO_2 absorption performance into aqueous, quasi-aqueous and non-aqueous MEA solutions," *International Journal of Greenhouse Gas Control*, vol. 63, pp. 281–288, Aug. 2017.

248. F. Liu, G. H. Jing, X. B. Zhou, B. H. Lv and Z. M. Zhou, "Performance and mechanisms of triethylene tetramine (TETA) and 2-amino-2-methyl-1-propanol (AMP) in aqueous and nonaqueous solutions for CO_2 capture," *ACS Sustainable Chemistry & Engineering*, vol. 6, no. 1, pp. 1352–1361, Jan. 2018.
249. S. Mukherjee and A. N. Samanta, "Kinetic study of CO_2 absorption in aqueous solutions of 2-((2-aminoethyl)amino)-ethanol using a stirred cell reaction calorimeter," *International Journal of Chemical Kinetics*, vol. 51, no. 12, pp. 943–957, Dec. 2019.
250. D. Nath and A. Henni, "Kinetics of carbon dioxide (CO_2) with 3-(Dimethylamino)-1-propylamine in water and methanol systems using the stopped-flow technique," *Industrial & Engineering Chemistry Research*, vol. 59, no. 33, pp. 14625–14635, Aug. 2020.
251. T. T. Ping, Y. Dong and S. F. Shen, "Energy-efficient CO_2 capture using nonaqueous absorbents of secondary alkanolamines with a 2-butoxyethanol cosolvent," *ACS Sustainable Chemistry & Engineering*, vol. 8, no. 49, pp. 18071–18082, Dec. 2020.
252. H. Rashidi, P. Valeh-e-Sheyda and S. Sahraie, "A multiobjective experimental based optimization to the CO_2 capture process using hybrid solvents of MEA-MeOH and MEA-water," *Energy*, vol. 190, Jan. 2020, Art. no. 116430.
253. F. J. Tamajon, E. Alvarez, F. Cerdeira and D. Gomez-Diaz, "CO_2 absorption into N-methyldiethanolamine aqueous-organic solvents," *Chemical Engineering Journal*, vol. 283, pp. 1069–1080, Jan. 2016.
254. I. von Harbou *et al.*, "Removal of carbon dioxide from flue gases with aqueous MEA solution containing ethanol," *Chemical Engineering and Processing-Process Intensification*, vol. 75, pp. 81–89, Jan. 2014.
255. R. R. Wanderley, D. Pinto and H. K. Knuutila, "Investigating opportunities for water-lean solvents in CO_2 capture: VLE and heat of absorption in water-lean solvents containing MEA," *Separation and Purification Technology*, vol. 231, Jan. 2020, Art. no. 115883.
256. F. Wang, L. M. Wang, S. Q. Wang, D. Fu and IOP, "Study on viscosity of MDEA-MeOH aqueous solutions," in *2nd International Conference on Advances in Energy Resources and Environment Engineering*, vol. 59, IOP Conference Series-Earth and Environmental Science, 2017.
257. Y. S. Yu, H. F. Lu, T. T. Zhang, Z. X. Zhang, G. X. Wang and V. Rudolph, "Determining the performance of an efficient nonaqueous CO_2 capture process at desorption temperatures below 373 K," *Industrial & Engineering Chemistry Research*, vol. 52, no. 35, pp. 12622–12634, Sept. 2013.
258. N. Zhong *et al.*, "Reaction kinetics of carbon dioxide (CO_2) with diethylenetriamine and 1-amino-2-propanol in nonaqueous solvents using stopped flow technique," *Industrial & Engineering Chemistry Research*, vol. 55, no. 27, pp. 7307–7317, July 2016.
259. N. Zhong *et al.*, "Kinetics of carbon dioxide (CO_2) with diethylenetriamine in non-aqueous solvents using stopped-flow technique," in *13th International Conference on Greenhouse Gas Control Technologies, GHGT-13*, vol. 114, T. Dixon, L. Laloui and S. Twinning, Eds. Energy Procedia, 2017, pp. 1869–1876.
260. R. R. Bhosale and V. V. Mahajani, "Kinetics of absorption of carbon dioxide in aqueous solution of ethylaminoethanol modified with N-methyl-2-pyrolidone," *Separation Science and Technology*, vol. 48, no. 15, pp. 2324–2337, Oct. 2013.
261. A. Blanco, A. Garcia-Abuin, D. Gomez-Diaz and J. M. Navaza, "Density, speed of sound, and viscosity of N-methyl-2-pyrrolidone plus ethanolamine plus water from T = (293.15 to 323.15) K," *Journal of Chemical and Engineering Data*, vol. 57, no. 11, pp. 3136–3141, Nov. 2012.
262. A. Garcia-Abuin, D. Gomez-Diaz, M. D. La Rubia, A. B. Lopez and J. M. Navaza, "Density, speed of sound, refractive index, viscosity, surface tension, and excess volume of N-Methyl-2-pyrrolidone+1-Amino-2-propanol (or Bis(2-hydroxypropyl)amine) from T = (293.15 to 323.15) K," *Journal of Chemical and Engineering Data*, vol. 56, no. 6, pp. 2904–2908, June 2011.

263. H. K. Karlsson, M. G. Sanku and H. Svensson, "Absorption of carbon dioxide in mixtures of N-methyl-2-pyrrolidone and 2-amino-2-methyl-1-propanol," *International Journal of Greenhouse Gas Control*, vol. 95, Apr. 2020, Art. no. 102952.

264. P. Pakzad, M. Mofarahi, A. A. Izadpanah, M. Afkhamipour and C. H. Lee, "An experimental and modeling study of CO_2 solubility in a 2-amino-2-methyl-1-propanol (AMP) + N-methyl-2-pyrrolidone (NMP) solution," *Chemical Engineering Science*, vol. 175, pp. 365–376, Jan. 2018.

265. H. Svensson, J. Edfeldt, V. Z. Velasco, C. Hulteberg and H. T. Karlsson, "Solubility of carbon dioxide in mixtures of 2-amino-2-methyl-1-propanol and organic solvents," *International Journal of Greenhouse Gas Control*, vol. 27, pp. 247–254, Aug. 2014.

266. H. Svensson, V. Z. Velasco, C. Hulteberg and H. T. Karlsson, "Heat of absorption of carbon dioxide in mixtures of 2-amino-2-methyl-1-propanol and organic solvents," *International Journal of Greenhouse Gas Control*, vol. 30, pp. 1–8, Nov. 2014.

267. L. S. Tan, A. M. Shariff, K. K. Lau and M. A. Bustam, "Impact of high pressure on high concentration carbon dioxide capture from natural gas by monoethanolamine/N-methyl-2-pyrrolidone solvent in absorption packed column," *International Journal of Greenhouse Gas Control*, vol. 34, pp. 25–30, Mar. 2015.

268. L. S. Tan, A. M. Shariff, K. K. Lau, M. S. Shaikh and G. Murshid, "High pressure rheology and viscosity of monoethanolamine with n-methyl-2-pyrrolidone and water hybrid solvent," in *Proceeding of 4th International Conference on Process Engineering and Advanced Materials*, vol. 148, M. A. Bustam *et al.*, Eds. Procedia Engineering, 2016, pp. 5–10.

269. L. S. Tan, A. M. Shariff, W. H. Tay, K. K. Lau, T. Tsuji and N. A. H. Hairul, "Impact of N-Methyl-2-pyrrolidone in monoethanolamine solution to the CO_2 absorption in packed column: Analysis via mathematical modeling," *Sains Malaysiana*, vol. 49, no. 11, pp. 2625–2635, Nov. 2020.

270. R. R. Wanderley, G. J. C. Ponce and H. K. Knuutila, "Solubility and heat of absorption of CO_2 into diisopropylamine and N,N-diethylethanolamine mixed with organic solvents," *Energy & Fuels*, vol. 34, no. 7, pp. 8552–8561, July 2020.

271. Y. Yuan and G. T. Rochelle, "CO_2 absorption rate in semi-aqueous monoethanolamine," *Chemical Engineering Science*, vol. 182, pp. 56–66, June 2018.

272. P. D. Mobley *et al.*, "CO_2 capture using fluorinated hydrophobic solvents" (in English), *Industrial & Engineering Chemistry Research*, vol. 56, no. 41, pp. 11958–11966, Oct. 18 2017.

273. R. F. Zheng *et al.*, "A single-component water-lean post-combustion CO_2 capture solvent with exceptionally low operational heat and total costs of capture – comprehensive experimental and theoretical evaluation" (in English), *Energy & Environmental Science*, vol. 13, no. 11, pp. 4106–4113, Nov. 1, 2020.

274. Q. Zhuang, B. Clements, J. Y. Dai and L. Carrigan, "Ten years of research on phase separation absorbents for carbon capture: Achievements and next steps," *International Journal of Greenhouse Gas Control*, vol. 52, pp. 449–460, Sept. 2016.

275. S. H. Zhang, Y. Shen, L. D. Wang, J. M. Chen and Y. Q. Lu, "Phase change solvents for post-combustion CO_2 capture: Principle, advances, and challenges," *Applied Energy*, vol. 239, pp. 876–897, Apr. 2019.

276. Y. Coulier, A. R. Lowe, A. Moreau, K. Ballerat-Busserolles and J. Y. Coxam, "Liquid-liquid phase separation of (amine-H_2O-CO_2) systems: New methods for key data," *Fluid Phase Equilibria*, vol. 431, pp. 1–7, Jan. 2017.

277. F. Liu *et al.*, "Carbon dioxide absorption in aqueous alkanolamine blends for biphasic solvents screening and evaluation," *Applied Energy*, vol. 233, pp. 468–477, Jan. 2019.

278. F. Liu, M. X. Fang, N. T. Yi and T. Wang, "Research on Alkanolamine-Based Physical-Chemical Solutions as Biphasic Solvents for CO_2 Capture," *Energy & Fuels*, vol. 33, no. 11, pp. 11389–11398, Nov. 2019.

279. F. Liu, M. X. Fang, N. T. Yi, T. Wang and Q. H. Wang, "Biphasic behaviors and regeneration energy of a 2-(diethylamino)-ethanol and 2-((2-aminoethyl)amino) ethanol blend for CO_2 capture," *Sustainable Energy & Fuels*, vol. 3, no. 12, pp. 3594–3602, Dec. 2019.

280. Y. J. Qiu, H. F. Lu, Y. M. Zhu, Y. Y. Liu, K. J. Wu and B. Liang, "Phase-Change CO_2 Absorption Using Novel 3-Dimethylaminopropylamine with Primary and Tertiary Amino Groups," *Industrial & Engineering Chemistry Research*, vol. 59, no. 19, pp. 8902–8910, May 2020.

281. Y. Shen *et al.*, "Two-stage interaction performance of CO_2 absorption into biphasic solvents: Mechanism analysis, quantum calculation and energy consumption," *Applied Energy*, vol. 260, Feb. 2020, Art. no. 114343.

282. Y. Shen, C. K. Jiang, S. H. Zhang, J. Chen, L. D. Wang and J. M. Chen, "Biphasic solvent for CO_2 capture: Amine property-performance and heat duty relationship," *Applied Energy*, vol. 230, pp. 726–733, Nov. 2018.

283. L. D. Wang, S. S. Liu, R. J. Wang, Q. W. Li and S. H. Zhang, "Regulating Phase Separation Behavior of a DEEA-TETA Biphasic Solvent Using Sulfolane for Energy-Saving CO_2 Capture," *Environmental Science & Technology*, vol. 53, no. 21, pp. 12873–12881, Nov. 2019.

284. M. M. Xu, S. J. Wang and L. Z. Xu, "Screening of physical-chemical biphasic solvents for CO_2 absorption," *International Journal of Greenhouse Gas Control*, vol. 85, pp. 199–205, June 2019.

285. M. M. Xu, S. J. Wang and L. Z. Xu, "N_2O Solubility in and Density and Viscosity of Novel Biphasic Solvents for CO_2 and Their Phase Separation Accelerators from 293.15 to 333.15 K," *Journal of Chemical and Engineering Data*, vol. 65, no. 2, pp. 598–608, Feb. 2020.

286. Z. C. Xu, S. J. Wang and C. H. Chen, "CO_2 absorption by biphasic solvents: Mixtures of 1,4-Butanediamine and 2-(Diethylamino)-ethanol," *International Journal of Greenhouse Gas Control*, vol. 16, pp. 107–115, Aug. 2013.

287. Z. C. Xu, S. J. Wang, G. J. Qi, J. Z. Liu, B. Zhao and C. H. Chen, "CO_2 Absorption by Biphasic Solvents: Comparison with Lower Phase Alone," *Oil & Gas Science and Technology-Revue d'IFP Energies Nouvelles*, vol. 69, no. 5, pp. 851–864, Sep-Oct 2014.

288. Z. C. Xu, S. J. Wang, B. Zhao and C. H. Chen, "Study on potential biphasic solvents: Absorption capacity, CO_2 loading and reaction rate," in *GHGT-11*, vol. 37, T. Dixon and K. Yamaji, Eds. Energy Procedia, 2013, pp. 494–498.

289. J. X. Ye *et al.*, "Novel Biphasic Solvent with Tunable Phase Separation for CO_2 Capture: Role of Water Content in Mechanism, Kinetics, and Energy Penalty," *Environmental Science & Technology*, vol. 53, no. 8, pp. 4470–4479, Apr. 2019.

290. Q. Ye, X. L. Wang and Y. Q. Lu, "Screening and evaluation of novel biphasic solvents for energy-efficient post-combustion CO_2 capture," *International Journal of Greenhouse Gas Control*, vol. 39, pp. 205–214, Aug. 2015.

291. W. D. Zhang, X. H. Jin, W. W. Tu, Q. Ma, M. L. Mao and C. H. Cui, "A Novel CO_2 Phase Change Absorbent: MEA/1-Propanol/H_2O," *Energy & Fuels*, vol. 31, no. 4, pp. 4273–4279, Apr. 2017.

292. X. B. Zhou, G. H. Jing, B. H. Lv, F. Liu and Z. M. Zhou, "Low-viscosity and efficient regeneration of carbon dioxide capture using a biphasic solvent regulated by 2-amino-2-methyl-1-propanol," *Applied Energy*, vol. 235, pp. 379–390, Feb. 2019.

293. X. B. Zhou, F. Liu, B. H. Lv, Z. M. Zhou and G. H. Jing, "Evaluation of the novel biphasic solvents for CO_2 capture: Performance and mechanism," *International Journal of Greenhouse Gas Control*, vol. 60, pp. 120–128, May 2017.

294. K. Zhu *et al.*, "Investigation on the Phase-Change Absorbent System MEA plus Solvent A (SA) + H_2O Used for the CO_2 Capture from Flue Gas," *Industrial & Engineering Chemistry Research*, vol. 58, no. 9, pp. 3811–3821, Mar. 2019.

295. J. F. Zhang, Y. Qiao and D. W. Agar, "Improvement of lipophilic-amine-based thermomorphic biphasic solvent for energy-efficient carbon capture," in *6th Trondheim Conference on CO_2 Capture, Transport and Storage*, vol. 23, N. A. Rokke, M. B. Hagg and M. J. Mazzetti, Eds. Energy Procedia, 2012, pp. 92–101.

296. J. F. Zhang, O. Nwani, Y. Tan and D. W. Agar, "Carbon dioxide absorption into biphasic amine solvent with solvent loss reduction," *Chemical Engineering Research & Design*, vol. 89, no. 8A, pp. 1190–1196, Aug. 2011.

297. J. F. Zhang, J. Chen, R. Misch and D. W. Agar, "Carbon dioxide absorption in biphasic amine solvents with enhanced low temperature solvent regeneration," in *Pres 2010: 13th International Conference on Process Integration, Modelling and Optimisation for Energy Saving and Pollution Reduction*, vol. 21, J. J. Klemes, H. L. Lam and P. S. Varbanov, Eds. Chemical Engineering Transactions, 2010, pp. 169–174.

298. N. Boulmal, R. Rivera-Tinoco and C. Bouallou, "Experimental assessment of CO_2 absorption rates for aqueous solutions of hexylamine, dimethylcyclohexylamine and their blends," *Scientific Study and Research-Chemistry and Chemical Engineering Biotechnology Food Industry*, vol. 19, no. 3, pp. 293–312, 2018.

299. D. J. Heldebrant et al., "Reversible zwitterionic liquids, the reaction of alkanol guanidines, alkanol amidines, and diamines with CO_2" (in English), *Green Chemistry*, vol. 12, no. 4, pp. 713–721, Apr. 2010.

300. D. Malhotra et al., "Reinventing Design Principles for Developing Low-Viscosity Carbon Dioxide-Binding Organic Liquids for Flue Gas Clean Up" (in English), *Chemsuschem*, vol. 10, no. 3, pp. 636–642, Feb. 8 2017.

301. P. M. Mathias et al., "Improving the regeneration of CO_2-binding organic liquids with a polarity change" (in English), *Energy & Environmental Science*, vol. 6, no. 7, pp. 2233–2242, July 2013.

302. N. Cihan, O. Y. Orhan and H. Y. Ersan, "Effect of non-aqueous solvents on kinetics of carbon dioxide absorption by (Bu3P)-Bu-t/B(C6F5)(3) frustrated Lewis pairs" (in English), *Separation and Purification Technology*, vol. 258, Mar. 1, 2021.

303. R. J. Liu, X. Liu, K. B. Ouyang and Q. Yan, "Catalyst-Free Click Polymerization of CO_2 and Lewis Monomers for Recyclable C1 Fixation and Release" (in English), *ACS Macro Letters*, vol. 8, no. 2, pp. 200–204, Feb. 2019.

304. O. Y. Orhan, N. Cihan, V. Sahin, A. Karabakan and E. Alper, "The development of reaction kinetics for CO_2 absorption into novel solvent Frustrated Lewis (FLPs)" (in English), *Separation and Purification Technology*, vol. 252, Dec. 1, 2020.

305. T. W. Yokley, H. Tupkar, N. D. Schley, N. J. DeYonker and T. P. Brewster, "CO_2 Capture by 2-(Methylamino)pyridine Ligated Aluminum Alkyl Complexes" (in English), *European Journal of Inorganic Chemistry*, vol. 2020, no. 31, pp. 2958–2967, Aug. 23, 2020.

3 Ionic Liquids in Carbon Capture

*George E. Romanos, Niki Vergadou, and
Ioannis G. Economou*

CONTENTS

3.1 INTRODUCTION TO IONIC LIQUIDS AND THEIR APPLICATIONS

Ionic liquids (ILs) is the generic term for a class of materials, consisting entirely of ions and being liquid below 373 K. If they are liquid at room temperature, they are also known as *room temperature ILs* (RTILs). Since the discovery of ethanolammonium nitrate, being the first IL ever reported by S. Gabriel and J. Weiner in 1888, and of ethylammonium nitrate, which was the first RTIL reported in 1914 by Paul Walden [1], nowadays approximately 1000 different ILs are known from literature and about one-third of them are commercially available. Compared to the number of 10^{18} different ILs predicted by some experts in this field, these numbers seem relatively small. The huge quantity of theoretically possible ILs that emerge from the plethora of anions and cations that can be combined to form an IL presumably offers an enormous potential for their involvement in various processes and state-of-the art technologies, as well as for industrial and academic research.

DOI: 10.1201/9781003162780-3

The anion and cation of an IL have a delocalized charge. The cation is usually organic and bulky, characterized by an asymmetric structure. These characteristics conclude, in comparison to conventional salts, to a weaker inter-ionic interaction and coordinating tendency of the ion pair as the result of a subtle balance between complex electrostatic interactions, polarizability and charge transfer phenomena, hydrogen bonds, and dispersion forces. Hence, by breaking the symmetry of the chemical structure and having the charge of the cation and/or the anion distributed over a larger volume of the molecule by resonance, the formation of a stable crystal lattice is prevented and the solidification of ILs takes place only at sub-zero temperatures. Stated alternatively, the bulky anion and cation of ILs and their high conformational flexibility result in small lattice enthalpies and large entropy changes that favor ILs to be in the liquid state at near ambient temperature conditions. Moreover, ILs exhibit a glassy-type dynamics at low temperatures and, in some cases, a glass transition is observed, especially if long aliphatic side chains are involved in the organic cation [2].

Due to their dual ionic and organic character, ILs are endowed with a prominent combination of properties and assets that render them attractive for a wide range of applications. They present huge chemical diversity and excellent electrochemical and thermal stability along with nonvolatility and noninflammability. ILs are liquid over a wide temperature range and can occur as isotropic fluids or form liquid crystalline structures [3]. Other important properties of ILs include their tunable miscibility with water and their thermomorphic behavior. The latter means that they can form monophasic systems with water or other solvents above an upper critical solution temperature and then separate again in two phases when the temperature drops. This is of high importance for separation applications and for achieving monophasic reaction systems, offering an easy recycling of catalyst [4]. Moreover, several types of ILs have paramagnetic properties, enhanced electric and ionic conductivity, wide electrochemical windows, high heat capacity, low corrosiveness, and high affinity to capture or separate CO_2 molecules from industrial gas streams in processes like absorption [5], adsorption [6], cryogenic separation [7], and membrane separation [8]. As such, ILs have improved performances in numerous areas such as electrochemistry, analytical chemistry, synthesis and catalysis, heat transport, and conversion, and they are used as functional fluids, additives, and process fluids for many separation processes in the gas and liquid phase.

Currently, there are many commercial uses of ILs in a wide spectrum of applications. However, what impedes the further expansion of ILs applicability is their high cost in conjunction with the shortage of conclusive data from all the necessary toxicity studies. Hence, the use of ILs has been commercialized only for the cases where ILs brought large process improvement, economic feasibility, and minimized environmental and human health hazards. For instance, due to their low isothermal compressibility combined with appropriate thermal expansion coefficients and viscosities, ILs are already used as effective hydraulic fluids and additives for extended lifetimes [9]. In the field of electrochemistry, and based upon their wide electrochemical windows and high conductivities, ILs found useful applications in sensors

[10] and supercapacitors development [11–12]. Notably, these applications required a small quantity of ILs that suffices the product requirement. Moreover, the inherent stability and degradation resistance resulted in extended life span and reduced the need for frequent replacement/replenishment. Other commercial applications include the synthesis of catalytic nanoparticles [13], organic synthesis [14], and the processes of protein crystallization [15], where ILs can be effectively recovered and reused many times. In the sector of analytics, ILs are used in Karl Fischer titration and as materials for gas chromatography (GC) [16], where their requirement in small amounts are sought after for creating immobilized phase in a GC column. The BASIL™ process uses ILs to produce the generic photo-initiator precursor alkoxyphenylphosphines [17]. In this process, the ILs have replaced triethylamine and achieved a similar enhanced space-time yield with much smaller reactor size. Another successful example of ILs application is the Difasol process [18]. Efficient scrubbing of mercury vapor from natural gas streams has also been demonstrated on an industrial scale, using chlorocuprate (II) ILs impregnated on porous solid supports with high surface area [19]. ILs are increasingly finding applications in pilot testing (immobilization of catalysts [SILCs], metal extraction, liquid–liquid [L/L] extraction, biomass conversion, dye-sensitized solar cells [DSSCs], sorption cooling, lubricants, surfactants). A high number of IL structures are already registered or pre-registered with the European Chemicals Agency (ECHA) [20] or other international chemical agencies, which indicates that many more applications will be brought forward by potential users, and some of them will be realized at commercial scale in the next couple of years.

There is also a multitude of other important processes that may gain great benefits and achieve innovative performance improvements by using ILs. Among them, the use of ILs as solvents for CO_2 capture is currently one of their most studied and propitious potential applications [21] and appears to get closer in satisfying and covering the stipulations and necessities of this very demanding process. Compared to other potential applications, the CO_2 capture demands for a diverse range of properties to be met, such as low volatility, low viscosity, chemical and thermal stability, fast CO_2 absorption and reaction rates, high CO_2 absorptivity and selectivity, positive or negligible effect of water on the CO_2 absorptivity, low energy demand for regeneration, miscibility with water, low toxicity and corrosiveness, and low cost. ILs can be tailored to fulfill the prerequisite of green CCS technologies that can substitute the current organic volatile solvents such as amines, which are widely used. Some of the aspects in which ILs perform better in comparison to the amines are in the reduction of parasitic energy load, reduced water use, the wide temperature range in which they can be utilized, and the almost zero release of solvents to the environment – facts that render ILs novel green solvents, in combination with high CO_2 capacities that are observed in many cases.

Besides their inherent properties and tuneability, which are highly attractive for a CO_2 capture process, ILs offer the additional benefit of high versatility. As such, ILs can be mixed with the currently used amine solvent formulations of post-combustion CO_2 capture, leading to reduction in energy consumption without affecting the absorption rate and CO_2 absorption capacity [22–23]; or they can be applied in their pure form as effective solvents for the pre-combustion CO_2 capture, eliminating cooling requirements for the conveyed flue gas stream. It has been recently proven that improving the CO_2 solubility in the IL $[HMIM]^+[Tf_2N]^-$ by 10% can lead to power use and separation cost of 34.6 MW and \$12 per ton of CO_2 separated, which are 35% and 29% less than the respective base case values [24]. Furthermore, due to their low volatility, high thermal stability, wide liquid range, and tunable surface wetting properties, ILs can be used as effective pore modifiers for a wide range of porous adsorbents and membranes [25–27], leading to composite materials with enhanced CO_2 selectivity. In addition, poly-(ionic liquids) (PILs), composed of covalently linked ILs, constitute a recently emerging class of IL-based absorbents and membranes for CO_2 capture and separation performance [28–29].

Although the structure tunability of ILs generates many prospects for their sustainable involvement in CO_2 capture, the multifactored process requirements for the CO_2 capture process poses a challenge for their technological development. Moreover, limited understanding of the microscopic mechanisms that govern the macroscopic behavior of ILs further hampers technological deployment. It was only in 1999 that Blanchard et al. [30] reported high CO_2 absorption by 1-butyl-3-methylimidazolium hexafluorophosphate ($[BMIM]^+[PF_6]^-$), showing that the IL is not transferred to the gas phase. However, numerous attempts to enhance specific properties of ILs for CO_2 capture have resulted in a decline of other equally important properties. For instance, the increasing degree of anion fluorination readily enhances both the CO_2 absorptivity [31] (due to enhanced free volume) and thermal stability of ILs [32] (tetrafluoroborate $[BF_4]^-$ < hexafluorophosphate $[PF_6]^-$ < bis(trifluoromethylsulfonyl)-imide $[Tf_2N]^-$ < tris(perfluoroalkyl)trifluorophosphate $[FAP]^-$), leading to less corrosivity and lower viscosity [32]. At the same time, and contrary to what required, highly fluorinated ILs are extremely hydrophobic (immiscible with water) [33], limited in their biodegradability [32], and costly. Similarly, longer alkyl chains on the cation increases the extent of free-volume and increases the CO_2 solubility [23], but the effect is negative on the thermal stability [34], the biodegradability [35] and the (oral) toxicity [36]. As a result, a compromise must be made between the cost, toxicity, and biodegradability of the IL and the need to ensure high CO_2 capture efficiency and selectivity, which hinders feasibility of a commercial application. In Figure 3.1, we provide an overview of the most common cations and anions examined to date as constituents of ILs for CO_2 capture applications.

Another important aspect is to devise and develop methods and processes that simultaneously fulfill all properties for the CO_2 capture process. An initial attempt was made in 1998 with the development of supported ionic liquid membranes (SILMs) for catalytic hydrogenation processes, and later in 2001 [37] and 2002 [38] with the use of ionic liquids as active separation layers in supported liquid membranes. A couple of years later, Scovazzo et al. [39] and Baltus et al. [40], published the first works on SILM membranes for gas separations. The concept behind these developments relies on the effective dispersion and stabilization of small amounts of ILs into the channels of highly porous substrates. This helps to reduce production cost, toxicity, and environmental hazards while imparting low viscosity and improving gas diffusivity. Similarly, the supported IL concept was also applied to solid support ionic liquid phase adsorbents (SILPs) for gas separation. The SILP concept helped in developing

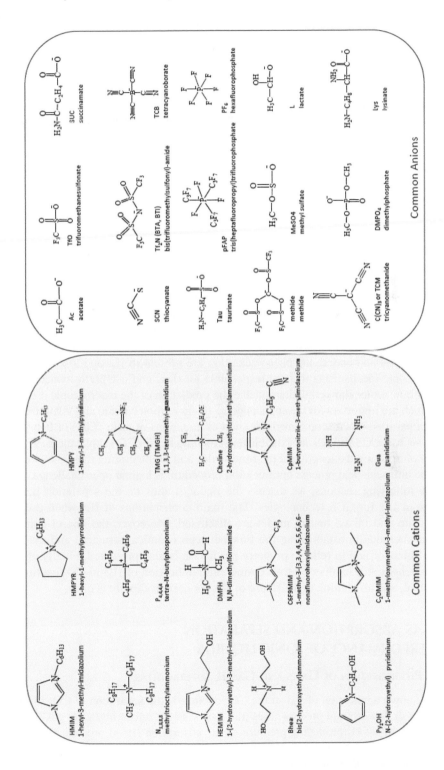

FIGURE 3.1 Cations and anions commonly used in ionic liquids for CO$_2$ capture.

efficient and recyclable catalysts for alkylation reactions, as described in the relevant pioneering works of DeCastro *et al.* [41] and Valkenberg *et al.* [42]. The first important reports on solid supported ionic liquid films for gas separation were reported by Baltus *et al.* [40] and Morgan *et al.* [43]. Since then, the progress on these IL-based composite and mixed matrix materials has followed exponential growth, with the involvement of many advanced porous adsorbents [44–47] (zeolites, MOFs, ZIFs, CNTs, MCM, SBA, CMS) and membranes [48–51] (polymeric, ZIF, zeolitic, carbon). The latter offer astonishingly high surface area and rich surface chemistry for dispersion and stabilization, in many cases through chemical grafting of ILs. Furthermore, in several solid supports, interaction of the deposited IL with the chemical functionalities of the pore surface (organic linkers in MOFs, acidic sites in zeolites, oxygenated functional groups in graphitic and carbon materials), or even the confinement into nanopores, endows ILs with distinctly modified and enhanced CO_2 absorption and selectivity properties when compared to the corresponding bulk liquid.

Computational methods significantly contribute in the efficient design of application-targeted IL materials and novel carbon capture and sequestration (CCS) processes. A multitude of computational methods are applied for the study of ILs and IL-based materials at various length and time scales and the field can greatly benefit from the advancements in efficiency of current methods and algorithms and the significant increase in computational power. Molecular simulations at the quantum mechanical, atomistic, or mesoscopic level are used for the investigation of the complex interactions and underlying phenomena that are present in ILs and to elucidate the microscopic mechanisms that are responsible for their end-use performance. At the same time, molecular simulations enable the prediction of the macroscopic properties, which are important for the application of ILs as carbon capture and separation media. In parallel, macroscopic methods such as equations of state (EoS) or hybrid methods such as COSMO-RS are valuable for screening ILs for CCS applications. The necessity of bridging the length and the time scales and of integrating these methods within the unit scale and process optimization procedure is a great open challenge.

In the following sections, we discuss the major studies on the separation performance of ILs for CCS technologies. The main contributions of ILs when used in composite and mixed matrix media are discussed. Moreover, the role of computational methods in establishing the link between chemical structure and macroscopic behavior and in reliable property prediction is also presented. Finally, the main challenges still hampering the commercial application of ILs in CO_2 capture are unveiled, and an outlook to the future of ILs for carbon capture is given.

3.2 GAS ABSORPTION AND SEPARATION PERFORMANCE OF IONIC LIQUIDS

3.2.1 Physisorption of Gases and Gas/IL Interactions

The first known application of ILs in the CO_2 capture is the separation of CO_2 from the flue gas or downstream process gases (tail gas of steam reforming/water gas shift reaction, tail gas of H_2 producing pressure-swing adsorption (PSA) process, tail gas from coal and biomass gasification, upgrade of biogas, sour natural gas, and shale gas). The main reason is the high solubility of CO_2 in ILs as compared to the other major gaseous components of these industrial streams, such as N_2, H_2, CO, CH_4, and O_2. ILs also offer other desired properties like negligible vapor pressure, high thermal stability, wide liquid range, nonflammability, and recyclability. Notably, acid gases like SO_2, H_2S, and NOx usually constitute part of the CO_2 capture challenge, as they occur concurrently (even in trace quantities) in the industrial streams producing CO_2. This brings further complexity to the CO_2 capture process due to competitive sorption, high corrosivity, and potential degradation of the capture medium. It is therefore not surprising to see many studies focusing on the mechanisms of acidic gas binding in ILs.

The main interactions responsible for the high CO_2 solubility in physisorbing ILs encompass electrostatic, van der Waals, and hydrogen-bonding forces [52]. It was initially proposed that the structure of the anions has a more profound effect on the CO_2 capture efficiency of the IL compared to the cation, following the order of $[BF_4]^-$ < $[TfO]^-$ < $[TfA]^-$< $[PF_6]^-$ < $[Tf_2N]^-$ < [methide]$^-$ < $[C_7F_{15}CO_2]^-$ < $[eFAP]^-$ < $[bFAP]^-$ [21]. This theory, however, could not explain why the Henry's law constants could differ as much as near to one order of magnitude for ILs with the same anion and various cations. For instance, the Henry's law constants of $[OMIM]^+[Tf_2N]^-$ and $[C_8F_{13}MIM]^+[Tf_2N]^-$ at 298.2 K are 30 and 4.5 bar, respectively [53], with the Henry's law constant expressing the pressure at which the mole fraction of the dissolved CO_2 in the IL equals unity. At the same time, another theory proposed that the enhanced CO_2 solubility in ILs should be attributed to interactions of the Lewis acid-base type, occurring between the acidic CO_2 and the usually basic anion of the IL. This theory was supported by experimental studies with ATR-IR spectroscopy [54], and although its origin emanated from experimentally verifiable observations, it completely failed in describing a universal trend for the CO_2 solubility in ILs. For example, it was contradictory to this theory that ILs composed of alkyl-methyl imidazolium cations and $[Tf_2N]$ anion exhibit lower Henry constants than the respective ones with $[BF_4]^-$ and $[PF_6]$ anions, when the ATR-IR spectroscopy concludes that the interaction of CO_2 with $[PF_6]^-$ and $[BF_4]^-$ is stronger than with $[Tf2N]^-$ [55]. Another inconsistency with this theory case is the basic anion IL $[BMIM]^+[DCA]^-$ (pK_a = 5.2), which exhibits a higher Henry constant than $[BMIM]^+[OAc]^-$, with the acetate anion being more acidic, having a pK_a value of 4.75. The study of macroscopic methods (EoS) or hybrid methods (COSMO-RS) convinced researchers to abandon the Lewis acid-base interaction. The macroscopic methods confirmed van der Waals forces as the main molecular interaction between CO_2 and IL instead of the Pearson's 'hard and soft acid/base' principle, which presumes that soft acids react faster and form stronger bonds with soft bases only, whereas hard acids react faster and form stronger bonds only with hard bases [52].

Since then the studies have progressed exponentially, and the number of IL structures examined for CO_2 capture increased tremendously, helping to better understand the IL structure/CO_2 solubility relation. Hence for physical absorption, the degree of anion and cation fluorination [56], the bromination on the anion [52], along with long alkyl chains with branching or ether linkages on the cation [57], and carbonyl or ester groups on the cation [58], endow ILs with enhanced CO_2 solubility relative to their alkyl analogues. Contrarily, site substitution of the C2 carbon on the cation with methyl [59], ether [60], hydroxyl, nitrile, and alkyne groups [61] is unfavorable for CO_2 solubility. In addition, most of the above empirical methods seem to converge with the theory of free volume, which is able to satisfactorily explain the trends of

CO_2 solubility in ILs relatively to their structure. The free volume effect, and specifically the concept of fractional free volume (FFV) defined by Shannon *et al.* [62] as $FFV = \dfrac{V_m - 1.3 V_{vdW}}{V_m}$ (with V_m being the molar volume and V_{vdW} being the van der Waals volume [63]), clarifies the unsolved issues and inadequacy of the anion effect and Lewis acid-base interaction, but it also explains the reduction in CO_2 solubility of ILs at high pressures (at a given temperature) and the increase in CO_2 solubility with increase in the molar volume of ILs, which follows an almost linear trend. If the alkyl chain length on the cations becomes long enough, the FFV of all ILs will converge asymptotically to a common value, and the solubility of CO_2 will be the same and independent of the anion type.

The most typical acid gas often existing in the stack gas of power plants is SO_2. Research shows that the effect of the alkyl chain length attached on the cation and of the type of the anion are not apparent for ILs interacting physically with SO_2, especially at the low-pressure region up to 1 bar [64]. The physical interaction of SO_2 with ILs composed of alkyl-methylimidazolium [RMIM]$^+$, alkyl-methylpyridinium [RMPy]$^+$, and tetramethylguanidinium [TMG]$^+$ cations and common fluorinated anions was evidenced by proton NMR and FT-IR spectroscopies [65]. In most cases and contrarily to CO_2, the SO_2 solubility declines with the increase of the carbon number in the alkyl chain while, surprisingly enough, the SO_2 solubility in [HMIM]$^+$[Tf$_2$N]$^-$ is even smaller than that in [HMIM]$^+$[BF$_4$]$^-$ [66–67]. Notably, the SO_2 solubility in ILs is always much higher than CO_2, which indicates the potential use of ILs for energy-efficient desulfurization processes. While the interaction between delocalized π electrons on the guanidine cation [68] and SO_2 through the van der Waals type of bonding is responsible for the high SO_2 solubility in TMG cation–based ILs, none of the reports explains the mechanism behind the high solubility of SO_2 in [RMIM]$^+$ and [HMPY]$^+$ cation–based ILs. There are also exceptional cases, such as N-butylpyridinium thiocyanate IL [C$_4$Py]$^+$[SCN]$^-$, and other ILs with specific functional groups. Interestingly, the functionalities (ether, nitrile) that are unfavorable for CO_2 solubility are favorable for SO_2 solubility, making such ILs the most efficient for selective SO_2/CO_2 separations [69].

As a general remark, up to 2013 SO_2 solubility was studied in marginally more than 20 different ILs, when at the same time more than 150 different IL structures had been investigated already for their CO_2 capture efficiency. The limited data related to SO_2 does not allow for the extraction of empirical conclusions on the IL structure/SO_2 solubility relation and the establishment of theories that could explain generalized trends. Currently, the studies on SO_2 are increasing rapidly compared to those for CO_2, and new unexplored anions and cations are added to the SO_2 list, such as the tetraalkyl-phosphonium and hydroxyl-ammonium cations and the tertazolate [70], pyridinolate [PyO]$^-$, pyridinecarboxylate [PyCOO]$^-$, pyridinesulfonate [PySO$_3$]$^-$ [71], phosphate, furoate [FA]$^-$, suberic acid [SUB]$^-$ [72–73], and poly(ethylene glycol) bis(carboxymethyl) ether [PBE]$^-$ anions [74]. Moreover, ether-linked diamine carboxylate ILs and caprolactam tetrabutyl ammonium bromide ILs (CPL-TBAB IL) have recently shown high SO_2 absorption capacity, satisfactory SO_2 removal efficiency, and excellent reusability [75], possessing good application prospects in flue gas desulfurization. In parallel, thermodynamic models based on UNIFAC-IL model extension [76] and a modified Sanchez-Lacombe EOS [77] are now capable of predicting quantitatively (with a <20% deviation from experimental data) the solubility of SO_2 in ILs, when a few years ago, in most cases the COSMO-RS models predicted SO_2 solubility values, which were by 60% lower as compared to the experimentally measured ones.

H_2S is the second acid gas of high industrial interest. The competitive absorption between CO_2 and H_2S constitutes one of the major challenges of the conventional amine scrubbing processes for natural gas sweetening [78]. Therefore, ILs are introduced as potential solvents to remove H_2S from these gases. In general, the single-gas solubility in ILs follows the order SO_2 (H_2S) > CO_2 ≈ N_2O > C_2H_4 > C_2H_6 > CH_4 > Ar > O_2 > N_2 > CO > H_2 at the same temperature and pressure. Therefore, H_2S shares common characteristics with SO_2, manifested by the higher solubility of both gases in ILs as compared to CO_2, but at the same time, the IL structure–H_2S solubility relation converges mostly to that of CO_2, meaning that the cation has a moderate effect on solubility. For alkyl-methylimidazolium cation-based ILs, the H_2S solubility increases with increasing alkyl chain length on the cation [79]. Similarly, with the solubility trend of CO_2 in ILs, the degree of anion fluorination improves the H_2S absorption capacity, which enhances with the increase of the number of trifluoromethyl (CF_3) groups on the anion [79]. Regarding the mechanism, several authors confirmed that strong hydrogen bonds are formed between H_2S and the anions [80]. The formation of hydrogen bonds (N–H–S) is also responsible for the high H_2S solubility in dual Lewis base functionalized ILs [81]. Other types of ILs include hydrophobic protic ILs and substituted benzoate-based ILs that exhibit not only high absorptivity but also extremely high H_2S/CO_2 selectivity, reaching values up to 40 [82]. Despite the confirmation of the H_2S/IL interaction mechanism, the literature to date also presents some controversial results on the effect of the anion, especially when examining ILs with different cations. For instance, for the 1-(2-hydroxyethyl)-3-methyl-imidazolium [HEMIM]⁺ cation we have the order [HEMIM]⁺[Tf₂N]⁻ > [HEMIM]⁺[TfO]⁻ > [HEMIM]⁺[PF₆]⁻ > [HEMIM]⁺[BF₄]⁻, while for the [HMIM]⁺ cation the H_2S solubility follows the order [BF₄]⁻ > [PF₆]⁻ ≈ [Tf₂N]⁻ [83]. There are also controversies in the results obtained for the same cation [BMIM]⁺, where some authors claim the order [Cl]⁻ >[BF₄]⁻ > [TfO]⁻ > [Tf₂N]⁻ >> [PF₆]⁻ [84], while others proposed that H_2S solubility follows the order [Tf₂N]⁻ > [BF₄]⁻ > [PF₆]⁻.

While the studies on acidic gases (H_2S and SO_2) are still scarce, NH_3 enjoys the performance of a high number of absorption tests at a wide range of temperatures and pressures, encompassing about 300 different IL structures, including hydroxyl-functionalized ILs such as the 1-2-(hydroxyethyl)-3 methylimidazolium tetrafluoroborate ([EtOHMIM]⁺[BF₄]⁻) and cholinium cation–based ILs such as the cholinium bis(trifluoromethylsulfonyl) imide ([choline]⁺[Tf₂N]⁻) [85]. Most of these studies concluded that NH_3 interacted more strongly with cations than anions *via* the formation of intermolecular complex interactions, which is different from the CO_2 absorption by ILs [86]. Since the most common cation of ILs encompasses heterocyclic structures, NH_3 solubility is mainly promoted by the formation of strong hydrogen bonds between the nitrogen atom in NH_3 and the ring hydrogen atoms of the heterocyclic cations. Especially, the occurrence of strong hydrogen bond interaction between NH_3 and the H atom of the hydroxyl group on the cation results in high NH_3 absorption capacity [87]. Protic ILs such as imidazolium cation–based ILs with side chain in position 1 (1-butyl-3-methylimidazolium bis(trifluoromethylsulfonyl) imide ([BMIM]$^+$[Tf$_2$N]$^-$)) exhibit a strong hydrogen bond donating ability for NH_3 absorption, and the solubility reaches up to 2.69 mol NH_3/mol IL at 313 K and 0.1 MPa [88].

Apart from the acidic and basic gases reported above, other gases of high industrial interest, such as N_2, H_2, CO, CH_4, and O_2, are also reported. It is promising that in the few existing studies relative to the solubility of these gases in ILs, the absorption capacity was negligible and often non detectable by the available experimental techniques. This means that CO_2 can be effectively separated from its mixtures with these gases. The gas solubility measurement techniques rely on three basic methods: gravimetric [89], isochoric saturation [90], and synthetic (bubble point) [91]. Each method has been applied by several research groups with slight modifications using variable setups and apparatuses. It is therefore important to note that readers should carefully identify the reliability of gas solubility data coming from different sources, since apart from the inaccuracies related to experimental techniques and complex calculations, there are also large discrepancies generated by the degree of IL purity and the initial regeneration of the samples used in the measurements. In this context, we do not intend to extensively focus on the gas solubility performance of ILs for CO, N_2, and O_2, since it is our belief that the controversy observed by several authors related to the temperature effect on the solubility of these gases emanates from the incapability of the experimental techniques to reliably detect the dissolution of extremely small quantities of gas into the IL. In fact, there are many controversial reports claiming enhancement of the solubility of these gases with the increase of temperature, and these controversies emphasize the need for more accurate experimental studies and deeper theoretical understanding based on molecular simulations. However, as a general trend, authors converge to the point that fluorination on the anion or the increase of alkyl chain length on the cation is favorable for increasing the solubility of CO, N_2, and O_2 [92]. As is the case with the aforementioned gases, the solubility of H_2 in ILs is remarkably lower than that of CO_2 (one to two orders of magnitude), and the effect of temperature on the H_2 solubility is not straightforward. In some cases, H_2 solubility shows an 'inverse' temperature effect, meaning that it presents a maximum at a temperature between 283 K and 343 K. This phenomenon was recently observed for the ILs [BMIM]$^+$[BF$_4$]$^-$ [93] and [EMIM]$^+$[EtSO$_4$]$^-$ [94], and although it can be counted among the many controversial results generated by the low solubility in conjunction with the inaccuracy of experimental methods, it also denotes that more work is still needed to explore the solubility mechanism and identify the structure-property relation.

Stepping over to the field of hydrocarbon gases, methane deserves the greatest focus since it constitutes the major component of natural gas, shale gas and biogas, and it must be purged from CO_2 and other constituents such as He, N_2, H_2S, SO_2, and some heavier hydrocarbons. In general, the CH_4 solubility in ILs follows similar trends with CO_2 [95]. What makes CO_2/CH_4 separation applications more challenging for ILs is that although CH_4 is less soluble in ILs as compared to CO_2, it exhibits much higher solubility than H_2, CO, N_2, and O_2. Therefore, the CO_2/CH_4 separation factor is not as high as in the case of CO_2 paired with other gases. Althuluth *et al.* [96] found that the solubility of CH_4 in various ILs slightly decreases in the order $[HMIM]^+[Tf_2N]^-$ > $[EMIM]^+[FAP]^-$ > $[BMIM]^+[Tf_2N]^-$, and since 1-ethyl-3 methylimidazolium tris(pentafluoroethyl)trifluorophosphate $[EMIM]^+[FAP]^-$ shows one of the highest carbon dioxide (CO_2) solubilities among all ILs studied so far, it shows great potential for a future CO_2/CH_4 separation application. However, it was only recently that a new class of ILs based on alkyl-methylimidazolium cation and tricyanomethanide anion ($[C(CN)_3]^-$ or $[TCM]^-$) became known, showing for the first time CO_2/CH_4 separation factors of above 10, along with significantly lower viscosities compared to ILs with fluorinated anions [97].

3.2.2 CHEMISORPTION IN TASK-SPECIFIC ILs (TSILs)

Chemical complexation of gases with ILs offers high and fast capture performances for tail gas applications at ambient pressure and temperature, containing a low percentage of the absorbate. Despite the benefits of low regeneration energy (the heat of CO_2 absorption of most physisorbing ILs is approx. -12 kJ/mol), and remarkably high CO_2 selectivity over N_2 and O_2, the physical solubility in ILs itself does not suffices for a feasible commercial application, especially when the target is the post-combustion CO_2 capture. The main reason is the demand for large IL circulation rates and compression/decompression cycles, which significantly increase the capture costs. It is a fact that physisorbing ILs would need a tenfold increase in their solubility to beat aqueous monoethanolamine (MEA) solvents. Contrarily, chemical complexation in ILs is strong enough to increase capacity and decrease IL circulation rates yet weak enough compared to MEA to keep regeneration energies and temperatures at a low level (40–50 kJ/mol compared to approx. 90 kJ/mol for MEA). The chemical interaction of gases with ILs is solely reported for CO_2 and SO_2, where the interaction is mostly built on amine chemistry while there are also reports of chemical interaction with task-specific ILs bearing an acetate moiety [98–104]. Most of the studies relevant to amine-functionalized ILs for CO_2 capture are focusing to go beyond the upper bound threshold of 1:2 CO_2 to IL molar uptake. This limit is imposed by the most plausible reaction mechanism, which means that for each amine moiety that binds one CO_2 molecule through the formation of a carbamate bond, it is necessary that one additional amine group is sacrificed (e.g., deactivated *via* protonation) to end up with the yield of a stable carbamate salt. Moreover, in most cases, the carbamate mechanism leads to a huge increase in the viscosity of the CO_2-IL system attributed to the formation of an extended hydrogen bonding network [98]. Thus, ILs having a limited number of free hydrogens able to participate in hydrogen bonding are required.

Mindrup and Schneider [105] showed that the tethering ion and tethering point of the amine group is decisive in controlling CO_2 reactions. Hence, local cation tethering favors 1:2 binding while local anion tethering opposes it, leading to CO_2 binding through the formation of carbamic acid instead of carbamate salt. For the challenge of viscosity increase, Goodrich *et al.* [106] and Gutowski *et al.* [99] studied TSILs with phosphonium cation (trihexyl(tetradecyl)phosphonium $[P_{6,6,6,14}]^+$) and several amino acid anions, including lysinate, isoleucinate, sarcosinate, methioninate, taurinate, glycinate, and prolinate. They found that ILs capturing CO_2 are readily subjected to an enhancement of their viscosity, which is intensified as we go from the prolinate to the lysinate anion in the above-described order and concluded to the point that prolinate anion, due to its ringed structure, has the least number of free hydrogens able to form hydrogen bonds. Furthermore, the reports of Gurkan *et al.* [107] and Seo *et al.* [108] showed that by retaining the amine moiety in ring structure, it is possible to further reduce free hydrogens while achieving CO_2 to IL molar uptake of about 1:1 at room temperature and pressure of 1 bar. In this context, aprotic heterocyclic anions (AHA) such as pyrrolides, pyrazolides, imidazolides, and triazolides combined with phosphonium-based cations have been used in these studies. Beyond showcasing the minor increase of the viscosity upon CO_2 binding, the studies proved that the reaction enthalpy (ΔH) of the IL with CO_2 depends on the AHA involved as anion, with ΔH values ranging between -54 kJ/mol and -41 kJ/mol for indazolide [In]$^-$ and SCH_3 substituted benzylimidazolide [2-SCH₃BnIm]$^-$ anions, respectively. Furthermore, phosphonium cations (due to the apparent acidity of the proton on the α-carbon [109]) contribute to the CO_2 uptake at high temperatures (e.g., 333 K) and form phosphonium ylide with a mechanism similar to the formation of carbene in imidazolium cation–based ILs [110–111].

Aprotic heterocyclic anions (AHA) help to tune the enthalpy of reaction between the IL and CO_2. They work with capability offered by tetra-alkylphosphonium–based cations to adjust the melting point of the IL *via* modification of the alkyl chain length. This has resulted in a new generation of phase change ILs (PCILs) for CO_2 capture, like tetraethylphosphonium benzimidazolide [P$_{2,2,2,2}$]$^+$[BnIm]$^-$. The melting point of the PCIL-CO_2 complex is below that of the pure PCIL [112]. Being solid at the normal flue gas processing temperatures (e.g., 40–80 °C), the PCIL reacts stoichiometrically and reversibly with CO_2 to form a liquid by absorbing heat, which is highly beneficial as it reduces the cooling duty in the absorber column. In the desorber column (stripper), the PCIL-CO_2 system goes from the liquid to the solid phase, releasing CO_2 and heat. Thus, the heat duty in the stripper is reduced by the heat of fusion of the phase change material. In another approach, triethylenetetramine (TETA) was used to react with several acids to synthesize ILs ([TETA]$^+$[Br]$^-$, [TETA]$^+$[BF$_4$]$^-$, or [TETA]$_3^+$[NO$_3$]$^-$). Then, these ILs were dissolved in various solvents to form phase change systems [113]. The concept behind the mixed solvents approach differs substantially from that of the AHA anion–based PCILs, as a mixed solvent system separates into one rich and one lean in CO_2 phases upon absorption of CO_2. Thus, the reduction of the energy required for regeneration stems primarily from the ability to separate the two phases. As such, the CO_2-containing stream is driven to the regenerator, resulting in significant reduction of both the sensible and latent heating requirements.

Amine-functionalized ILs are the most promising candidates for post-combustion CO_2 capture, both as direct solvents and phase-change materials (usually blended with other solvents to form phase change systems). Although the technical benchmark is to exceed the technical capture performance and energy penalty of aqueous MEA solvent, efforts should also focus on improving the TSIL synthesis approach in terms of cost. For instance, the usual way to produce

TSILs is to displace the halide from an organic halide containing the functional group by a parent imidazole or phosphine. Although this is simple, it requires complete removal of the halide, which is time-consuming, costly, and leads to depressed yields. Moreover, in some applications, halide contamination may not be acceptable, even in ppm quantities. Especially for the application in CO_2 capture, halide-free ILs are desirable since halides can increase viscosities and corrosion rate (together with free amines and free acids). Alternatively, TSILs with amino-functionalized anions can be synthesized by a neutralization reaction of commercially available amino acids with a base such as [RMIM]+OH− or alkyl-methylmorpholine hydroxide [RMMorph]+OH−. This synthetic route involves three steps. As an example, for the TSIL [EMIM]+[Lys]−, the route starts with the reaction between 1-ethyl-3-methylimidazolium methyl carbonate and H_2SO_4, continues with the production of 1-ethyl-3-methylimidazolium hydroxide through reaction between [EMIM]+[HSO$_4$]− and barium hydroxide octahydrate, and ends with the neutralization reaction between [EMIM]+[OH]− and the amino acid lysine. The benefits of this synthetic approach, which completely avoids the involvement of halides, are counterbalanced by the drawback of a final product containing large amounts of water that should be evaporated. Thus, the process is highly energy-consuming and increases the cost of the TSIL. It can be realized that based on their CO_2 capture performance, TSILs present great potential for a commercial application. However, there are still certain impediments towards such an achievement, mainly related to the high cost of TSILs and the difficulty to obtain them in pure form.

Sulfur dioxide (SO$_2$) constitutes the second acidic gas, which can be chemisorbed by ILs and TSILs. In TSILs composed of tetramethylguanidinium [TMG]+ and monoethanolamine [MEA]+ cations and lactate [L]− anion, the chemical interaction of SO_2 with the amine group on the cation has been verified by FTIR spectroscopy [114–116]. This was also showcased in the studies of Yuan et al. [117], who evidenced that chemical reactions between the amine groups and SO_2 are mainly responsible for the high SO_2 absorptivity of hydroxyl ammonium cation-based ILs. Using the gravimetric method, Shiflett et al. [118] showed that hydroxyl ammonium-based ILs can achieve a mole ratio of IL to SO_2 equal to 1. In general, SO_2 dissolution by TSILs including ILs with acetate [OAc]− and methyl sulfate [MeSO$_4$]− anions is based on both chemical and physical interaction. Comparing experimental solubility data with predicted results by the COSMO-RS model constitutes further proof on this statement. Indeed, the convergence at high pressures and the deviation at low pressures between the predicted and experimental results indicates that absorption based on a physical mechanism is predominant in the high-pressure region, whereas in the low-pressure region SO_2 dissolution is mainly based on chemical interaction.

It should be noticed that CO_2 solubilities are mostly reported in literature on a mole fraction basis and that the Henry constant is the property frequently used to compare the CO_2 capture performance of ILs. Carvalho and Coutinho [119] claimed that physical CO_2 absorption in ILs is dominated by entropic effects, hence CO_2 solubilities when plotted as molality (e.g., mmol CO_2/g IL) versus pressure are solvent independent and fall on a common universal curve. Although there are exceptions to this general trend, such as in the case of acetates (m-2-HEAA) and

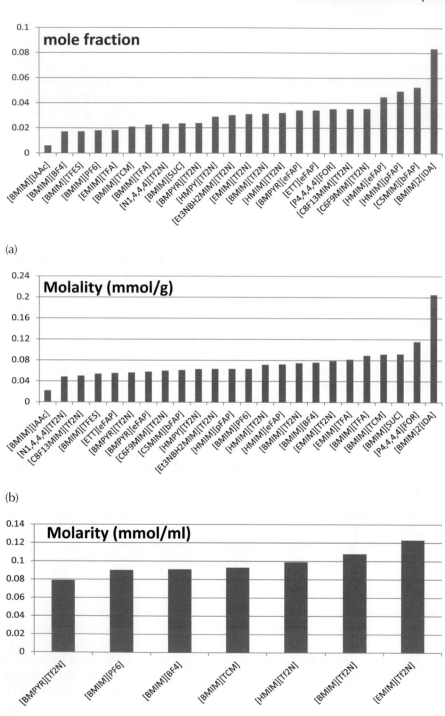

FIGURE 3.2 Ranking of ILs regarding their performance in capturing CO_2 based on: (a) the mole fraction, (b) the molality, and (c) the molarity of CO_2.

formates (m-2HEAF) where the solubility is dominated by the formation of electron donor-acceptor (EDA) complexes, we believe that comparing the CO_2 capture performance of ILs based on the molality – or better on the molarity (mol CO_2/ m^3 solvent) – is more valuable from a practical point of view. It is a fact that the demand for large solvent volumes brings up a cascade of adverse effects to the performance of a CO_2 capture process, such as the requirement for huge installations and the concomitant energy expense to heat large volumes of inert materials. Thus, ILs exhibiting the highest CO_2 molarity should be the preferred choice for an industrial application, with the prerequisites that their viscosity is kept within reasonable limits and further side effects in the capture performance should be reduced. CO_2 capture performance of 25 different ILs when ranked on the basis of the CO_2 mole fraction, molarity, and molality is stressed in the bar charts presented in Figure 3.2. To create these plots, CO_2 mole fractions found in the literature were based at a temperature of 298 K and pressure of 1 bar [120–121], while the mole fractions were converted to molarity and molality by knowing the molecular weight and the density of the ILs.

3.3 MICROSCOPIC MECHANISMS AND PROPERTY PREDICTIONS USING COMPUTATIONAL METHODS

The key characteristic of ILs is their chemical tunability allowing targeted application, as their properties can largely be controlled by slightly modifying or changing their chemical structure. Hence, tailor-made ILs can be developed with a controlled separation performance while being environmentally benign and energy efficient for carbon capture technologies [21, 30, 56–57, 122–127]. In this direction, there is a set of important features that ILs need to possess for their industrial scale application as separation media in terms of their transport properties such as viscosity and dynamical properties, gas permeability as well as selectivity in the presence of more than one gas species. Computational methods are a powerful and valuable tool for the molecular design of materials in general [128], and especially in the case of ILs and IL-based materials. They enable the investigation and unravelling of the microscopic mechanisms that govern the macroscopic properties of this diverse class of materials and the prediction of a wide range of properties. Molecular simulation methods are extensively applied at various length and time scales [129], including ab initio and quantum mechanical (QM) methods [130], QM methods in combination with molecular mechanics (MM), atomistic simulations such as molecular dynamics (MD) and Monte Carlo (MC) methods, mesoscopic coarse-grained modeling, and multi-scale molecular modeling [131–132]. Thermodynamic modeling using equations of state (EoS), correlative methods such as QSPR [133–136], and hybrid QM/continuum models [137–139] are also widely employed for the screening of IL systems [140] at a macroscopic scale. The existing literature on computational methods is enormous and rapidly expanding, and an exhaustive analysis is not feasible within the scope of the present chapter. Thus, this section will only involve a representative discussion, primarily focusing on briefly identifying the role of computational methods in CCS and on analyzing some indicative gas/IL-related computational approaches.

ILs are highly structured materials [150] compared to conventional fluids and exhibit a complex dynamical and spatiotemporal behavior [151–161], even resembling attributes of supercooled liquids at lower temperatures. The intricate underlying characteristics are the outcome of an interplay of a number of coexisting factors that emanate from the simultaneous ionic and organic nature of the ions and their complex interactions. These include the strong electrostatic interactions, dispersion forces, the anisotropic formation of hydrogen bonds, polarizability and charge transfer phenomena that are present. All these factors directly affect their macroscopic behavior such as transport properties, especially diffusivities and viscosities, hence determining their end-use performance. The intense Coulombic, h-bond, and van der Waals interactions in IL systems affect for instance, the viscosities that are much higher compared to molecular fluids. Investigation of the influence of the anion, for example, in the highly studied imidazolium-based ILs category reveals that the fluorinated anions are more viscous than the cyano-based ones following the order $[PF_6]^- >$ $[BF_4]^- > [Tf_2N]^- > [N(CN)_2]^- > [B(CN)_4]^- > [C(CN)_3]^-$ (or $[TCM]^-$), as shown in Figure 3.3. The TCM-based ILs exhibit low viscosities, as determined both experimentally [141, 147] and computationally [141, 153, 162] due to charge delocalization effects and a weaker interaction between the ions, while $[PF_6]^-$ possesses the highest viscosity as expected from the strong interionic interactions. For the fluorinated anions, the viscosity decreases as the size of the anion increases [163] due to the reduced electrostatic interactions and the enhanced

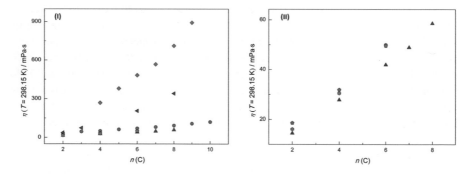

FIGURE 3.3 Plots of: (I) Viscosity at 298.15 K as a function of the number of carbons on the alkyl chain of the cation, n(C), for several IL series; (II) zoom in on the viscosities of the cyano-based ILs.
Abbreviations: ▲, $[C_nmim]^+[TCM]^-$ (n = 2, 4, 6–8) [141]
■, $[C_nmim]^+[Tf_2N]^-$ (n = 2–10) [142–143]
♦, $[C_nmim]^+[PF_6]^-$ (n = 4–9) [144]
◄, $[C_nmim]^+[BF_4]^-$ (n = 2, 3, 6, 8) [145–146]
♠, $[C_nmim]^+[B(CN)_4]^-$ (n = 2, 4, 6) [147–148]
●, $[C_nmim]^+[N(CN)_2]^-$ (n = 2, 4, 6) [148–149]
Source: Reproduced from [141].

flexibility of the anions, a fact that is not detected in cyano-based cases, where the TCM-ILs exhibit the lowest viscosity [148]. The effect of the alkyl chain length increase has a much greater impact on the highly viscous fluorinated anion ILs in comparison to the cyano-based ones, as shown in Figure 3.3. It generally results in a reduction of the melting point of the ILs [164], while for long alkyl tails self-assembles of nanostructures have also been observed [150]. The correlation of the macroscopic behavior with the ion structures is not straight forward to determine and understand and this fact is enhanced as the complexity of the ions escalates. The implementation of molecular simulation methods is pivotal, in shedding light in this direction by investigating the microscopic complexity and by property predictions at conditions where experimental measurements are difficult to be performed or cannot be carried out reliably, such as, for example, for very light gases, for which inconsistent experimental findings are often reported. Computational methods can also be used for decoupling the numerous underlying and interconnected factors that control the macroscopic properties and to isolate specific concepts at a time by applying targeted modifications on the simulated systems, which in many cases can also be artificial in order to determine a specific trend [165].

The subatomic and ionic interactions and structures [166–174] of ILs are being widely investigated by applying ab initio and QM methods [130]. QM approaches are very computationally intensive and are therefore restricted to small systems of single ions, ion pairs, or small ion clusters; in some cases, hybrid QM/MM are utilized to extend the system sizes and study reaction mechanisms. Charge distributions and charge scaling effects [153, 167, 175–176], which result in a reduced total ionic charge, are detected. Moreover, the polarizability phenomena, which are rooted in the distortion of the electron cloud of the ions due to local strong interactions, are also examined [176–179]. Hydrogen bond interactions [180–185], often present in ILs as well as their coupling with the electrostatic forces, are also a very important aspect, especially when multiple ion hydrogen bonds may exist within an IL pair resulting in chelated structures, as all electronegative atoms and not only oxygen are able form hydrogen bonds. QM methods have been applied for the determination of CO_2-IL ion interactions, mostly focusing on imidazolium-based ILs. Density functional theory (DFT) was applied by Bhargava et al. [186] for the calculation of CO_2 with different anions, and a Lewis acid-base interaction was revealed as dominant, while the anion-CO_2 binding energies exhibited an inverse trend to the CO_2 solubility in the ILs under study. Similar findings [187] on a Lewis acid-base interaction were found for CO_2 and [Ac]$^-$ anions, both from DFT calculations and Raman spectroscopy, identifying that this interaction may be one significant factor of some CO_2-IL systems among other important factors, while reporting no perturbation of the IL structure by the CO_2 presence. Kirchner and co-workers [188] have implemented QM and ab initio MD (AIMD) and detected dispersion interactions between CO_2 and the cation for the [EMIM]$^+$[OAc]$^-$ IL. Firaha and Kirchner [189] also applied AIMD of ethylammonium nitrate, reporting no effects on the IL structure due to the solvation of CO_2. An effect of the cation in CO_2 solubility was also reported from DFT calculations of brominated ions pairs via the formation of halogen bonds [190]. QM interaction descriptors based on Koopmans' theorem and DFT were calculated [191] for imidazolium and pyridinium-based cations paired with various anions, showing a dominant role of the anion in CO_2 solubility. DFT methods, in combination with symmetry adapted perturbation theory (SAPT), were implemented [192] for the study on CO_2 sorption in various protic and aprotic ILs, identifying the ions with a strong dispersion component, such as with bulky groups (such as Tf_2N, carboxylic and sulfonic groups) and groups with the π delocalization, as the best candidates for physical absorption of CO_2. QM studies also exist on the investigation of simultaneous capture of CO_2 and SO_2 gases by ILs [193].

Solubility of solutes in a solvent can be determined from atomistic simulations by calculating the excess chemical potential of the solutes, implementing free energy schemes such as Widom test particle insertions [194], Bennett's acceptance ratio [195], thermodynamic integration [196], and expanded ensemble methods [197]. These methods allow determination of the solubility in the infinite dilution regime and have been used for the calculation of Henry's law constants of various gases in ILs based on trajectories obtained from MC or MD simulations. Deschamp *et al.* [198] used thermodynamic integration to calculate infinite dilution solubility of CO_2, CH_4, Ar, O_2, and N_2 in $[BMIM]^+[PF_6]^-$ and $[BMIM]^+[BF_4]^-$ ILs, extracting the correct relative solubility of the various gases and correctly predicting an exothermic solvation for CO_2, but they were unable to reproduce the temperature trend for the nonpolar gases. Shah *et al.* [199] calculated the CO_2 Henry's law constant also in $[BMIM]^+[PF_6]^-$ using the Widom test particle insertion method, overestimating the solubility in comparison with experiments. They subsequently used an expanded ensemble MC scheme [200], achieving a better agreement for CO_2 and qualitative agreement for other penetrants such as methane and oxygen gases. Paschek and co-workers [201] determined CO_2 infinite dilution properties in $[RMIM]^+[TF_2N]^-$ ILs (with R= ethyl, butyl, hexyl, octyl), in the temperature range 300–500 K, using both Bennett's overlapping method and the Widom test particle method; implementing appropriate sampling they extracted results in good agreement between the two techniques and also with experimental data. They predicted a decreasing CO_2 solubility trend with increasing temperature, which was connected to the negative solvation entropy of CO_2 due to the attractive gas-IL interactions. Various additional gases (noble gases, H_2, N_2, O_2, CH_4) were also studied in the same system [202] and in ILs with imidazolium-based cations [203] paired to $[BF_4]^-$ and $[PF_6]^-$; infinite dilution solubility was determined and enthalpy-entropy compensation effects were reported. Ghobadi *et al.* [204] calculated CO_2 and SO_2 Henry's law constants in imidazolium-based ILs. Liu *et al.* [205] implemented an alchemical free energy scheme and Bennett's acceptance ratio analysis for the study of CO_2 solubility in $[EMIM]^+$-based ILs paired to $[BF_4]^-$, $[TF_2N]^-$ and $[B(CN)_4]^-$ anions. They underestimated the Henry law constants for $[BF_4]^-$ and $[B(CN)_4]^-$ and strongly overestimated the permeability properties of $[B(CN)_4]^-$, but the overall expected solubility trend for the ILs under study was extracted. The study of various gases in $[EMIM]^+[TF_2N]^-$ IL [205] led to the correct temperature dependence of solubility for both polar and nonpolar gases, such as N_2, which exhibited an endothermic solvation. Vergadou *et al.* [206–207] studied the infinite dilution solubility of a number of gases such as Ar, CO_2, N_2, CH_4, O_2, SO_2, and H_2S in the [RMIM] $[C(CN)_3]$ IL family incorporating an optimized and validated force field [141, 153] and in a wide temperature range using Widom test particle insertion method, with their results being in very good agreement with experimental measurements [121, 208] available for CO_2 and N_2. A very good separation ability was extracted for the CO_2/N_2 separation, with N_2 solubility being more than two orders of magnitude less than the one of CO_2, while SO_2 solubility was determined as more than one and a half orders of magnitude higher than CO_2. The temperature dependence of N_2 solubility using a relatively simple two-site model for N_2 showed an endothermic trend, while the inverse was observed for CO_2 and most of the other gases.

Molecular simulations on the calculation of gas solubility at higher pressures and the determination of the sorption isotherms are quite challenging and scarcer. Maurer *et al.* [209–211] were the first to calculate sorption isotherms by applying the Gibbs ensemble MC (GEMC) at constant pressure and temperature to calculate the solubility of CO_2, O_2, CO, and H_2 in [BMIM]$^+$[PF$_6$]$^-$ IL in a united-atom representation. Shi *et al.* [212] carried out a continuous fractional component MC scheme (CFC MC), which enables the gradual insertion or deletion of solute molecules in combination with a hybrid MC/MD scheme [213], and predicted CO_2 sorption isotherms in [HMIM]$^+$[TF$_2$N]$^-$ IL and reported a negative excess molar volume. In subsequent studies [214–215] they included more gases such as N_2 and O_2, reporting an exothermic behavior for N_2, and they also implemented GEMC for the determination of mixed gas sorption, presenting a competing behavior for CO_2/SO_2 at high pressures and near-ideal CO_2/O_2 and SO_2/N_2 selectivities [214]. In imidazolium-based [TF$_2$N]$^-$ ILs, GEMC was used [216–217] to study the pure and binary behavior of CO_2 and H_2, and CH_4 and CO_2 and a slightly nonideal CH_4/CO_2 solubility selectivity was reported. Pure and mixed gas solubility for CO_2 and CH_4 was also recently investigated in binary ionic liquid mixtures of [BMIM]$^+$[TF$_2$N]$^-$ and [BMIM]$^+$[Cl]$^-$ [218]. Zhang *et al.* [219] used a CFC MC scheme to determine the sorption isotherms of CO_2 in [HMIM]$^+$[FEP]$^-$ at two temperatures and up to 20 MPa and captured the experimentally observed trends satisfactorily. The above MC methods require additional MD simulations in order to study the mechanisms and dynamic properties of the IL-gas mixtures, which is an important contributing factor, besides solubility, in the permeability performance and selectivity of separation media. Recently, Karanasiou *et al.* [220–221] implemented a multistage iterative scheme [222] that incorporated a series of MD simulations in the NPT ensemble and the Widom particle insertion method until convergence in pressure is achieved, to determine the sorption isotherms and associated volumetric effects. They applied the method for the study of CO_2 sorption up to 10 MPa in [BMIM]$^+$[C(CN)$_3$]$^-$ in the temperature range 298–398K and predicted a negative molar volume change and sorption isotherms in very good agreement with experimental measurements [223–224] of the same IL/CO_2 system and at similar thermodynamic conditions. Simultaneously, transport properties and local dynamics were quantified, with the ion dynamics being enhanced as the CO_2 concentration is increased, while the viscosity decreases in a self-consistent manner. Diffusivity of CO_2 in the [BMIM]$^+$[C(CN)$_3$]$^-$ was calculated and determined to be higher than in other common ILs such as [BMIM]$^+$[PF$_6$]$^-$ and [BMIM]$^+$[BF$_4$]$^-$, due to the higher viscosity of the latter ILs, and activation energies were also reported. No significant effect on the local structure properties of the IL was detected due to the CO_2 presence for gas mole fractions up to 0.4.

The effect of CO_2 presence on the structure of ILs has been studied by several atomistic simulations [205, 212, 220–221, 225–232] of IL-CO_2 mixtures and various trends for the ILs solvation ability have been examined. In agreement with experimental findings, the IL structure was largely detected as almost unperturbed by the presence of CO_2 up to moderate concentrations, indicating that CO_2 is accommodated in pre-existing cavities in the IL, as CO_2 exhibits a very small partial molar volume when dissolved in ILs than the one of the bulk supercritical CO_2 at the same thermodynamic conditions [231]. Therefore, the free volume was also considered as an additional important characteristic for improved CO_2 sorption in ILs, along with the CO_2-ion interactions. Consequently, factors that enhance the free volume in the IL structure, like weaker anion-cation interactions or longer cation alkyl chains or incorporation of fluorinated ions [219], also have been considered as favorable, although the interplay between all these factors is not totally unambiguous. Several molecular simulation studies analyze the relation between the strength of interaction between the ions and CO_2 absorption capability. A dominant anion-CO_2 interaction was initially reported [227] with the cation playing a secondary role. Babarao *et al.* [229] performed MD simulations of imidazolium-based ILs with nitrile-containing anions and observed a higher CO_2 solubility for [EMIM]$^+$ paired to [B(CN)$_4$]$^-$ than when paired to [TF$_2$N]$^-$ due to the weaker anion-cation interactions. In a subsequent study [233], they reported a highest solubility for [B(CN)$_4$]$^-$ anion, by increasing the nitrile groups in anions paired to [BMIM]$^+$, attributed to weaker anion-cation interactions that were extracted though using QM methods on isolated ion pairs. Tong *et al.* [232] investigated 12 amino acid ionic liquids and suggested that high CO_2 solvation capacity in ILs is linked to strong CO_2-cation van der Waals interactions as well as strong CO_2-anion electrostatic interactions. Huang *et al.* [231] performed MD simulations of [BMIM]$^+$[PF$_6$]$^-$ in mixture with CO_2 and showed that although the radial distribution function that characterizes the spatial organization of the anions remained almost unchanged, the size of the calculated pre-existing cavities was not sufficient to host the CO_2 molecules, and an orientation rearrangement of the anions was necessary for the accommodation of CO_2. Local rearrangement phenomena in order to form larger cavities for the CO_2 accommodation were also observed in molecular simulations of hexamethylguanidinium lactate ILs [80]. A similar conclusion for the IL reorganization was drawn by Klähn *et al.* [234], who performed MD simulations of 10 ILs in the imidazolium-based family in pure and in mixture with CO_2 at experimentally predetermined saturated concentrations at 10 bar and 298 K. They presented a strong relationship between the unoccupied space in pure ILs, which results from a weaker ion cohesion, and the IL CO_2 solvation ability. An opposite finding was reported in the work of Lourenço *et al.* [235], in which a weak relationship between fractional free volume and CO_2 solubility was found. Recently, Liu *et al.* [236] simulated a series of multivalent ILs with imidazolium-based cations paired to sulfonate and sulfonamide anions and concluded that the anion choice is significant in the ILs with larger fractional free volume (FFV), while the free volume effects become dominant in the ILs that possess small FFV, indicating also that the electrostatic potential can provide a manifestation of the available FFV for adsorption (see Figure 3.4a and 3.4b). They also presented a strong correlation of CO_2 solubility with respect to FFV but solely if restricted to a common anion as shown in Figure 3.4(c).

Molecular simulations also have been conducted to investigate the effect of the presence of other gases in ILs, such as SO_2 [193, 204, 239–244]. For the SO_2 case, anion-SO_2 interactions have been reported as dominant in an ab initio study of

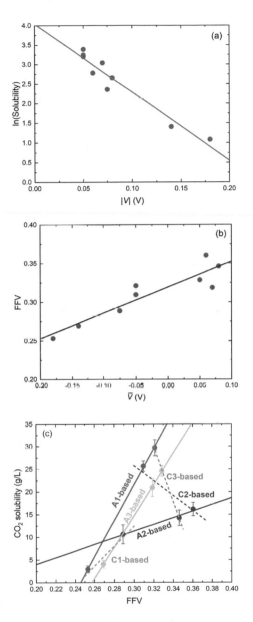

FIGURE 3.4 (a) The logarithmic regression of electrostatic complementarity $|\overline{V}|$ [237–238] between a solute and a solvent with the CO_2 solubility in ILs ($R^2 = 0.934$); (b) the linear regression between average electrostatic potential $\overline{V}$ and the FFV of the IL solvent ($R^2 = 0.843$); (c) correlation between FFV of ILs with CO_2 solubility. Red: A1-based ionic liquids (e.g., A1C1, A1C2, A1C3) ($R^2 = 0.9996$); blue: A2-based ILs ($R^2 = 0.9751$); orange: A3-based ILs ($R^2 = 0.9998$); green: C1-based ILs ($R^2 = 0.8953$); purple: C2-based ILs ($R^2 = 0.8817$); gray: C3-based ILs ($R^2 = 0.9984$).

Abbreviations: A1: [NpO$_2$]$^{2-}$; A2: [Np(TfNO)$_2$]$^{2-}$; A3: [BzO$_3$]$^{3-}$; C1: [Bzmim$_2$]$^{2+}$; C2: [Bzmim$_3$]$^{3+}$; C3: [Bzmim$_4$]$^{4+}$.

Source: Reproduced from [236].

1,3-dimethylimidazolium with several different anions [242]. Apart from anion-SO_2 interactions, anion-cation interaction also has been shown as important in DFT studies of 55 ILs [245]. The significant influence of both anions and cations interactions with the solute gas on the IL, SO_2 absorption behavior has been also reported [240] by a combined DFT and MD study. Zhang *et al.* [244] have also performed molecular simulations (ab initio and MD) of various ILs and their mixtures with SO_2 at experimentally measured saturated concentrations at 298 K and 0.1 MPa. In their analysis, high SO_2 absorption is reported as favored when SO_2 is associated with strong van der Waals interactions with the cation and strong electrostatic interactions with anion. Charge transfer phenomena from the IL to the gas were detected and an important enhancement of the ions' mobility at high SO_2 content. The counterion interactions and the free volume are not significant in this mechanism, while the latter may play a more important role in the solvation of nonpolar gases such as N_2 and O_2. Considering the chemical diversity of the ions of current and potential ILs, it is evident that simplified conclusions cannot be made for the underlying mechanisms of gas-IL systems, and further and deeper understanding is undoubtedly necessary.

ILs functionalized with reactive groups on their ions also have been explored using molecular simulation strategies [99, 107, 246–248]. CO_2-reactive ILs suffer from the disadvantage of high viscosities, and the issue of gas kinetics on the surface and of the selective reactive absorption of CO_2 over other gases are still to be studied. A comprehensive review on the topic of molecular modeling of CO_2-reactive ILs can be found in [249]. Molecular simulation has been applied for the study of the effect of confinement on ILs properties [250–257], which is important for the design and utilization of ILs in SILMs and in composite and mixed matrix membranes, aiming at the the fundamental understanding and the prediction of the separation performance of ILs at interfaces [258–262] or in various composite materials [263–271] using QM methods and atomistic simulations. Examples of molecular simulation studies of IL composites and confined ILs are also discussed in Section 3.5.

Macroscopic methods are valuable in the direction of screening ILs for specific applications such as in CO_2 capture and in separation processes. Correlative methods such as QSPR [272–274] are utilized to detect a statistical relationship between a number of molecular descriptors and a specific property such as gas solubilities, viscosities, melting points, and densities. The approach involves the extraction of correlations based on a training set of experimental data and is in some cases able to make predictions beyond the regressed range and in relation to the latter it may significantly benefit from the current advancements in the field of machine learning [275]. It is generally fast but becomes more computationally demanding if QM calculations are necessary to place quantitative structure-property correlations. Hybrid statistical thermodynamics/QM methods are also used for the calculation of gas solubility and the high throughput screening of ILs. COSMO-RS [276–280] is a commonly used method for performing such implicit solvent QM calculations. For instance, Zhang *et al.* [281] implemented the COSMO-RS method to calculate CO_2 Henry's law constants in 408 ILs and identified a dominant anion influence on CO_2 solubility, with the [FEP]⁻-based ILs having the highest solubility. They extracted Henry's law constants that were in good agreement with experiments at room temperature but could not reproduce the temperature dependence. The work of Gonzalez-Miquel *et al.* [277] is also an example of a COSMO-RS implementation to estimate CO_2/N_2 selectivity in 224 ILs, in which [SCN]−-based ILs were determined as the best candidate ILs. Such methods, although being very useful in the direction of screening, suffer from important limitations as far as their quantitative predictive ability is concerned. An extensive review on solvation studies based on the COSMO-RS approach can be found in [279]. Various thermodynamic models in the form of EoS [120, 282–283] also have been applied for the study of the phase behavior of gases in ILs. Cubic EoS, such as Peng-Robinson-type [284–286], Redlich-Kwong [282], Soave-Redlich-Kwong [285] and Cubic-Plus-Association (CPA) [287–290], have been used to model solubility of gas in ILs and correlate experimental data. Statistical Association Fluid Theory (SAFT) EoS have been applied in several variational schemes, such as Soft-SAFT [291–293], ePC-SAFT [223, 294–295], tPC-SAFT [296–297], and Heterosegmented-SAFT [298] for the characterization of the thermodynamic behavior of gas/IL mixtures. SAFT-based EoS provide a stronger theoretical basis and a more reliable screening means in comparison to cubic EoS, however, they are also constrained by the requirement of various adjustable parameters in each of the schemes. In that respect, the predictive ability of such EoS is relatively limited.

Molecular simulation and computational modeling offer the significant advantage of exploring various length and time scales. Apart from addressing the open important challenges that each methodological category faces in unraveling the structure-property-performance in the fascinating and diverse area of ILs, another significant challenge is still to be faced: the design of novel materials, devices, and processes from the molecular scale to the level of industrial applications demands the development and wide use of hierarchical and systematic multi-scale simulation strategies that need to be coupled [299–303] in a rigorous manner with unit scale simulations and process optimization methods. Synergistic advancements in this direction will greatly facilitate the efficient design and development of novel green technologies that can be applied at large scales and address current crucial environmental and sustainability issues.

3.4 ILS AS CARBON CAPTURE MEDIA IN CCS TECHNOLOGIES

3.4.1 CCS TECHNOLOGIES

Broadly, ILs and TSILs can be applied in all different configurations of the existing technologies for CO_2 capture from a certain large point source. These are presented in Figure 3.5. First, the post-combustion capture where the fuel is fully combusted with air generating a flue gas stream from which CO_2 must be separated mainly from N_2 and O_2 (the latter existing in the flue gas of power plants burning natural gas [NG]). Second, the pre-combustion capture where CO_2 must be separated from H_2 contained in the tail gas of a gasification or reforming process, in which a solid, liquid, or gas fuel is converted into syngas (a mixture of CO and H_2) and further, in the case of methane steam reforming, into a mixture of H_2 and CO_2 via the water-gas shift reaction. Pre-combustion also covers the case of CO_2/CH_4 separation from a sour NG source or from biogas (NG or biogas sweetening/upgrade). Third, the oxyfuel combustion, which is the simpler one since the fuel is combusted in pure

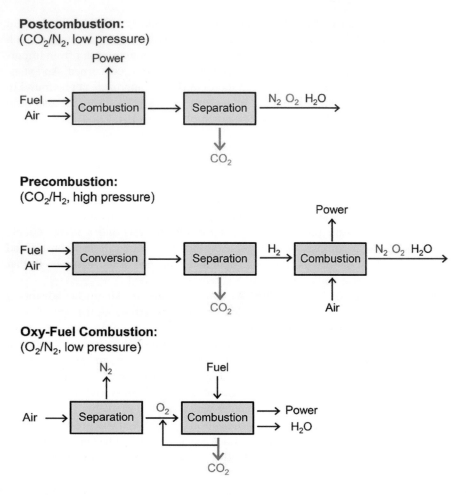

FIGURE 3.5 The three basic CO_2 capture technologies and the main gas separation required in each technology.

Source: **Obtained from [304].**

oxygen, generating a flue gas stream that contains solely water vapor and CO_2. Each configuration caries out a different separation task and is therefore constrained to certain conditions and gas concentrations, as presented in Table 3.1. One of the challenges of gas separation is the small difference in the kinetic diameters of the gases contained in the several streams of interest.

However, the differences in the electronic properties of gases (e.g., dipole and quadrupole moment and polarization) are more pronounced and combined with the great variety of chemical and physicochemical properties of the materials used in gas separation, assist the selective absorption or adsorption of certain gases over others by a solvent or solid adsorbent.

TABLE 3.1

Typical Operating Conditions and Gas Composition in Post-Combustion and Pre-Combustion CO_2 Capture Processes and the Kinetic Diameter of the Gas Molecules

Gas	Post-combustion	Pre-combustion	Kinetic diameter (Å)
	Composition/wt%	Composition/wt%	
CO_2	15–16	35.5	3.3
H_2O	5–7	0.2	2.65
H_2		61.5	2.89
O_2	3–4		3.45
CO	0.002	1.1	3.75
N_2	70–75	0.25	3.64
SO_x	0.08		4.11 (SO_2)
NO_x	0.05		4.01–5.02
H_2S		1.1	3.6
Conditions			
Temperature (K)	323–348	313	
Pressure (bar)	1	30	

The benchmark technology for post-combustion CO_2 capture is the chemical absorption using aqueous solutions of monoethanolamine (MEA) [305]. The main drawback of this process is the huge energy expense for regeneration. The regeneration of amine solvents is highly energy demanding for two reasons. The first one is the low concentration of CO_2 in the flue gas and the operation at atmospheric pressure conditions which bring up the necessity for a large amount of amine and fast circulation rates of the aqueous amine solvent. The second one is the corrosive strength of amines which requires dilution with large amounts of water. Consequently, relatively large equipment size is required. Hence, the regeneration of the CO_2-rich solvent in the stripping tower entails three components of thermal energy. Sensible heat to bring the temperature of a large amount of solvent from 323 K up to its boiling point (about 383 K) and to heat the huge mass of inert materials used in the construction of the capture plant (parasitic loss). Latent heat to evaporate water and generate the vapor required to flow upwards the stripper column for generating CO_2-lean conditions. Thermal energy to break the carbamate bond formed between the amine group and CO_2 is approx. 100 kJ/mol [306]. Further implications also include the use of inhibitors to control the oxidative degradation of amines and avoid corrosion and foaming. Strategies to enhance the separation efficiency and lowering the regeneration energy have led to sterically hindered amines such as the 2-amino-2-methyl-1-propanol (APM). Moreover, sterically hindered amines can achieve CO_2 loadings higher than 0.5 equivalents and higher regeneration rates [307].

Pre-combustion capture has certain advantages over post-combustion capture, emanating from the higher partial pressure of CO_2 in the flue gas and the higher operating pressure. Under these conditions, the strength of physical absorbents suffices for an efficient CO_2 capture process, and due to the weaker interactions with the CO_2 molecules, the heat of regeneration associated with pre-combustion is lower than that of the post-combustion. Indeed, the energy consumption (penalty) is reduced to 10%–16%, almost half of that of the post-combustion processes [308].

The standard, commercially available solvents to remove acid gases (CO_2/H_2S) from the syngas stream (which also contains $H_2/CO/CH_4/N_2$) or from other high-pressure CO_2-containing streams related to pre-combustion are Selexol® (Union Carbide, Houston, Texas, US) and Rectisol® (Lurgi AG, Frankfurt am Main, Germany). The Selexol® process involves a mixture of dimethyl ethers of polyethylene glycol, while the Rectisol® process uses methanol at low temperature (chilled methanol). The hydrophilicity and high vapor pressure of both solvents along with their corrosive strength at elevated temperatures raises the demand for lowering the capture temperature to sub-ambient conditions (283 K for Selexol and 263 K for Rectisol) and then bring the purified gas stream to about 473 K for combustion. Hence the processes are inefficient from an energy expense perspective, and this large CO_2 capture energy penalty results to high cost and significant reduction of the power plant's efficiency. Other challenges facing the pre-combustion technology are related to the fuel treatment, the large-scale application, the continuous and dynamic operation, and the combustion of H_2-rich syngas in the gas turbines.

Oxyfuel combustion provides the easiest separation task, since the CO_2 must be isolated from water vapor by condensing the latter. However, combustion of low-quality fuels results in high concentrations of SO_2 and NO_x in the flue gas, which further complicates the flue gas cleaning [309]. Other implications of the process arise due to the high energy demand of the air separation stage and the need to redesign the turbines, which constitutes the retrofitting of existing plants economically unfeasible.

3.4.2 CO₂ CAPTURE MATERIALS

Several materials and physicochemical processes can be combined with the three basic configurations of technologies for CO_2 capture. The main processes are absorption, adsorption, membrane separation, and chemical looping. The active materials of these processes (absorbents, adsorbents, membranes, and metal-oxides) interact differently with CO_2 and demand for specific conditions of temperature and pressure to augment their interaction and achieve enhanced capture and separation performance. The challenge to be met is to develop new innovative materials that are effectively regenerated with low energy demand while being resistant to chemical or thermal degradation and unsusceptible to erosion and attrition.

Absorption of gas molecules is a physicochemical process occurring in the bulk of a fluid (solvent) or polymer *via* chemical or physical interaction. To date, chemical or physical absorption with solvents constitutes the most widely applied technology in the chemical, natural gas and biogas treatment, petroleum, and power generation industries to separate acid gases such as CO_2 and H_2S. For low CO_2 concentrations, chemical adsorption with primary, secondary, tertiary, and sterically hindered amines today accounts for 70%–90% of the CO_2 captured and yields high purity CO_2 (>99%) [40, 310]. Despite the progress achieved so far with the use of novel solvent formulations (mixtures of amines) and additives to limit corrosion, solvent degradation, and foaming, recent cost estimates for post-combustion capture in a coal-fired power plant extend to the point where the total costs of an amine scrubbing system, including the addition of fresh solvent and other operating and maintenance costs, increase the cost of electricity by $0.06/kWh and the 'avoided cost of capture' is in the range of $57–60/ton CO_2. The energy penalty of the absorption process, excluding operation and maintenance costs, is approximately 0.34 kWh/kg CO_2 [40], and the solvent losses account to 1.6 kg for each ton of captured CO_2 [311].

For high CO_2 concentrations and high operating pressures, physical solvents such as Selexol® and Rectisol® solvents or a mixture of N-formyl morpholine and N-acetylmorpholine developed by Udhe, bind CO_2 based on physical interaction (dispersive forces). Notably, the N-formyl morpholine/N-acetylmorpholine solvent exhibited high absorption capacity for acidic gases, low solubility for short chain hydrocarbons (C1-C3) and capability to simultaneously remove mercaptans. The Kwoen gas plant in Canada was the first commercial application of this process. In general, lower heats of absorption are associated with physical solvents, and regeneration can occur by lowering the pressure or heating the mixture or by applying both driving forces. Physical absorption, however, is competitive solely for high CO_2 concentrations and when contaminants or other gases with high affinity for the solvent are absent or only present in trace quantities in the gas stream under treatment.

Adsorption differs from absorption in the sense that the CO_2 interactions mainly involve chemisorption or physisorption on the surface and not in the bulk of an adsorbent, the latter being usually a solid nanoporous material of large specific surface area. Adsorbent materials used in CO_2 capture include activated carbons, carbon molecular sieves, zeolites, porous polymers [312], and recently metal organic frameworks (MOFs) and zeolitic imidazolate frameworks (ZIFs) [304, 313]. Chemical looping can be also classified as an adsorption process since it proceeds with chemisorption of CO_2 on metal-oxide solids at elevated temperatures followed by regeneration at slightly higher temperature. There are two types of chemical looping systems. Type I systems utilize oxygen carrier particles to perform the reduction-oxidation cycles, while Type II utilize CO_2 carrier particles to conduct carbonation-calcination cycles [314]. The adsorbent materials, usually in the form of pellets, pearls, beads (and lately in the form of bespoke structured microporous monoliths prepared by 3D printing [315]) are placed in the adsorber, which is usually a packed or fluidized or even a moving trickle bed, and the flue gas passes through, allowing for adsorption to take place. Adsorption and desorption cycles undergo either pressure swing, temperature swing, vacuum swing (Figure 3.6), or electric swing. Solid adsorbents, even the chemisorbing ones, are more energy efficient than amine-based solvents for three reasons. First, the heat capacity of most solids is far less than that of water. Second, the expense of latent heat vanishes, since just a small quantity of water adsorbed from the gas phase on the adsorbent must be evaporated during regeneration. Third, interactions of the pore walls with the amine functionalities grafted on the surface of the adsorbent result in a lower energy requirement for breaking the carbamate bond. Despite these benefits, solid handling, upscaling and achievement of the same heat exchange efficiency in the cycle between the cold and hot sections of the capture process (e.g., ability to fluidize) is still challenging [316].

Membrane technology is the most prominent regarding the energetic cost of gas separation since there is not any requirement for regeneration. Endowed also with

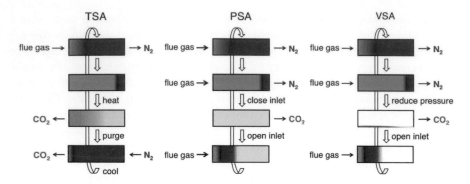

FIGURE 3.6 Schematic diagrams of idealized temperature swing adsorption (TSA), pressure swing adsorption (PSA), and vacuum swing adsorption (VSA) processes for regenerating solid adsorbent in a fixed-bed column.

Source: **Obtained from [304].**

the capacity to selectively extract CO_2 from other gas constituents of a mixed gas stream and with high flexibility in their possible configurations, membranes are expected to achieve high CO_2 separation efficiency relative to the adsorption-desorption cycles with solid adsorbents [317]. In polymeric membranes, which are usually shaped in the form of hollow fibers with the purpose to attain a high aspect ratio and reduce the footprint of the process, the most prominent mechanism of separation is solution-diffusion. Lately, poly (ether-block-amide) copolymer (Pebax) hollow fiber membranes appear as the most efficient ones for CO_2/N_2 and CO_2/CH_4 separation [318–319]. The performance of Pebax membranes is further augmented *via* inclusion of selective diffusion paths into the bulk of the membrane with the incorporation of MOF nanoparticles as fillers of the polymer matrix. This approach has led to the next generation of CO_2 separation membranes known as MOF-based mixed matrix hollow fiber membranes, or MMM-HFs [320]. There are also porous polymeric membranes made of a class of polymers known as polymers with intrinsic micro porosity (PIMs). PIM-1 is the polycondensation product of ultra-high-purity monomers of 5,5′,6,6′-tetrahydroxyl-3,3,3′,3′-tetramethyl-1,1′-spirobisindane (TTSBI) and 2,3,5,6-tetra fluoroterephthalonitrile (TFTPN), and is endowed with high perm-selectivity performance due to the very high fractional free volume and favorable interconnectivity between micro-cavities, assets induced by the spirobisindane moiety, which creates rigid ladder-type polymeric chain structures, with significant steric hindrance preventing chain rotation and limiting chain packing [321–322]. In this type of membranes as well as in zeolitic, MOF, ZIF, and other microporous ceramic (SiO_2, TiO_2) membranes, sieving is the most prominent mechanism of separation, and microporous diffusion governs the transport of gas molecules through the membrane. Microporous diffusion is an activated process, meaning that the membrane permeance increases significantly with the temperature, since the intensified thermal motion of gas molecules enhances the probability to be directly transferred (adsorbed) from the gas phase to the cavity of the pore (the number of strikes with the pore aperture increases due to enhanced thermal motion). Moreover, the passage through the membrane is implemented *via* a hopping mechanism of the gas molecules between active sites on the pore walls (adsorption sites), meaning that the gas molecules must overcome an energetic barrier to move across the membrane. Another approach, which involves porous and nonporous polymeric membranes and aims to the enhancement of the selectivity, is the inclusion into the pores or the polymer matrix of moving or immobilized gas molecule carriers with high specificity for the gas to be separated. For instance, immobilized amino-group carriers with high affinity for CO_2 can be either attached on the backbone of the polymeric chains or grafted onto the pore surface of polymeric or inorganic membranes. It is also possible to attach immobilized carriers on the surface of the nanoparticles used as fillers of mixed matrix membranes.

Alternatively, moving carriers are usually contained in liquid formulations entrapped and stabilized into the pore cavities. This type of facilitated transport membranes [323] is ideal for CO_2 separation at low partial pressures and therefore for post-combustion CO_2 capture, since at higher concentrations the carriers are saturated and do not contribute to the enhancement of the facilitated transport mechanism. Despite the high potential, industrial scale application of membranes for post-combustion CO_2 capture is so far not feasible due to the low CO_2 content, which rises the demand for multiple stages to achieve a high degree of separation, along with other issues complicating the application of membranes such as their fouling and degradation, especially of the polymeric ones.

3.4.3 ILs as Carbon Capture Media

The chemical versatility and the capacity to fine-tune the structure towards the desired properties, constitutes ILs as promising candidates to be applied and bring step changes in the performance of almost all the above-described CO_2 capture technologies. Apart from the case of high temperature CO_2 capture processes, such as the calcium looping process, ILs can be applied as chemisorbing or physisorbing absorbents (solvents), as immobilized or moving carriers in facilitated transport membranes, and as pore fillers or pore-wall modifiers in nanoporous solid adsorbents and membranes.

In post-combustion CO_2 capture, water–IL–amine ternary systems can be used to facilitate the regeneration at relatively low temperatures (60 °C) and decrease the viscosity [324]. Other benefits arising by using ILs in mixtures with aqueous amines include anti-corrosion properties [325] and less amine loss by evaporation during desorption or by the formation of sulfate and heat-stable salts [326]. A recent study on the use of ILs based on alkyl-methylimidazolium cations and tricyanomethanide [TCM]⁻ anions in mixture with aqueous diethanolamine (DEA) solvent has concluded to beneficial effect of the ILs on the capture efficiency, toxicity, and corrosiveness of the standard amine solution [327]. This study concluded that solvent formulations containing the tricyanomethanide anion-based ILs and about 10 vol% of primary or secondary amines exhibit the same CO_2 capture performance as the 25 vol% standard amine solutions.

Another strategy that is highly relevant to the use of ILs as effective solvents for pre-combustion CO_2 capture application is the mixing between ILs or the mixing of

ILs with conventional solvents. The concept behind this strategy is to combine synergistically ILs of high capture efficiency, which are usually of high viscosity (e.g., fluorinated ILs or TSILs) with ILs of moderate CO_2 absorptivity but significantly lower viscosity (e.g., nonfluorinated ILs) so that their properties are complemented towards an improved solvent formulation for CO_2 capture. The same holds for the mixing of ILs with organic solvents. In this case, the main target is to effectively control the viscosity of the IL while combining the individual advantages of organic solvents and ILs. This is readily attainable since the viscosity of ILs can decrease sharply by mixing them with just a small amount of water or organic solvent [328–330]. Therefore, the mixing of ILs or the mixing of ILs and organic solvents is proposed as a new solvent formulation, endowed with the desired assets of low viscosity, high solubility, and high selectivity and having the capacity to bind CO_2 by both chemical and physical interactions. It is important to note that even in the case that a chemisorbing TSIL of high viscosity is mixed with a physisorbing, non-fluorinated IL with the target to depress the viscosity, it is also possible to predict the CO_2 solubility of the mixed solvent by applying a lever rule, knowing the mole fractions of individual ILs in the mixture and the solubility of CO_2 in pure ILs [60, 331–332]. Regarding the factors that are important in determining the CO_2 solubility in the mixture of organic solvent and IL, the consensus now is that these are the free volume in the IL/organic solvent mixture and the interaction between the solute and IL. Indeed, the effect of mixing ILs with alcohols and ketones on the CO_2 solubility in ILs has been studied by Kühne et al. [333], who showed that ketones should be the preferable solute. Their experimental results on phase equilibria for ternary mixtures of a common IL ([BMIM]$^+$[BF$_4$]$^-$) with conventional solvents (4-isobutylacetophenone, 1-phenylethanol, acetophenone, and 1-(4-isobutylphenyl)ethanol) concluded to lower solubility of CO_2 when using an alcohol as solute, due to the presence of hydroxyl groups that may form hydrogen bonds with [BF$_4$]$^-$ and thus reduce the CO_2-IL interaction.

For judging whether ILs have the potential to become a viable alternative of the conventional Selexol, Purisol, Rectisol, and Fluor solvents currently applied in precombustion CO_2 capture, it is necessary to define the exceptional properties and the concomitant advantages they could bring to the capture process, along with ensuring that other equally important assets are maintained. In this context Ramdin et al. [334] showed that the ideal CO_2/CH_4 selectivities, the stability, and heat capacity of several ILs are in the range of the conventional solvents. These included ILs with very common cations (phoshonium, ammonium, pyridinium, sulfonium) and anions (dicyanamide, diethylphosphate and bis(trifluoromethylsulfonyl)imide) [334], which due to their low vapor pressure and low corrosive strength do not demand for the feed stream to be cooled below ambient conditions, thus constituting economically viable replacements for Selexol and Rectisol in pre-combustion capture of CO_2. In addition, IL-based solvents can be regenerated using waste or low-grade heat, which allows net electrical energy savings at a power plant or industrial facility equipped with pre-combustion CO_2 capture.

At this point, and before shifting to the discussion on the broad category of composite and mixed matrix, IL-based adsorbents and membranes in Section 3.5, it is essential to make a short notice relevant to the practical use of ILs as solvents for CO_2

capture. In this context, a real gas stream is in most of the cases water saturated or contains a certain amount of moisture and other gases that may act competitively or degrade the CO_2 absorption capacity of ILs. Although solubility data of mixed gases in ILs is important for developing real processes, it is a small amount compared to the solubility data of pure gases. This is a major shortcoming, since the gas solubility may be significantly influenced by the presence of another gas, and the real CO_2 selectivity may be much lower than the ideal selectivity calculated from single phase gas solubility measurements or predicted from thermodynamic models. Moreover, in a real capture process that involves ILs as pure solvents or even as modifiers of porous adsorbents and membranes, it is inevitable that several among the used ILs will contain various amounts of water due to their hygroscopic nature.

In general, studies to date show that the single-gas solubility in physisorbing ILs follows the order SO_2 (H_2S) > CO_2 ≈ N_2O > C_2H_4 > C_2H_6 > CH_4 > Ar > O_2 > N_2 > CO > H_2 at the same temperature and pressure. Although these gases are of high industrial interest due to their frequent coexistence with CO_2 in almost all the flue and tail gas streams produced in industry, references on their effect on the CO_2 absorption capacity of ILs are very scarce, especially under conditions of temperature, pressure, and concentration like those met in a real industrial stream. In fact, the existing studies can be broadly divided into three categories: (1) gases are only sparsely soluble in the IL; (2) one gas is well soluble, but the other one is sparsely soluble in the IL; and (3) both gases are well soluble in the IL. The cases of interest for CO_2 capture pertain mostly to the categories (2) and (3).

Hence, Yokozeki et al. [335] proved experimentally the efficiency of [BMIM]$^+$[PF$_6$]$^-$ for CO_2/H$_2$ separation. Mixed gas streams were involved in their experiments and the operating conditions simulated the real hydrogen purification process. However, Solinas et al. [336], conducting high-pressure NMR measurements, had showcased co-solubilization effects of pre-absorbed CO_2 in [EMIM]$^+$[Tf$_2$N]$^-$, which enhanced the H_2 solubility and affected adversely the CO_2/H$_2$ selectivity. In another study on the effect of temperature on the CO_2 co-solubilization capacity for H_2 in ILs, it was found that at temperatures below 330 K, the addition of CO_2 in the binary system of H_2-[BMIM]$^+$[BF$_4$]$^-$ increased the H_2 solubility, while at temperatures above 340 K, the effect was the opposite [337]. Quite similar conclusions were drawn for the cases of CO_2 mixtures with CH_4 and O_2 by Hert et al. [338], who reported that the presence of CO_2 increases the solubility of O_2 and CH_4 in [HMIM]$^+$[Tf$_2$N]$^-$ due to CO_2 co-solubilizing the sparsely soluble gas via favorable dispersion interactions.

Another topic of great industrial interest relates to the simultaneous capture of SO_2, H_2S, and CO_2 with ILs. These three gases frequently coexist in the stack emissions of the power and petroleum industry and in many natural sources of gaseous fuels. Shiflett et al. [53, 118, 339–340] reported a series of simultaneous solubility data of CO_2 and H_2S in [BMIM]$^+$[MeSO$_4$]$^-$ and [BMIM]$^+$[PF$_6$]$^-$, and of CO_2 and SO_2 in [OMIM]$^+$[PF$_6$]$^-$ $^+$and [BMIM]$^+$[MeSO$_4$]$^-$. They observed an anti-solvent effect of H_2S and SO_2 on CO_2 solubility. In parallel, they developed a ternary RK (Redlich-Kwong) equation of state for describing the CO_2/SO$_2$/IL systems, taking into consideration the interaction between SO_2 and CO_2. They showed that the experimental and modeling results converged to the conclusion that the CO_2/SO$_2$ selectivity in the gas phase is significantly enhanced with the addition of ILs.

As a general trend, it seems that CO_2, H_2S and SO_2, which are highly soluble in ILs, have the capacity to act as co-solvents when paired with CH_4, H_2, O_2 and N_2, enhancing the solubility of the latter and reducing their own solubility. On the other hand, when both gases are highly soluble in IL (CO_2/SO_2, CO_2/H_2S), the outcome is not always straightforward, although there is indication that the more soluble gas acts as anti-solvent for the other, meaning that H_2S and SO_2 dissolution in the IL may have an adverse effect on the solubility of CO_2. However, molecular simulation results of the systems $O_2/CO_2/[HMIM]^+[Tf_2N]^-$ and $N_2/SO_2/[HMIM]^+[Tf_2N]^-$ do not agree with the positive effect of CO_2 and SO_2 on the solubility of O_2 and N_2 [214]. These contradictory results emphasize the need for deeper theoretical works towards understanding of the solubility mechanism and specific interaction between mixed gases and ILs [214].

It should be stated that the second topic of high practical interest, relevant to the influence of water on the CO_2 solubility in ILs, is already studied adequately and encompasses three major effects: enhancement, degradation, and dilution. The degradation effect is frequent in ILs with fluorinated anions, especially in ILs composed of small fluorinated anions, such as $[PF_6]^-$ and $[BF_4]^-$. These ILs are hygroscopic in nature, contrarily to ILs containing bulkier fluorinated anions such as $[Tf_2N]^-$, which are hydrophobic. It is currently well known that $[PF_6]^-$ and $[BF_4]^-$ are unstable and easy to hydrolyze in the presence of water at moderate temperatures (e.g., 353–373K), with the fluoride-based impurities being formed. In the case that such an incident takes place during the CO_2 capture process, it will cause severe corrosion problems and damage several points and sub-equipment of the capture plant.

In most of the other cases, water dilutes the physisorbing ILs and inhibits the dissolution of CO_2 as expected, because the affinity between water and CO_2 is very weak. Therefore, the effect is always negative, and the only discrepancies between the results reported in literature concern the intensity of this negative impact. Some of the authors claim that the influence of degradation and dilution effects on gas solubility in ILs is significant and connect the existence of different amounts of water in ILs with the deviations among the CO_2 solubility data coming from different sources [341–342]. Other authors however report that the influence of water on the CO_2 solubility in ILs is not so intense, and as an example, we can refer to a recent work showing that the deviation of CO_2 solubility in $[BMIM]^+[PF_6]^-$ with water contents up to 1.6 wt% was less than 6.7% at the same temperature and pressure [343]. In another study relative to the system $CO_2/[OMIM]^+[Tf_2N]^-$, the conclusions on the minor effect of water are more convincing, since the Henry's law constant of CO_2 remained almost unchanged with a relative humidity of about 40% [344]. Contrarily, the reports declaring positive effect of water on the CO_2 solubility in physisorbing ILs are scarce, and among them there is one that encompasses a systematic study based on gravimetric analysis together with in-situ Raman spectroscopy and excess molar volume and viscosity deviation measurements in a series of ILs with alkyl-methylimidazolium cations and tricyanomethanide anions [345]. The verified four-fold and tenfold enhancement of the CO_2 solubility and diffusivity, respectively, in the binary $[RMIM]^+[TCM]^-/H_2O$ system was explained *via* a molecular exchange mechanism between CO_2 in the gas phase and H_2O in the liquid phase and the subtle competition between the TCM-H_2O and TCM-CO_2 interactions. Further to this, the enhancement effect of water is common in almost all chemisorbing, amino-functionalized or acetate anion TSILs, since it promotes complexation of CO_2 with amines to form carbamates [98, 346] or with [OAc]$^-$ to form AB or AB2 types of chemical complexation [347], thus increasing the solubility of CO_2 in ILs. In addition to the existence of trace impurities and moisture in the industrial streams, the performance of a real CO_2 capture process can be adversely affected by the migration of metal corrosion products into the solvent or by the use of corrosion and foaming inhibitors that interact with the solvent and degrade the CO_2 absorptivity. To elucidate the effect of corrosion products on the CO_2 absorptivity of ILs, our group has conducted a systematic study of CO_2 absorption in 3-alkyl-1-methylimidazolium cation, tricyanomethanide anion ILs ([RMIM]$^+$[TCM]$^-$ with R=ethyl, butyl, and hexyl) and in an IL with butyrolactam cation and fluorinated anion ([BHC]$^+$[Tf$_2$N]$^-$), after having been in contact with mild steel (MS) for several days under elevated temperatures [348]. Concerning the corrosion mechanism, it was concluded that with the 1-alkyl-3-methylimidazolium TCM ILs, corrosion initiated with the dissolution of MnS inclusions, which are present on the steel surface and subsequently, depending on the alkyl chain length (mostly for ethyl and butyl), magnetite and maghemite ferrites have been formed as corrosion products around the inclusion sites. Contrarily, the corrosive strength of the IL with the fluorinated anion [BHC]$_2$$^+$[Tf$_2$N]$_2$$^-$ was much higher, promoting general etching over the macroscopic surface of the alloy and generating a plethora of corrosion products including ferrites (mainly hematite), zinc oxide, sulfates, and carbonates. Most importantly, this study unveiled that the transfer of MnS inclusions and corrosion products to the IL phase had no effect on the CO_2 absorption capacity and kinetics. This was achieved by conducting gravimetric CO_2 absorption measurements on the selected ILs before and after their contact with MS. It is also notifiable that the dispersion of sodium molybdate as a corrosion inhibitor in [BHC]$^+$[Tf$_2$N]$^-$ also had no effect on the CO_2 solubility and absorption rate while significantly limiting the etching rate. In general, this study verified that the CO_2 capture performance of [RMIM]$^+$[TCM]$^-$ and [BHC]$^+$[Tf$_2$N]$^-$ ILs remains unaffected by the dispersion of corrosion products and corrosion inhibitors into the liquid phase, which is a highly important asset for a real application.

3.5 COMPOSITE AND MIXED MATRIX IL MEDIA FOR CO$_2$ CAPTURE

The concept of composite and mixed matrix IL media for CO_2 capture, also known as supported IL phase adsorbents (SILPs) and membranes (SILMs), had been conceived and materialized contemporarily with the surging progress of the studies relatively to the CO_2 solubility in ILs. Researchers had realized early that the drawbacks of high viscosity, high cost, and unknown toxicity of ILs could be addressed solely if the ILs are finely dispersed in the form of tiny droplets enveloped by a solid matrix. Alternatively, ILs could be encapsulated in a gel matrix or stabilized as ultra-thin layers on the pore surface of adsorbents and membranes. As such, it would be attainable to use effectively the less possible amount of ILs and have the capacity to recover them, along with avoiding their accidental spillage to the environment, thus overcoming the problems of high cost and unknown toxicity. Moreover, spreading an ultra-thin layer of IL onto the surface of a solid support drastically increases the gas/liquid contact area, leading to much faster sorption rates than in the case of bulk IL. Hence, a technical solution was devised to overcome the mass transfer limitations of ILs in gas-liquid reaction [349] and separation [350] processes, as imposed by their high viscosity. Recent reviews on the performance of SILPs [351–352] and SILMs [353–354] for CO_2 capture applications show that the inclusion of ILs readily boosts the CO_2 selectivity while slightly depressing the CO_2 diffusivity and permeability. As such, SILMs have already achieved overcoming the limitation of the trade-off relationship between membrane permeability and selectivity, the latter meaning that the increase of permeability comes from the sacrifice of selectivity and vice versa. Many of the recently reported SILMs exhibit performances that are above the upper bound threshold of Robeson plots [355] for CO_2/CH_4 and CO_2/N_2 separation, with the most prominent example being an IL-based composite membrane (ILCM) [356]. The membrane was prepared *via* hydrogen bonding interactions between 1-(3-aminopropyl)-3-methylimidazolium bromide modified graphene oxide and poly(ether-block-amide) (Pebax 1657) and exhibited permeance up to 900 GPU and CO_2/N_2 selectivity of 45, which exceeded the 2008 Robeson's upper bound line.

In general, the composite and mixed matrix IL media for CO_2 capture can be divided into three broad categories. The first category encompasses all adsorbents and membranes that are developed by filling the pores or by immobilizing a thin IL layer on the pore surface of a solid substrate that can be either inorganic, carbon, or polymeric. The second category consists of the systems that are developed *in situ*, meaning that the IL phase is involved in the reaction sol that produces the solid substrate. This approach has been successfully applied to produce ILs encapsulated in silica gel matrices and for the synthesis of MOFs in ILs with the iono-thermal method. The third category encompasses IL-based materials in the form of tiny IL droplets on the scale of a few microns, encapsulated in a solid matrix that in most of the cases consists of metal oxide nanoparticles. ILs encapsulation is usually achieved *via* sol-gel (or else iono-gel) synthesis of an IL within a silica-like network [357] or *via* a phase inversion process of a silica/solvent/IL system [358], while thin layer deposition is attained with physical imbibition and chemical grafting techniques under vacuum or with the assistance of excess pressure. Apart from the method of materials preparation, the major difference between these two approaches relates to the amount of IL loading into the solid matrix. With physical imbibition and chemical grafting techniques, the IL solvent uploaded on to the support is generally lower than 20% w/w [359], reaching up to 50% w/w in some cases [360], and the main impediments towards higher loadings are imposed by the porosity, the pore size, the tortuosity, and the pore surface properties of the solid support. Though ILs are regarded as universal pore wetters, it is in some cases extremely difficult to completely fill the pore structure with the IL phase, especially when the pore surface is hydrophilic, the size of the pore mouth does not exceed 2–3 times the molecular diameter of the IL and when the pore structure is highly tortuous and characterized by pore constrictions (bottle neck pores or pore cavities constricted by narrow pore mouths). On the other hand, encapsulation can conclude to IL loadings up to 85% w/w [361]. Although the limited amount of IL loaded into a porous material *via* a physical imbibition approach may lead to much lower absorption capacity of the derived SILP as compared to that of the bulk IL and the pristine porous support, the impacts of the solid interface and nanoconfinement on the properties of ILs should also be considered. Hence, in most of the cases, confinement of the IL phase into pores with size smaller than 50 nm results in higher CO_2 solubility and diffusivity when compared to values observed in the unconfined IL [350]. Analytical methods combined with atomistic and molecular simulation studies explain the improved performance, which is attributed to the enhancement of the self-diffusivity of the IL due to weaker ion-ion interactions upon confinement [250, 362]. For instance, when the IL $[HMIM]^+[Tf_2N]^-$ nests under extreme confinement into the lumen space of carbon nanotubes (CNTs), atomistic simulations indicate that the IL undergoes nanoscale organization within the CNT, forming highly ordered structures with the cations oriented towards the bore walls and the anions towards the core of the tube [362]. Exhibiting this organized conformation, the IL undergoes real enhancement of its self-diffusivity and its capacity to faster absorb higher amounts of CO_2 as compared to the bulk IL phase. Moreover, nanoconfinement in the CNT resulted also in an increase of the CO_2/H_2 selectivity.

In general, the dynamics of ILs under confinement in nanostructured carbon- and silica-based porous materials and more recently in MOFs, especially the translational dynamics (self-diffusion), are highly relevant to the gas diffusivity and permeation properties of the SILPs and SILMs and are theoretically evaluated by MD simulations and experimentally defined and understood with a series of advanced techniques, the most prominent among them being the pulsed-field gradient NMR (PFG NMR) [363]. The dynamic properties of ILs confined to silica-based nanoporous materials either increase or decrease, with the changing trend and extent, depending on the molecular details of the confining space and IL and on the IL loading. When the hydrophilicity of the silica surface is altered by silanization and a hydrophilic IL such as the $[BMIM]^+[BF_4]^-$ is confined into pores with size smaller than 10 nm, the diffusion coefficients exhibit an Arrhenius-like thermal activation behavior. Explanatively, at high temperatures, the mobility of the IL under confinement (self-diffusivity of the ions) is of the same magnitude with that of the bulk IL phase. However, as the temperature decreases, a remarkable enhancement of the mobility is observed, which is more pronounced under confinement into pores of quite small mean diameter (7.5 nm). As such, the measured diffusion coefficients of the ions are more than two orders of magnitude higher than the corresponding bulk IL values. In consistency with the results of simulations [364], the enhancement at lower temperatures is attributed to the reduction in the density of ILs in small pores and the concomitant changes in ion packing.

Regarding the confinement of ILs into nanostructured carbons, the most studied case is that of CNTs, and the reports on diffusivity enhancement are much more frequent compared to those related to nanoporous silica. Usually, ILs that have a more pronounced tendency to be self-organized in the bulk state, such as the $[OMIM]^+[BF_4]^-$ as compared to $[BMIM]^+[Tf_2N]^-$, exhibit an increase of the self-diffusion coefficient under confinement, likely due to a frustration of IL self-organization at the nanoscale [365]. MD simulations also predict the significant intensification of the ILs mobility when confined into the bore of CNTs. This unusual gain is explained by the CNTs wall/IL interaction, which inhibits the formation of a hydrogen-bonded and electrostatically driven network in the bulk IL (e.g., $[EMIM]^+[Cl]^-$), and by the very smooth bore surface of CNTs that promotes the frictionless movement of ions near the CNTs sidewall. It is also interesting that the mobility increases more intensively into the CNTs with the narrower bore size. For instance, the mobility increased by a factor of 2 and more than 10 when $[EMIM]^+[Cl]^-$ was confined in 3-nm and 1.4-nm diameter CNTs, respectively [366–367].

MOFs are a new generation of porous crystalline materials that are lately examined as efficient adsorbents for CO_2 capture due to their astonishing high surface area and large pore volume. Initially, the concept behind ILs/MOFs synergy emanated from the capacity and benefits obtained *via* the synthesis of MOFs in ILs with the iono-thermal approach. However, very recently MOFs are considered as a unique platform for the direct nanoconfinement of ILs, mostly due to the versatility of pore shape and chemistry they offer, which can in turn allow for tunable interactions of MOFs with the guest ILs, along with effective control over the structure and dynamics of ILs under nanoconfinement [368, 369]. Similarly to what happens in nanoporous silica and CNTs, ILs confined in MOFs (e.g., IRMOF-1) are packed more orderly than in the bulk phase [267]. A distinguishing feature of MOFs as compared to CNTs and nanoporous silicas is that they offer two centers for promoting the IL's ordering. One center is located at the open cage and the other on the MOF framework. The open cage center is the organic linker, which is usually the preferred site for the bulky imidazolium cations to reside in, due to configuration entropy effects. The framework center is the metal node, which is the preferred site of residence for the anions of the ILs due to strong interactions. As regards the dynamics of ILs confined into MOFs, these are not so extensively studied to date, but most of the reports conclude the point that the mobility of the IL in the IL@MOF composite is much lower than that in the bulk phase. Therefore, no enhancement effects are concluded, and the mobility of both cations and anions reduces with an increase in the IL loading in the composite, mainly due to the enhanced confinement effect [267].

In conclusion, the confinement of ILs into nanopores results in a novel class of composite and mixed matrix materials with great potential for gas separation applications. In most of the cases, the interaction of pore walls with the entrapped IL phase acts synergistically to the CO_2 absorptivity and diffusivity, resulting in composite absorbents or membranes, where the CO_2 capture performance of the confined IL is augmented as compared to the bulk IL phase. Moreover, through the SILP and SILM approach, major drawbacks of bulk ILs are overwhelmed, such as their high viscosity and cost, the latter due to the small amount of IL that suffices for a certain scale of application. Interestingly enough, the effect of spatial confinement and dominant surface forces at such small length scales results also in the improvement of several other features that eventually become more attractive for a CO_2 capture application, such as thermal stability, phase transition behavior and stability of the IL phase when subjected to continuous pressure swings. However, the most prominent asset of IL-based composite and mixed matrix absorbents and membranes is the capability they offer to finely tune properties important for the CO_2 capture by varying a multitude of factors, such as the structure and chemical composition of the ILs, the extent of pore loading, the pore geometry and size, the surface chemistry and architecture (roughness of pores), the density of electrical charges in the pore walls, and the external conditions (temperature and pressure).

Due to the great potential of these materials, it is important to consider their possible configurations and relevant performance in the CO_2 capture process. Hence, the focus will be mostly on systems derived by the post-impregnation (physical imbibition), method because it is the most widely used and highly adaptable for nearly all porous supports and membranes. Moreover, a suggestion on the most promising configuration relatively to the targeted capture technology will be provided. Hence, as illustrated in Figure 3.7, there are four possible conformations of a SILP system depending on the way and the extent at which the IL fills the pores of the solid support.

The conformation A is attainable for larger nanopores (i.e., more than three times the molecular diameter of the IL) and with the requirement that the IL is first diluted in a common solvent (ethanol or methanol) before been introduced into the pores. Then, by solvent evaporation under mild conditions (vacuum and temperature up to 353–363 K) it is possible to obtain an ultra-thin and continuous layer of IL on the pore surface. Alternative, grafting of ILs having appropriate functionalities on the cation or the anion allows for chemical bonding between the functional groups of the IL and those existing on the pore surface and leads to the formation of a monomolecular IL layer on the pore walls. Our group has proposed the silanization of 3-alkyl-1-methylimidazolium cation-based ILs, using chloropropyl-trialkoxysilanes instead of haloalkanes in the starting alkylation reaction of 1-methylimidazolium, followed by anion metathesis of the halide with a salt bearing the desired anion [370]. This is a versatile process, as it can be adapted to silanize several types of cations (imidazoles, pyridines, pyrrolidines, phosphines, trialkylamines) and produce the respective silylated ILs via their combination with a multitude of common [BF_4^-, PF_6^-, Tf_2N^-, TfO^-, OAc^-] and task specific [lysinate, prolinate] anions. The produced silylated ILs can be chemically attached on any siliceous porous substrate (zeolites, MCM, SBA, KIT-6) via condensation reactions between the surface silanol groups, and the alkoxysilane and the synthetic approach can proceed either after having synthesized the silylated IL or by involving sequential surface chemistry reactions that initiate with the silylation of the pore surface followed by reaction with the 1-methyl imidazolium and anion metathesis with the appropriate salt. The grafting procedure always concludes to a composite material with the conformation D (Figure 3.7) because the

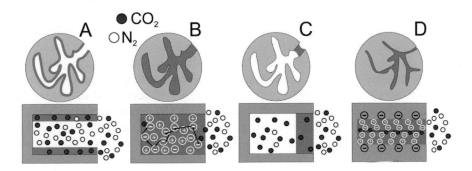

FIGURE 3.7   Possible conformations of SILP and SILM systems.

Source: **Obtained from [358].**

silylated cations are chemically attached on the pore surface and the anions are oriented towards the bulk of the pore. A conformation of type D can be also obtained with the physical impregnation method owing to the interactions between the pore surface and ILs. As evidenced by atomistic MD simulations of a series of ILs based on the 1-butyl-3-methyl imidazolium cation [BMIM]$^+$ and fluorinated anions ([BF$_4$]$^-$, [PF$_6$]$^-$, [TfO]$^-$, and [Tf$_2$N]$^-$) of different size and shape, when these ILs are interacting with siliceous (quartz) surfaces [371] there is a strong dependence of their orientational preferences on the size and shape of the anions and the surface charge (positive or negative). This is because the interfacial IL/surface interactions consist mainly of strong electrostatic forces and/or the formation of hydrogen bonds between the confined ionic species and the interfacial groups of the quartz surface. Hence, the [BMIM]$^+$ is always attracted by the negatively charged Si(OH)$_2$ surface, and the orientational preference is such that the imidazolium ring lies perpendicular to the siliceous surface, whereas the methyl and butyl chains are oriented towards and elongated along the surface. On the contrary, on a positively charged surface such as this of alumina, the main adsorbed species are anions, instead of cations, because of strong electrostatic interactions [371]. The main differentiation between the methods of chemical and physical SILP development is that chemical grafting always achieves the oriented conformation of IL independently of the pore size, while with the physical impregnation method, only the IL molecules nesting close to the pore wall exhibit a specified orientation to the surface. Those located in the center region are randomly orientated or layered. Washing with solvents is the only way to get rid of the randomly orientated IL phase, but this is not easily attainable, as washing may also leach out the IL layer near the pore wall. Hence, the physical method can achieve conformation D, with the prerequisite that the pore size equals two to three molecular diameters of the IL. Soaking a previously prepared host network in an IL may also yield composite and mixed matrix IL media with the conformations B and D.

Conformation A is ideal for heterogeneous catalysis since the open core of the pores ensures the accessibility of gaseous reactants and the unhindered removal of gaseous products from the huge surface area of the deposited IL's diffusion layer, which is used to disperse and stabilize metallic nanoparticles, metal complexes, or homogeneous organocatalysts. In such cases, the incorporated ILs are fully exploited, offering an excellent platform for the selective absorption of the reactants and have the potential to change the effective concentrations of adducts and intermediates but can also exert ligand effects and improve the catalytic efficiency ('co-catalytic effect') in hydrogenation, hydroformylation, oxidation, and carbonylation reactions. This conformation, however, is not suitable for CO$_2$ absorption and separation because the small amount of IL does not suffice to endow the composite absorbent with the CO$_2$ absorptivity and selectivity properties of the IL. Hence, the open pore space, especially if belonging to the nanoscale size, has the capacity to host large amounts of gas and consequently define to a large extent the sorption and selectivity properties of the composite. Conformations B and D are appropriate for absorption processes involving pressure or temperature swings, since the large amount of IL entrapped into the pores ensures the stability of SILP performance under repeated absorption/desorption cycles as compared to conformation C, which is mostly suitable for membrane technology (SILM). In the case of membranes, the IL should block only a very thin section of the pores near to the gas/solid interface, so that the gas permeation is not significantly degraded, and the selectivity properties are totally controlled by the dissolution/diffusion mechanism through the immobilized IL phase. Nanoconfined IL phases of enhanced ion mobility and the concomitant enhanced gas diffusivity, having a dominant role on the gas absorption and selectivity properties of the composite absorbents are desirable for conformations B and D. While enhanced translational dynamics is a common asset of both cases, conformation D constitutes a more straightforward way to achieve enhanced CO_2 diffusivity.

This is a bit contradictory to the general feature of nanoconfined ILs, the detailed local structure (ionic layer structure and orientation) and dynamics (ionic mobility) of which are related to the distance of the ions from the pore walls. In this context, ILs near the pore wall are expected to form a layered structure with solid properties and lowered dynamics because of the 'template' effect of the solid surface and the strong interaction between ILs and the pore walls. As shown in Figure 3.7, in a SILP absorbent that presents the conformation D, the confined IL phase consists entirely of molecules that are in close proximity to the pore walls. However, our group has showed and validated enhanced CO_2 diffusivity in both SILPs and SILMs that exhibited this conformation [372]. This extraordinary asset has been attributed to the orientation of the anions towards the core of the pore, which leads to the formation of straight channels for the CO_2 molecules that diffuse through the confined IL phase *via* a hopping mechanism between the anions (see Figure 3.8).

It can be comprehended that the knowledge on the possible conformation achieved *via* the post-impregnation (physical imbibition) method of an IL into a porous support is decisive for extracting valid conclusions on the effects of nanoconfinement and pore walls/IL interactions on the solubility and diffusivity properties of the confined

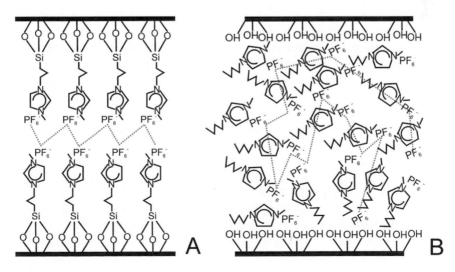

FIGURE 3.8 Straight diffusion channels for CO_2 achieved with the grafting method.

Source: **Obtained from [373].**

IL, especially when attempting to confirm enhancement effects in comparison to the bulk IL phase. It is also important to be aware of the loading fraction of the IL when trying to understand the effect of confinement on the rotational and translational motions of ILs in porous materials by NMR relaxation measurements for 1H spin-lattice relaxation times (T1) and 1H pulsed-field gradient NMR (PFG NMR), since the IL loading exerts a profound effect on the dynamic behavior of ILs confined in nanopores. Due to the exceptionally low solubility of nitrogen in ILs and the slow diffusivity at extremely low temperatures, nitrogen adsorption at 77 K (liquid nitrogen porosimetry [LN_2]) constitutes the simplest and most widely applied method to conclude on the extent of pore filling by the IL phase. When the LN_2 porosimetry of the SILP shows zero N_2 adsorption (Figure 3.9a), it can be assumed that the IL phase has effectively filled the entire pore space.

Then, by knowing the pore volume of the pristine solid substrate (Figure 3.9a) and the density of the IL, and having obtained experimentally the CO_2 absorption isotherm of the bulk IL at near ambient temperature (Figure 3.9b), it is possible to predict the CO_2 absorption capacity of the composite absorbent. Direct comparison of the predicted absorption isotherm with that obtained experimentally provides a valuable indication on the positive or negative impact of IL nanoconfinement on the CO_2 absorptivity. The isotherms presented in Figure 3.9b (unpublished data) unveil an almost fivefold enhancement of the CO_2 absorptivity of the [BMIM]$^+$[PF$_6$]$^-$ under confinement into the 4-nm and 7.3-nm pores of the MCM-41 and SBA-15, respectively. Hence, the presence of the pore surface appears to change the physical properties of [BMIM]$^+$[PF$_6$]$^-$, perhaps through a reorganization of the cations and anions at the interface, increasing the free volume available for CO_2 absorption. There are also many reports showing that not only the solubility but also the diffusivity of CO_2 in ILs are readily augmented as a result of IL nanoconfinement [350]. Nonetheless, extracting conclusions based solely on the results of LN_2 porosimetry is a bit vague, and the reason is that LN_2 porosimetry is more sensitive in defining pore constrictions rather than in providing information for the bulk of the pore. As an example, LN_2 porosimetry might produce the misleading result of zero pore volume for a SILP absorbent having the conformations A or C (Figure 3.7), even though the core of the pore is not filled with the IL phase. The reason is obvious for a material of type C, but even in the case that a thin film of IL is attached on the pore walls (Figure 3.7, case A), the pore window may be of smaller size than the kinetic diameter of N_2 molecules (0.36 nm), thus inhibiting their passage towards the empty core space.

The situation may become more complex when subjecting such types of materials to a CO_2 absorption experiment. The first reason is the smaller kinetic diameter of CO_2 molecules (0.33 nm) as compared to N_2; the second relates to the capacity of CO_2 to be dissolved into the IL layer near to the pore mouth and eventually find its way towards the empty core of the pore *via* diffusion through the IL layer. Hence, the CO_2 absorptivity of the SILP absorbent will be contributed by both the thin IL layer and the empty pore space. Under these circumstances, it is not correct to end up with conclusions on the effect of nanoconfinement on the CO_2 absorption capacity of the IL. To address this challenge and confirm the positive effect of IL nanoconfinement on the CO_2 solubility (Figure 3.9), there is the requirement to develop analytical techniques for investigating whether the IL has been partially or completely filled within the pores of the

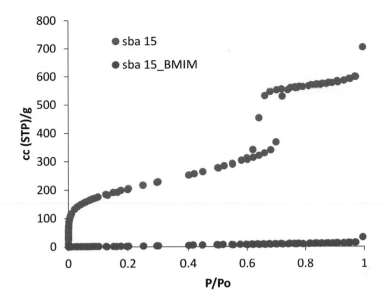

(a)

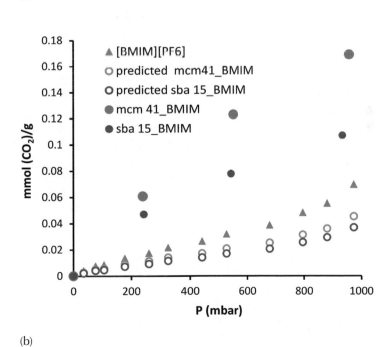

(b)

FIGURE 3.9 (a) Liquid nitrogen porosimetry (LN_2) results of SBA-15 and of the respective SILP prepared with the IL $[BMIM]^+[PF_6]^-$; (b) CO_2 absorption isotherm of $[BMIM]^+[PF_6]^-$ at 308 K (triangles); CO_2 absorption isotherms of the composite absorbents at 308 K (filled circles); predicted CO_2 absorption isotherms of the composite absorbents at 308 K (open circles).

solid matrix. As an example of such a powerful technique that has been developed and applied to elucidate the pore-filling extent of SILPs consisting of [BMIM]⁺[PF₆]⁻ and ordered mesoporous silicas (SBA-15 and MCM-41), we can refer to small angle neutron scattering (SANS) measurements [374–375]. SANS have significant advantages over the conventional gas adsorption methods because neutrons can provide information for the bulk of the pores, meaning that they can also 'see' pores that are inaccessible to gas molecules, such as pores that are blocked due to deposition of the IL near the pore aperture. The technique is based on the scattering theory, according to which the intensities of the Bragg reflections for a two-phase system (e.g., SiO_2-empty pore and SiO_2-IL) are related to the square of the contrast, defined as the difference of the scattering length density (SLD) between the silica matrix and the pore content. When the pores are filled with IL, the intensity ratio of the reflections of the pristine to the IL-loaded matrices is equal to the squared SLD ratio. In this way, it is possible to conclude whether a complete pore filling with IL has been achieved.

3.6 CHALLENGES AND OUTLOOK

The major challenges to be addressed towards the effective removal of carbon dioxide (CO_2) from industrial flue gas and tail gas streams are primarily related to technical and economic issues. The currently available CO_2 capture processes, which are mostly based on chemical or physical absorption with solvents, physisorption, or chemisorption with solid adsorbents and separation with membranes, are highly energy intensive and costly, features that render them unattractive for large-scale applications. Among the several new materials examined for their potentiality towards addressing these problems, ILs appear as the most promising ones, not only due to their inherent properties of low volatility, high selectivity for CO_2 absorption, and tunable structure, but also because of the versatility they offer to be applied as solvents or in mixtures with other solvents, as pore modifiers of nanoporous materials, and as solubility enhancers encapsulated into mixed matrix and IL-composite adsorbents and membranes.

Studies to date have proven that conventional ILs are inadequate to compete with the standard amine process in the conditions of post-combustion CO_2 capture. The reason is the high viscosity of ILs and the concomitant slow CO_2 diffusivity of the physical sorption mechanism in combination with the low content of CO_2 in the flue gas effluent of post combustion. Contrarily, ILs have the potential to bring step changes in the performance of the pre-combustion CO_2 capture, where they can be effectively applied due to the high CO_2 concentration in the flue gas. Hence, significant energy savings can be achieved because of the low vapor pressure, lack of affinity for water, and low corrosiveness of ILs, properties that allow for operation at higher temperatures than the conventional hydrophilic solvents. Thus, the high energy-consuming stage of cooling the feed stream below ambient conditions is avoided.

Conclusively, when dealing with the post-combustion CO_2 capture, conventional (physisorbing) ILs can be applied in mixtures with the standard aqueous amine solutions to assist overcoming the drawbacks of amine volatility, degradation susceptibility, and corrosiveness, along with participating in the capture mechanism, by abstracting protons during the mechanism of stable carbamate salt formation.

Thus, the involvement of ILs assists in going beyond the upper bound threshold of 1:2 CO_2 to amine molar uptake because less amine molecules are sacrificed (protonated) during carbamate bonding, while mitigating the energy expense for regeneration due to the contribution of physisorption in the overall CO_2 capture mechanism. Task specific ILs (TSILs) with an amine moiety attached on the anion, especially those with aprotic heterocyclic anions (AHAs; pyrazolides, imidazolides and triazolide) can be also applied for post-combustion capture, but again in mixtures with conventional amines. Although ILs with AHA anions have been showcased to undergo minor increases of their viscosity upon CO_2 binding, the viscosity is still high for an application without mixing with another solvent. The benefits obtained are mostly related to the enhanced absorptivity and mitigation of the energy expense for regeneration, since this type of ILs can achieve CO_2 to IL molar uptake of about 1:1 at room temperature and the reaction enthalpy (ΔH) of the IL with CO_2 is limited to values of up to 50 kJ/mol. However, an issue of great concern that should be also considered is the high cost of TSILs, which are 2–4 times more expensive than the conventional ILs.

Concerning the application of ILs in the pre-combustion CO_2 capture, the main efforts have as a target to control the problem of high ILs viscosity. This is usually achieved by mixing between ILs or by mixing of ILs with conventional solvents. Thus, complements of high viscosity (e.g., with fluorinated IL) and low viscosity (e.g., with nonfluorinated IL), high solubility and high selectivity, or chemical and physical interactions can be achieved. In addition to the capability for effective control of the viscosity, the strategy of mixing between ILs offers a straightforward way to predict and control the CO_2 solubility in the IL mixture. Many studies have proven that this is possible and can be achieved according to a lever rule, knowing the mole fractions of individual ILs in the mixture and the mole fractions (i.e., solubility) of CO_2 in pure ILs.

The broad field of composite and mixed matrix IL media presents great potential for accelerated deployment towards an industrial application. A novel area in membrane technology is the further development of polymer-IL composite membranes [376–378] that will be able to combine these two powerful categories of materials. For absorbents (SILPs) and membranes (SILMs) consisting of a solid nanoporous substrate that hosts the IL phase under extreme confinement into the nanoscale pores, the most prominent approaches from a practical point of view are those entailing the chemical attachment (grafting) of the IL phase onto the pore walls. Despite the low volatility of ILs, the stability of the liquid phase under extreme conditions of pressure and temperature swings or under high differential pressures across the membrane is doubtful, so there is still a strong need for lifetime and recyclability studies on this type of composite IL/solid materials. In this context, since the reports on the analysis of long-term stability of ILs confined into pores are scarce, the chemical attachment of ILs constitutes a head start towards the achievement of robust SILP and SILM systems. ILs encapsulation, achieved *via* iono-gel synthesis of an IL within a silica-like network or *via* the addition of ILs into the polymeric solutions used as precursors of flat sheet and hollow fiber membranes produced by a phase inversion process, constitutes another synthetic approach with great potential for the development of absorbents and membranes characterized by extended lifetime. Apart from the stability, another issue of great concern for both SILPs and SILMs is the effect of confinement and pore wall/IL interactions on the CO_2 diffusivity into the IL phase that nests inside the pores. While most of the recent studies report on the beneficial effect of ILs confinement on the CO_2 solubility and selectivity, the conclusions on the kinetics of absorption are controversial, and this can become a major roadblock towards further deployment. In fact, one of the major targets behind the concept of SILPs and SILMs is to confront the problem of high viscosity and the concomitant slow mobility. Hence, it may be a great impediment if scientists eventually conclude that IL confinement acts in the opposite direction of what expected. Hence, it is about time for research studies to focus more intensively on elucidating the CO_2 diffusion mechanism through confined or encapsulated ILs in relation to a great variety of factors such as the structure and chemical composition of the ILs; the filling ratio into the pores; the pore geometry and size; the surface chemistry, charge, and architecture of the pore walls; and the external conditions (temperature, pressure). The importance of these studies is underlined by the fact that the results to date on the most promising SILP absorbents consisting of MOFs and ILs have showcased significant attenuation of the CO_2 diffusivity in the confined IL phase as compared to the bulk IL. Also considering the high cost of MOFs and ILs, the current results suggest that it is not worth to continue making further efforts in this direction.

Computer-aided molecular design can assist in overcoming such predicaments. Combining molecular simulation with experimental characterizations will provide an efficient method to reveal the structure-property relationship and CO_2 absorption mechanisms in ILs and understand interfacial effects and the effects of confinement, enabling the rational design and development of novel IL-based composite absorbents with high CO_2 gravimetric capacity, selectivity, and diffusivity. The development of transferable accurate interaction potentials, also incorporating powerful machine learning techniques [379–383], will enable the implementation of systematic multiscale molecular simulation strategies and the elucidation of the complex microscopic phenomena and reliable property predictions in a wide range of length and time scales.

Concluding, we consider it essential to highlight a few roadblocks that should be surmounted soon to validate the sustainability of a CO_2 capture process based on ILs.

- *Scarcity of thermophysical properties data*: Considering that from a practical point of view, the CO_2 solubility in ILs should be evaluated in terms of molality, and that most of the results in the recent literature are provided as CO_2 mole fraction or Henry constants, it is essential to have at least more data on the density of ILs. Thus, it will be possible to convert the CO_2 mole fraction to molality. Then, the most promising among the highly absorbing ILs can be selected based on their viscosity, diffusion coefficients, surface tension, specific heat, thermal stability, water solubility, corrosiveness, and so forth. These properties of ILs have scarcely been reported in the literature, and accurate property predicting tools are still missing.
- More data are needed as an output of safety, health, and environmental studies. Biodegradability and toxicity of the most promising ILs should

be assessed thoroughly to prevent environmental pollution, and more ILs should be registered with the European Chemicals Agency or other international chemicals agencies.

- *High price of ILs.* The current price of small-scale ILs synthesis is extremely high (~1000 €/kg), which is 100–1000 times more expensive than conventional amine solvents. Of course, the price may drop for a large-scale production (40 €/kg), but even this cost is much higher than the cost for conventional solvents. A price level typical for conventional solvents should not be expected soon, because ILs are complex molecules requiring advanced synthesis and purification steps. In this context, efforts have already been made to produce ILs in continuous flow microreactors that minimize the expense of solvents and confront the problem of hot spots occurring in traditional batch processes. Continuous flow synthesis of ILs is already achieved for the alkylation reaction and the anion metathesis reaction, including the post-treatment purification stages, leading to significant cost reduction (200 €/kg at the 100-kg scale). It is also prominent that many studies suggest that the IL purity does not have a significant effect on CO_2 solubility and diffusivity.
- *Scarcity of engineering and scale-up studies.* Systematic process engineering studies can assist towards reducing the synthesis and process costs of ILs. The reason behind the scarcity of such studies is mostly due to the absence of physicochemical properties of ILs. Moreover, lab-scale processes need to be scaled up to a pilot-plant scale to assess the feasibility of ILs on industrial scale and environment. Industry and investors almost always demand results from experimental campaigns on a pilot scale for a long period before commercialization.

REFERENCES

1. P. Walden, "Ueber die Molekulargrösse und elektrische Leitfähigkeit einiger geschmolzenen Salze," *Bull. Acad. Impér. Sci. St. Pétersbourg*, vol. 8, no. 6, pp. 405–422, 1914.
2. S. A. Mirkhani, F. Gharagheizi, P. Ilani-Kashkouli and N. Farahani, "Determination of the glass transition temperature of ionic liquids: A molecular approach," *Thermochimica Acta*, vol. 543, pp. 88–95, 2012, doi:10.1016/j.tca.2012.05.009.
3. S. Zhang, J. Zhang, Y. Zhang and Y. Deng, "Nanoconfined Ionic Liquids," *Chemical Reviews*, vol. 117, no. 10, pp. 6755–6833, 2017, doi:10.1021/acs.chemrev.6b00509.
4. M. V. Quental *et al.*, "Enhanced separation performance of aqueous biphasic systems formed by carbohydrates and tetraalkylphosphonium- or tetraalkylammonium-based ionic liquids," *Green Chemistry*, vol. 20, no. 13, pp. 2978–2983, 2018, doi:10.1039/C8GC00622A.
5. S. Stevanovic, A. Podgorsek, L. Moura, C. C. Santini, A. A. H. Padua and M. F. Costa Gomes, "Absorption of carbon dioxide by ionic liquids with carboxylate anions," *International Journal of Greenhouse Gas Control*, vol. 17, pp. 78–88, 2013, doi:10.1016/j.ijggc.2013.04.017.
6. A. A. Abd, S. Z. Naji, A. S. Hashim and M. R. Othman, "Carbon dioxide removal through physical adsorption using carbonaceous and non-carbonaceous adsorbents: A review," *Journal of Environmental Chemical Engineering*, vol. 8, no. 5, p. 104142, 2020, doi:10.1016/j.jece.2020.104142.

7. E. Knapik, P. Kosowski and J. Stopa, "Cryogenic liquefaction and separation of CO_2 using nitrogen removal unit cold energy," *Chemical Engineering Research and Design*, vol. 131, pp. 66–79, 2018, doi:10.1016/j.cherd.2017.12.027.

8. A. Brunetti, F. Scura, G. Barbieri and E. Drioli, "Membrane technologies for CO_2 separation," *Journal of Membrane Science*, vol. 359, no. 1, pp. 115–125, 2010, doi:10.1016/j.memsci.2009.11.040.

9. T. Regueira, L. Lugo and J. Fernández, "Ionic liquids as hydraulic fluids: Comparison of several properties with those of conventional oils," *Lubrication Science*, vol. 26, no. 7–8, pp. 488–499, 2014.

10. D. Wei and A. Ivaska, "Applications of ionic liquids in electrochemical sensors," *Analytica Chimica Acta*, vol. 607, no. 2, pp. 126–135, 2008, doi:10.1016/j.aca.2007.12.011.

11. T. Abdallah, D. Lemordant and B. Claude-Montigny, "Are room temperature ionic liquids able to improve the safety of supercapacitors organic electrolytes without degrading the performances?" *Journal of Power Sources*, vol. 201, pp. 353–359, 2012, doi:10.1016/j.jpowsour.2011.10.115.

12. A. Eftekhari, "Supercapacitors utilising ionic liquids," *Energy Storage Materials*, vol. 9, pp. 47–69, 2017, doi:10.1016/j.ensm.2017.06.009.

13. E. Weiss, B. Dutta, A. Kirschning and R. Abu-Reziq, "BMIm-PF6@SiO2 Microcapsules: Particulated Ionic Liquid as A New Material for the Heterogenization of Catalysts," *Chemistry of Materials*, vol. 26, no. 16, pp. 4781–4787, 2014, doi:10.1021/cm501840d.

14. S. Subbiah, I. C. Cathy and C. Yen-Ho, "Ionic Liquids for Green Organic Synthesis," *Current Organic Synthesis*, vol. 9, no. 1, pp. 74–95, 2012.

15. R. A. Judge, S. Takahashi, K. L. Longenecker, E. H. Fry, C. Abad-Zapatero and M. L. Chiu, "The Effect of Ionic Liquids on Protein Crystallization and X-ray Diffraction Resolution," *Crystal Growth & Design*, vol. 9, no. 8, pp. 3463–3469, 2009, doi:10.1021/cg900140b.

16. C. Cagliero, C. Bicchi, C. Cordero, E. Liberto, P. Rubiolo and B. Sgorbini, "Ionic liquids as water-compatible GC stationary phases for the analysis of fragrances and essential oils," *Analytical and Bioanalytical Chemistry*, vol. 410, no. 19, pp. 4657–4668, 2018, doi:10.1007/s00216-018-0922-0.

17. R. D. Rogers and K. R. Seddon, "Chemistry. Ionic liquids – solvents of the future?" *Science*, vol. 302, no. 5646, pp. 792–793, 2003, doi:10.1126/science.1090313.

18. Y. Chauvin, S. Einloft and H. Olivier, "Catalytic Dimerization of Propene by Nickel-Phosphine Complexes in 1-Butyl-3-methylimidazolium Chloride/AlEtxCl3-x (x = 0, 1) Ionic Liquids," *Industrial & Engineering Chemistry Research*, vol. 34, no. 4, pp. 1149–1155, 1995, doi:10.1021/ie00043a017.

19. M. Abai *et al.*, "An ionic liquid process for mercury removal from natural gas," *Dalton Transactions*, vol. 44, no. 18, pp. 8617–8624, 2015, doi:10.1039/C4DT03273J.

20. A. C. Leitch *et al.*, "The toxicity of the methylimidazolium ionic liquids, with a focus on M8OI and hepatic effects," *Food and Chemical Toxicology*, vol. 136, 2020, Art no. 111069, doi:10.1016/j.fct.2019.111069.

21. M. Ramdin, T. W. de Loos and T. J. H. Vlugt, "State-of-the-Art of CO_2 Capture with Ionic Liquids," *Industrial & Engineering Chemistry Research*, vol. 51, no. 24, pp. 8149–8177, 2012, doi:10.1021/ie3003705.

22. F. Xu *et al.*, "Solubility of CO_2 in aqueous mixtures of monoethanolamine and dicyanamide-based ionic liquids," *Fluid Phase Equilibria*, vol. 365, pp. 80–87, 2014, doi:10.1016/j.fluid.2013.12.020.

23. M. Aghaie, N. Rezaei and S. Zendehboudi, "A systematic review on CO_2 capture with ionic liquids: Current status and future prospects," *Renewable and Sustainable Energy Reviews*, vol. 96, pp. 502–525, 2018, doi:10.1016/j.rser.2018.07.004.

24. H. Zhai and E. S. Rubin, "Systems Analysis of Physical Absorption of CO_2 in Ionic Liquids for Pre-Combustion Carbon Capture," *Environmental Science & Technology*, vol. 52, no. 8, pp. 4996–5004, 2018, doi:10.1021/acs.est.8b00411.

25. K. Suresh Kumar Reddy, B. Rubahamya, A. A. Shoaibi and C. Srinivasakannan, "Solid support ionic liquid (SSIL) adsorbents for mercury removal from natural gas," *International Journal of Environmental Science and Technology*, vol. 16, no. 2, pp. 1103–1110, 2019, doi:10.1007/s13762-018-1781-0.

26. K. N. Ruckart, R. A. O'Brien, S. M. Woodard, K. N. West and T. G. Glover, "Porous Solids Impregnated with Task-Specific Ionic Liquids as Composite Sorbents," *The Journal of Physical Chemistry C*, vol. 119, no. 35, pp. 20681–20697, 2015, doi:10.1021/acs.jpcc.5b04646.

27. L. J. Lozano, C. Godínez, A. P. de los Ríos, F. J. Hernández-Fernández, S. Sánchez-Segado and F. J. Alguacil, "Recent advances in supported ionic liquid membrane technology," *Journal of Membrane Science*, vol. 376, no. 1, pp. 1–14, 2011, doi:10.1016/j.memsci.2011.03.036.

28. J. Yuan, D. Mecerreyes and M. Antonietti, "Poly(ionic liquid)s: An update," *Progress in Polymer Science*, vol. 38, no. 7, pp. 1009–1036, 2013, doi:10.1016/j.progpolymsci.2013.04.002.

29. X. Zhou, J. Weber and J. Yuan, "Poly(ionic liquid)s: Platform for CO_2 capture and catalysis," *Current Opinion in Green and Sustainable Chemistry*, vol. 16, pp. 39–46, 2019, doi:10.1016/j.cogsc.2018.11.014.

30. L. A. Blanchard, D. Hancu, E. J. Beckman and J. F. Brennecke, "Green processing using ionic liquids and CO_2," *Nature*, vol. 399, no. 6731, pp. 28–29, 1999.

31. M. Farsi and E. Soroush, "Chapter 4 – CO_2 absorption by ionic liquids and deep eutectic solvents," in *Advances in Carbon Capture*, M. R. Rahimpour, M. Farsi and M. A. Makarem Eds. Woodhead Publishing, 2020, pp. 89–105.

32. A. B. Pereiro *et al.*, "Fluorinated Ionic Liquids: Properties and Applications," *ACS Sustainable Chemistry & Engineering*, vol. 1, no. 4, pp. 427–439, 2013, doi:10.1021/sc300163n.

33. J. J. Tindale and P. J. Ragogna, "Highly fluorinated phosphonium ionic liquids: Novel media for the generation of superhydrophobic coatings," *Chemical Communications*, no. 14, pp. 1831–1833, 2009, doi:10.1039/b821174d.

34. R. Markiewicz *et al.*, "Influence of Alkyl Chain Length on Thermal Properties, Structure, and Self-Diffusion Coefficients of Alkyltriethylammonium-Based Ionic Liquids," *International Journal of Molecular Sciences*, vol. 22, no. 11, 2021, doi:10.3390/ijms22115935.

35. A. Jordan and N. Gathergood, "Biodegradation of ionic liquids – a critical review," *Chemical Society Reviews*, vol. 44, no. 22, pp. 8200–8237, 2015, doi:10.1039/C5CS00444F.

36. D. O. Hartmann and C. S. Pereira, "Chapter 13 – toxicity of ionic liquids: Past, present, and future," in *Ionic Liquids in Lipid Processing and Analysis*, X. Xu, Z. Guo and L. Z. Cheong, Eds. AOCS Press, 2016, pp. 403–421.

37. M. Medved, P. Wasserscheid and T. Melin, "Ionic Liquids as Active Separation Layer in Supported Liquid Membranes," *Chemie Ingenieur Technik*, vol. 73, no. 6, pp. 715–715, 2001, doi:10.1002/1522-2640(200106)73:6<715::AID-CITE7152222>3.0.CO;2-O.

38. P. Scovazzo *et al.*, "Supported Ionic Liquid Membranes and Facilitated Ionic Liquid Membranes," in *Ionic Liquids*, vol. 818, (ACS Symposium Series, no. 818): American Chemical Society, 2002, ch. 6, pp. 69–87.

39. P. Scovazzo, J. Kieft, D. A. Finan, C. Koval, D. DuBois and R. Noble, "Gas separations using non-hexafluorophosphate [PF6](-) anion supported ionic liquid membranes," *Journal of Membrane Science*, vol. 238, no. 1–2, pp. 57–63, 2004, doi:10.1016/j.memsci.2004.02.033.

40. R. E. Baltus *et al.*, "Examination of the potential of ionic liquids for gas separations," *Separation Science and Technology*, vol. 40, no. 1–3, pp. 525–541, 2005, doi:10.1081/Ss-200042513.

41. C. DeCastro, E. Sauvage, M. H. Valkenberg and W. F. Hölderich, "Immobilised Ionic Liquids as Lewis Acid Catalysts for the Alkylation of Aromatic Compounds with Dodecene," *Journal of Catalysis*, vol. 196, no. 1, pp. 86–94, 2000, doi:10.1006/jcat.2000.3004.

42. M. H. Valkenberg, C. deCastro and W. F. Hölderich, "Friedel-Crafts acylation of aromatics catalysed by supported ionic liquids," *Applied Catalysis A: General*, vol. 215, no. 1, pp. 185–190, 2001, doi:10.1016/S0926-860X(01)00531-2.

43. D. Morgan, L. Ferguson and P. Scovazzo, "Diffusivities of Gases in Room-Temperature Ionic Liquids: Data and Correlations Obtained Using a Lag-Time Technique," *Industrial & Engineering Chemistry Research*, vol. 44, no. 13, pp. 4815–4823, 2005, doi:10.1021/ie048825v.

44. J. M. Vicent-Luna, J. J. Gutiérrez-Sevillano, S. Hamad, J. Anta and S. Calero, "Role of Ionic Liquid [EMIM]+[SCN]− in the Adsorption and Diffusion of Gases in Metal-Organic Frameworks," *ACS Applied Materials & Interfaces*, vol. 10, no. 35, pp. 29694–29704, 2018, doi:10.1021/acsami.8b11842.

45. M. Zeeshan, H. C. Gulbalkan, Z. P. Haslak, S. Keskin and A. Uzun, "Doubling CO_2/N_2 separation performance of CuBTC by incorporation of 1-n-ethyl-3-methylimidazolium diethyl phosphate," *Microporous and Mesoporous Materials*, vol. 316, 2021, Art no. 110947, doi:10.1016/j.micromeso.2021.110947.

46. M. Zeeshan, S. Keskin and A. Uzun, "Enhancing CO_2/CH_4 and CO_2/N_2 separation performances of ZIF-8 by post-synthesis modification with [BMIM][SCN]," *Polyhedron*, vol. 155, pp. 485–492, 2018, doi:10.1016/j.poly.2018.08.073.

47. D. S. Karousos *et al.*, "Physically bound and chemically grafted activated carbon supported 1-hexyl-3-methylimidazolium bis(trifluoromethylsulfonyl)imide and 1-ethyl-3-methylimidazolium acetate ionic liquid absorbents for SO_2/CO_2 gas separation," *Chemical Engineering Journal*, vol. 306, pp. 146–154, 2016, doi:10.1016/j.cej.2016.07.040.

48. Y. Huang *et al.*, "Ionic liquid functionalized multi-walled carbon nanotubes/zeolitic imidazolate framework hybrid membranes for efficient H_2/CO_2 separation," *Chemical Communications*, vol. 51, no. 97, pp. 17281–17284, 2015, doi:10.1039/C5CC05061H.

49. O. Tzialla *et al.*, "Zeolite Imidazolate Framework – Ionic Liquid Hybrid Membranes for Highly Selective CO_2 Separation," *The Journal of Physical Chemistry C*, vol. 117, no. 36, pp. 18434–18440, 2013, doi:10.1021/jp4051287.

50. B. Liu *et al.*, "Room-temperature ionic liquids modified zeolite SSZ-13 membranes for CO_2/CH_4 separation," *Journal of Membrane Science*, vol. 524, pp. 12–19, 2017, doi:10.1016/j.memsci.2016.11.004.

51. A. Ilyas, N. Muhammad, M. A. Gilani, I. F. J. Vankelecom and A. L. Khan, "Effect of zeolite surface modification with ionic liquid [APTMS][Ac] on gas separation performance of mixed matrix membranes," *Separation and Purification Technology*, vol. 205, pp. 176–183, 2018, doi:10.1016/j.seppur.2018.05.040.

52. J. Palomar, M. Gonzalez-Miquel, A. Polo and F. Rodriguez, "Understanding the Physical Absorption of CO_2 in Ionic Liquids Using the COSMO-RS Method," *Industrial & Engineering Chemistry Research*, vol. 50, no. 6, pp. 3452–3463, 2011, doi:10.1021/ie101572m.

53. A. Yokozeki and M. B. Shiflett, "Separation of carbon dioxide and sulfur dioxide gases using room-temperature ionic liquid [hmim][Tf2N]," *Energy Fuel*, vol. 23, pp. 4701–4708, 2009.

54. S. G. Kazarian, B. J. Briscoe, and T. Welton, "Combining ionic liquids and supercritical fluids: ATR-IR study of CO dissolved in two ionic liquids at high pressures," *Chemical Communications*, no. 20, pp. 2047–2048, 2000, doi:10.1039/b005514j.

55. J. L. Anthony, J. L. Anderson, E. J. Maginn and J. F. Brennecke, "Anion Effects on Gas Solubility in Ionic Liquids," *The Journal of Physical Chemistry B*, vol. 109, no. 13, pp. 6366–6374, 2005, doi:10.1021/jp046404l.

56. J. L. Anderson, J. K. Dixon and J. F. Brennecke, "Solubility of CO_2, CH_4, C_2H_6, C_2H_4, O_2, and N_2 in 1-Hexyl-3-methylpyridinium Bis(trifluoromethylsulfonyl)imide: Comparison to Other Ionic Liquids," *Accounts of Chemical Research*, vol. 40, no. 11, pp. 1208–1216, 2007, doi:10.1021/ar7001649.

57. J. E. Bara *et al.*, "Guide to CO_2 Separations in Imidazolium-Based Room-Temperature Ionic Liquids," *Industrial & Engineering Chemistry Research*, vol. 48, no. 6, pp. 2739–2751, 2009, doi:10.1021/ie8016237.

58. M. J. Muldoon, S. N. Aki, J. L. Anderson, J. K. Dixon and J. F. Brennecke, "Improving carbon dioxide solubility in ionic liquids," *The Journal of Physical Chemistry B*, vol. 111, no. 30, pp. 9001–9009, 2007, doi:10.1021/jp071897q.

59. E. J. Beckman, "A challenge for green chemistry: Designing molecules that readily dissolve in carbon dioxide," *Chemical Communications*, no. 17, pp. 1885–1888, 2004, doi:10.1039/B404406C.

60. Z. Lei, J. Han, B. Zhang, Q. Li, J. Zhu and B. Chen, "Solubility of CO_2 in Binary Mixtures of Room-Temperature Ionic Liquids at High Pressures," *Journal of Chemical & Engineering Data*, vol. 57, no. 8, pp. 2153–2159, 2012, doi:10.1021/je300016q.

61. T. K. Carlisle, J. E. Bara, C. J. Gabriel, R. D. Noble and D. L. Gin, "Interpretation of CO_2 Solubility and Selectivity in Nitrile-Functionalized Room-Temperature Ionic Liquids Using a Group Contribution Approach," *Industrial & Engineering Chemistry Research*, vol. 47, no. 18, pp. 7005–7012, 2008, doi:10.1021/ie8001217.

62. M. S. Shannon, J. M. Tedstone, S. P. O. Danielsen, M. S. Hindman, A. C. Irvin and J. E. Bara, "Free Volume as the Basis of Gas Solubility and Selectivity in Imidazolium-Based Ionic Liquids," *Industrial & Engineering Chemistry Research*, vol. 51, no. 15, pp. 5565–5576, 2012, doi:10.1021/ie202916e.

63. A. Bondi, *Physical Properties of Molecular Crystals, Liquids and Gases*. John Wiley and Sons, 1968.

64. J. L. Anderson, J. K. Dixon, E. J. Maginn and J. F. Brennecke, "Measurement of SO_2 solubility in ionic liquids," *The Journal of Physical Chemistry B*, vol. 110, no. 31, pp. 15059–15062, 2006, doi:10.1021/jp063547u.

65. J. Huang, A. Riisager, P. Wasserscheid and R. Fehrmann, "Reversible physical absorption of SO_2 by ionic liquids," *Chemical Communications*, no. 38, pp. 4027–4029, 2006, doi:10.1039/b609714f.

66. N. A. Manan, C. Hardacre, J. Jacquemin, D. W. Rooney and T. G. A. Youngs, "Evaluation of Gas Solubility Prediction in Ionic Liquids using COSMOthermX," *Journal of Chemical & Engineering Data*, vol. 54, no. 7, pp. 2005–2022, 2009, doi:10.1021/je800857x.

67. Y. Y. Jiang, Z. Zhou, Z. Jiao, L. Li, Y. T. Wu and Z. B. Zhang, "SO_2 gas separation using supported ionic liquid membranes," *Journal of Physical Chemistry B*, vol. 111, no. 19, pp. 5058–5061, 2007, doi:10.1021/jp071742i.

68. J. Huang, A. Riisager, R. W. Berg and R. Fehrmann, "Tuning ionic liquids for high gas solubility and reversible gas sorption," *Journal of Molecular Catalysis A: Chemical*, vol. 279, no. 2, pp. 170–176, 2008, doi:10.1016/j.molcata.2007.07.036.

69. S. J. Zeng *et al.*, "Improving SO_2 capture by tuning functional groups on the cation of pyridinium-based ionic liquids," *RSC Advances*, vol. 5, no. 4, pp. 2470–2478, 2015, doi:10.1039/c4ra13469a.

70. S. Ren, Y. Hou, K. Zhang and W. Wu, "Ionic liquids: Functionalization and absorption of SO_2," *Green Energy & Environment*, vol. 3, no. 3, pp. 179–190, 2018, doi:10.1016/j.gee.2017.11.003.

71. G. Cui *et al.*, "Tuning Ionic Liquids with Functional Anions for SO_2 Capture through Simultaneous Cooperation of N and O Chemical Active Sites with SO_2," *Industrial & Engineering Chemistry Research*, vol. 59, no. 49, pp. 21522–21529, 2020, doi:10.1021/acs.iecr.0c05190.

72. Y. Jiang, X. Liu and D. Deng, "Absorption of SO_2 in Furoate Ionic Liquids/PEG200 Mixtures and Thermodynamic Analysis," *Journal of Chemical & Engineering Data*, vol. 63, no. 2, pp. 259–268, 2018, doi:10.1021/acs.jced.7b00306.

73. X. Meng, J. Wang, H. Jiang, X. Zhang, S. Liu and Y. Hu, "Guanidinium-based dicarboxylic acid ionic liquids for SO_2 capture," *Journal of Chemical Technology & Biotechnology*, vol. 92, no. 4, pp. 767–774, 2017, doi:10.1002/jctb.5052.

74. L. Wang, Y. Zhang, Y. Liu, H. Xie, Y. Xu and J. Wei, "SO_2 absorption in pure ionic liquids: Solubility and functionalization," *Journal of hazardous materials*, vol. 392, 2020, Art no. 122504, doi:10.1016/j.jhazmat.2020.122504.

75. E. Duan, B. Guo, M. Zhang, Y. Guan, H. Sun and J. Han, "Efficient capture of SO_2 by a binary mixture of caprolactam tetrabutyl ammonium bromide ionic liquid and water," *Journal of Hazardous Materials*, vol. 194, pp. 48–52, 2011, doi:10.1016/j.jhazmat.2011.07.059.

76. Y. Chen, X. Liu, J. M. Woodley and G. M. Kontogeorgis, "Gas Solubility in Ionic Liquids: UNIFAC-IL Model Extension," *Industrial & Engineering Chemistry Research*, vol. 59, no. 38, pp. 16805–16821, 2020, doi:10.1021/acs.iecr.0c02769.

77. F. Gholizadeh, A. Kamgar, M. Roostaei and M. R. Rahimpour, "Determination of SO_2 solubility in ionic liquids: COSMO-RS and modified Sanchez-Lacombe EOS," *Journal of Molecular Liquids*, vol. 272, pp. 878–884, 2018, doi:10.1016/j.molliq.2018.09.137.

78. C. Cleeton, O. Kvam, R. Rea, L. Sarkisov and M. G. De Angelis, "Competitive H_2S-CO_2 absorption in reactive aqueous methyldiethanolamine solution: Prediction with ePC-SAFT," *Fluid Phase Equilibria*, vol. 511, 2020, Art no. 112453, doi:10.1016/j.fluid.2019.112453.

79. H. Sakhaeinia, V. Taghikhani, A. H. Jalili, A. Mehdizadeh and A. A. Safekordi, "Solubility of H_2S in 1-(2-hydroxyethyl)-3-methylimidazolium ionic liquids with different anions," *Fluid Phase Equilibria*, vol. 298, no. 2, pp. 303–309, 2010, doi:10.1016/j.fluid.2010.08.027.

80. S. Aparicio and M. Atilhan, "Computational Study of Hexamethylguanidinium Lactate Ionic Liquid: A Candidate for Natural Gas Sweetening," *Energy & Fuels*, vol. 24, no. 9, pp. 4989–5001, 2010, doi:10.1021/ef1005258.

81. K. Huang, D.-N. Cai, Y.-L. Chen, Y.-T. Wu, X.-B. Hu and Z.-B. Zhang, "Dual Lewis Base Functionalization of Ionic Liquids for Highly Efficient and Selective Capture of H_2S," *ChemPlusChem*, vol. 79, no. 2, pp. 241–249, 2014, doi:10.1002/cplu.201300365.

82. K. Huang, X.-M. Zhang, X.-B. Hu and Y.-T. Wu, "Hydrophobic protic ionic liquids tethered with tertiary amine group for highly efficient and selective absorption of H_2S from CO_2," *AIChE Journal*, vol. 62, no. 12, pp. 4480–4490, 2016, doi:10.1002/aic.15363.

83. M. Rahmati-Rostami, C. Ghotbi, M. Hosseini-Jenab, A. N. Ahmadi and A. H. Jalili, "Solubility of H_2S in ionic liquids [hmim][PF6], [hmim][BF4], and [hmim][Tf2N]," *J Chem Thermodyn*, vol. 41, no. 9, pp. 1052–1055, 2009, doi:10.1016/j.jct.2009.04.014.

84. C. S. Pomelli, C. Chiappe, A. Vidis, G. Laurenczy and P. J. Dyson, "Influence of the interaction between hydrogen sulfide and ionic liquids on solubility: Experimental and theoretical investigation," *Journal of Physical Chemistry B*, vol. 111, no. 45, pp. 13014–13019, 2007, doi:10.1021/jp076129d.

85. J. Bedia, J. Palomar, M. Gonzalez-Miquel, F. Rodriguez and J. J. Rodriguez, "Screening ionic liquids as suitable ammonia absorbents on the basis of thermodynamic and kinetic analysis," *Separation and Purification Technology*, vol. 95, pp. 188–195, 2012, doi:10.1016/j.seppur.2012.05.006.

86. A. Yokozeki and M. B. Shiflett, "Vapor-liquid equilibria of ammonia + ionic liquid mixtures," *Applied Energy*, vol. 84, no. 12, pp. 1258–1273, 2007, doi:10.1016/j.apenergy.2007.02.005.
87. Z. Li *et al.*, "Efficient absorption of ammonia with hydroxyl-functionalized ionic liquids," *RSC Advances*, vol. 5, no. 99, pp. 81362–81370, 2015, doi:10.1039/C5RA13730F.
88. D. Shang *et al.*, "Protic ionic liquid [Bim][NTf2] with strong hydrogen bond donating ability for highly efficient ammonia absorption," *Green Chemistry*, vol. 19, no. 4, pp. 937–945, 2017, doi:10.1039/C6GC03026B.
89. M. B. Shiflett and A. Yokozeki, "Solubilities and diffusivities of carbon dioxide in ionic liquids: [bmim][PF6] and [bmim][BF4]," *Industrial & Engineering Chemistry Research*, vol. 44, no. 12, pp. 4453–4464, 2005, doi:10.1021/ie058003d.
90. Y. S. Kim, W. Y. Choi, J. H. Jang, K. P. Yoo and C. S. Lee, "Solubility measurement and prediction of carbon dioxide in ionic liquids," *Fluid Phase Equilibria*, vol. 228, pp. 439–445, 2005, doi:10.1016/j.fluid.2004.09.006.
91. J. Y. Ahn, B. C. Lee, J. S. Lim, K. P. Yoo and J. W. Kang, "High-pressure phase behavior of binary and ternary mixtures containing ionic liquid [C6-mim][Tf2N], dimethyl carbonate and carbon dioxide," *Fluid Phase Equilibria*, vol. 290, pp. 75–79, 2010.
92. C. A. Ohlin, P. J. Dyson and G. Laurenczy, "Carbon monoxide solubility in ionic liquids: Determination, prediction and relevance to hydroformylation," *Chemical Communications*, pp. 1070–1071, 2004.
93. J. Jacquemin, M. F. C. Gomes, P. Husson and V. Majer, "Solubility of carbon dioxide, ethane, methane, oxygen, nitrogen, hydrogen, argon, and carbon monoxide in 1-butyl-3-methylimidazolium tetrafluoroborate between temperatures 283 K and 343 K and at pressures close to atmospheric," *J Chem Thermodyn*, vol. 38, no. 4, pp. 490–502, 2006, doi:10.1016/j.jct.2005.07.002.
94. J. Jacquemin, P. Husson, V. Majer, A. A. H. Padua and M. F. C. Gomes, "Thermophysical properties, low pressure solubilities and thermodynamics of solvation of carbon dioxide and hydrogen in two ionic liquids based on the alkylsulfate anion," *Green Chemistry*, vol. 10, no. 9, pp. 944–950, 2008, doi:10.1039/b802761g.
95. D. Camper, J. Bara, C. Koval and R. Noble, "Bulk-fluid solubility and membrane feasibility of Rmim-based room-temperature ionic liquids," *Industrial & Engineering Chemistry Research*, vol. 45, no. 18, pp. 6279–6283, 2006, doi:10.1021/ie060177n.
96. M. Althuluth, M. C. Kroon and C. J. Peters, "Solubility of Methane in the Ionic Liquid 1-Ethyl-3-methylimidazolium Tris(pentafluoroethyl)trifluorophosphate," *Industrial & Engineering Chemistry Research*, vol. 51, no. 51, pp. 16709–16712, 2012, doi:10.1021/ie302472t.
97. M. Althuluth, M. C. Kroon and C. J. Peters, "High pressure solubility of methane in the ionic liquid 1-hexyl-3-methylimidazolium tricyanomethanide," *The Journal of Supercritical Fluids*, vol. 128, pp. 145–148, 2017, doi:10.1016/j.supflu.2017.05.021.
98. E. D. Bates, R. D. Mayton, I. Ntai and J. H. Davis, "CO_2 capture by a task-specific ionic liquid," *Journal of the American Chemical Society*, vol. 124, no. 6, pp. 926–927, 2002, doi:10.1021/ja017593d.
99. K. E. Gutowski and E. J. Maginn, "Amine-Functionalized Task-Specific Ionic Liquids: A Mechanistic Explanation for the Dramatic Increase in Viscosity upon Complexation with CO_2 from Molecular Simulation," *Journal of the American Chemical Society*, vol. 130, no. 44, pp. 14690–14704, 2008, doi:10.1021/ja804654b.
100. B. E. Gurkan *et al.*, "Equimolar CO_2 Absorption by Anion-Functionalized Ionic Liquids," *Journal of the American Chemical Society*, vol. 132, no. 7, pp. 2116–2117, 2010, doi:10.1021/ja909305t.
101. D. Camper, J. E. Bara, D. L. Gin and R. D. Noble, "Room-Temperature Ionic Liquid-Amine Solutions: Tunable Solvents for Efficient and Reversible Capture Of CO_2," *Industrial & Engineering Chemistry Research*, vol. 47, no. 21, pp. 8496–8498, 2008, doi:10.1021/ie801002m.

102. C. Wang, H. Luo, D. E. Jiang, H. Li and S. Dai, "Carbon dioxide capture by superbase-derived protic ionic liquids," *Angewandte Chemie*, vol. 49, no. 34, pp. 5978–81, 2010, doi:10.1002/anie.201002641.

103. G. R. Yu *et al.*, "Design of task-specific ionic liquids for capturing CO_2: A molecular orbital study," *Industrial & Engineering Chemistry Research*, vol. 45, no. 8, pp. 2875–2880, 2006, doi:10.1021/ie050975y.

104. J. J. H. Davis, "Task-Specific Ionic Liquids," *Chemistry Letters*, vol. 33, no. 9, pp. 1072–1077, 2004, doi:10.1246/cl.2004.1072.

105. E. M. Mindrup and W. F. Schneider, "Computational Comparison of Tethering Strategies for Amine Functionalised Ionic Liquids," in *Ionic Liquids: From Knowledge to Application*, vol. 1030, (ACS Symposium Series, no. 1030): American Chemical Society, 2009, ch. 27, pp. 419–430.

106. B. F. Goodrich *et al.*, "Effect of water and temperature on absorption of CO_2 by amine-functionalized anion-tethered ionic liquids," *The Journal of Physical Chemistry B*, vol. 115, no. 29, pp. 9140–9150, 2011, doi:10.1021/jp2015534.

107. B. Gurkan *et al.*, "Molecular Design of High Capacity, Low Viscosity, Chemically Tunable Ionic Liquids for CO_2 Capture," *Journal of Physical Chemistry Letters*, vol. 1, no. 24, pp. 3494–3499, 2010, doi:10.1021/jz101533k.

108. S. Seo, M. A. DeSilva and J. F. Brennecke, "Physical Properties and CO_2 Reaction Pathway of 1-Ethyl-3-Methylimidazolium Ionic Liquids with Aprotic Heterocyclic Anions," *The Journal of Physical Chemistry B*, vol. 118, no. 51, pp. 14870–14879, 2014, doi:10.1021/jp509583c.

109. T. Ramnial *et al.*, "Carbon-Centered Strong Bases in Phosphonium Ionic Liquids," *The Journal of Organic Chemistry*, vol. 73, no. 3, pp. 801–812, 2008, doi:10.1021/jo701289d.

110. M. Besnard *et al.*, "On the spontaneous carboxylation of 1-butyl-3-methylimidazolium acetate by carbon dioxide," *Chemical Communications*, vol. 48, no. 9, pp. 1245–1247, 2012, doi:10.1039/C1CC16702B.

111. G. Gurau, H. Rodriguez, S. P. Kelley, P. Janiczek, R. S. Kalb and R. D. Rogers, "Demonstration of chemisorption of carbon dioxide in 1,3-dialkylimidazolium acetate ionic liquids," *Angewandte Chemie*, vol. 50, no. 50, pp. 12024–12026, 2011, doi:10.1002/anie.201105198.

112. S. Seo *et al.*, "Phase-Change Ionic Liquids for Postcombustion CO_2 Capture," *Energy & Fuels*, vol. 28, no. 9, pp. 5968–5977, 2014, doi:10.1021/ef501374x.

113. H. Zhou, X. Xu, X. Chen and G. Yu, "Novel ionic liquids phase change solvents for CO_2 capture," *International Journal of Greenhouse Gas Control*, vol. 98, 2020, Art no. 103068, doi:10.1016/j.ijggc.2020.103068.

114. S. Ren, Y. Hou, W. Wu, Q. Liu, Y. Xiao and X. Chen, "Properties of Ionic Liquids Absorbing SO_2 and the Mechanism of the Absorption," *The Journal of Physical Chemistry B*, vol. 114, no. 6, pp. 2175–2179, 2010.

115. M. Jin *et al.*, "Solubilities and thermodynamic properties of SO_2 in ionic liquids," *The Journal of Physical Chemistry B*, vol. 115, no. 20, pp. 6585–6591, 2011, doi:10.1021/jp1124074.

116. W. Z. Wu, B. X. Han, H. X. Gao, Z. M. Liu, T. Jiang and J. Huang, "Desulfurization of flue gas: SO_2 absorption by an ionic liquid," *Angewandte Chemie*, vol. 43, no. 18, pp. 2415–2417, 2004, doi:10.1002/anie.200353437.

117. X. L. Yuan, S. J. Zhang and X. M. Lu, "Hydroxyl ammonium ionic liquids: Synthesis, properties, and solubility of SO_2," *Journal of Chemical & Engineering Data* vol. 52, no. 2, pp. 596–599, 2007.

118. M. B. Shiflett and A. Yokozeki, "Chemical Absorption of Sulfur Dioxide in Room-Temperature Ionic Liquids," *Industrial & Engineering Chemistry Research*, vol. 49, no. 3, pp. 1370–1377, 2010, doi:10.1021/ie901254f.

119. P. J. Carvalho and J. A. P. Coutinho, "On the Nonideality of CO_2 Solutions in Ionic Liquids and Other Low Volatile Solvents," *Journal of Physical Chemistry Letters*, vol. 1, no. 4, pp. 774–780, 2010, doi:10.1021/jz100009c.

120. Z. Lei, C. Dai and B. Chen, "Gas Solubility in Ionic Liquids," *Chemical Reviews*, vol. 114, no. 2, pp. 1289–1326, 2014, doi:10.1021/cr300497a.

121. I. F. Zubeir, G. E. Romanos, W. M. A. Weggemans, B. Iliev, T. J. S. Schubert and M. C. Kroon, "Solubility and Diffusivity of CO_2 in the Ionic Liquid 1-Butyl-3-methylimidazolium Tricyanomethanide within a Large Pressure Range (0.01 MPa to 10 MPa)," *Journal of Chemical & Engineering Data*, vol. 60, no. 6, pp. 1544–1562, 2015, doi:10.1021/je500765m.

122. J. Huang and T. Rüther, "Why are Ionic Liquids Attractive for CO_2 Absorption? An Overview," *Australian Journal of Chemistry*, vol. 62, no. 4, pp. 298–308, 2009, doi:10.1071/CH08559.

123. J. F. Brennecke and B. E. Gurkan, "Ionic Liquids for CO_2 Capture and Emission Reduction," *Journal of Physical Chemistry Letters*, vol. 1, no. 24, pp. 3459–3464, 2010, doi:10.1021/jz1014828.

124. F. Karadas, M. Atilhan and S. Aparicio, "Review on the Use of Ionic Liquids (ILs) as Alternative Fluids for CO_2 Capture and Natural Gas Sweetening," *Energy & Fuels*, vol. 24, no. 11, pp. 5817–5828, 2010, doi:10.1021/ef1011337.

125. K. R. Seddon, "Ionic liquids for clean technology," *J Chem Technol Biot*, vol. 68, no. 4, pp. 351–356, 1997.

126. A. Finotello, J. E. Bara, D. Camper and R. D. Noble, "Room-Temperature Ionic Liquids: Temperature Dependence of Gas Solubility Selectivity," *Industrial & Engineering Chemistry Research*, vol. 47, no. 10, pp. 3453–3459, 2007, doi:10.1021/ie0704142.

127. S. Zhang, Y. Chen, F. Li, X. Lu, W. Dai and R. Mori, "Fixation and conversion of CO_2 using ionic liquids," *Catalysis Today*, vol. 115, no. 1–4, pp. 61–69, 2006.

128. I. G. Economou, P. Krokidas, V. K. Michalis, O. A. Moultos, I. N. Tsimpanogiannis and N. Vergadou, "The role of molecular thermodynamics in developing industrial processes and novel products that meet the needs for a sustainable future," in *The Water-Food-Energy Nexus: Processes, Technologies, and Challenges*, I. M. Mujtaba, R. Srinivasan and N. O. Elbashir, Eds. CRC Press, 2017.

129. K. Dong, X. Liu, H. Dong, X. Zhang and S. Zhang, "Multiscale Studies on Ionic Liquids," *Chemical Reviews*, vol. 117, no. 10, pp. 6636–6695, 2017, doi:10.1021/acs.chemrev.6b00776.

130. E. I. Izgorodina, Z. L. Seeger, D. L. A. Scarborough and S. Y. S. Tan, "Quantum Chemical Methods for the Prediction of Energetic, Physical, and Spectroscopic Properties of Ionic Liquids," *Chemical Reviews*, vol. 117, no. 10, pp. 6696–6754, 2017, doi:10.1021/acs.chemrev.6b00528.

131. F. Dommert *et al.*, "Towards multiscale modeling of ionic liquids: From electronic structure to bulk properties," *Journal of Molecular Liquids*, vol. 152, no. 1–3, pp. 2–8, 2010, doi:10.1016/j.molliq.2009.06.014.

132. Y. L. Wang, S. Sarman, B. Li and A. Laaksonen, "Multiscale modeling of the tri-hexyltetradecylphosphonium chloride ionic liquid," *Physical Chemistry Chemical Physics*, vol. 17, no. 34, pp. 22125–22135, 2015, doi:10.1039/c5cp02586a.

133. M. Karelson, V. S. Lobanov and A. R. Katritzky, "Quantum-Chemical Descriptors in QSAR/QSPR Studies," *Chemical Reviews*, vol. 96, no. 3, pp. 1027–1044, 1996, doi:10.1021/cr950202r.

134. A. R. Katritzky *et al.*, "Quantitative Structure- Property Relationship Studies on Ostwald Solubility and Partition Coefficients of Organic Solutes in Ionic Liquids," *Journal of Chemical & Engineering Data*, vol. 53, no. 5, pp. 1085–1092, 2008, doi:10.1021/je700607b.

135. K. Tämm and P. Burk, "QSPR analysis for infinite dilution activity coefficients of organic compounds," *Journal of Molecular Modeling*, vol. 12, no. 4, pp. 417–421, 2006, doi:10.1007/s00894-005-0062-2.

136. D. M. Eike, J. F. Brennecke and E. J. Maginn, "Predicting Infinite-Dilution Activity Coefficients of Organic Solutes in Ionic Liquids," *Industrial & Engineering Chemistry Research*, vol. 43, no. 4, pp. 1039–1048, 2004, doi:10.1021/ie034152p.

137. V. S. Bernales, A. V. Marenich, R. Contreras, C. J. Cramer and D. G. Truhlar, "Quantum Mechanical Continuum Solvation Models for Ionic Liquids," *The Journal of Physical Chemistry B*, vol. 116, no. 30, pp. 9122–9129, 2012, doi:10.1021/jp304365v.

138. M. Diedenhofen, F. Eckert and A. Klamt, "Prediction of Infinite Dilution Activity Coefficients of Organic Compounds in Ionic Liquids Using COSMO-RS," *Journal of Chemical & Engineering Data*, vol. 48, no. 3, pp. 475–479, 2003, doi:10.1021/je025626e.

139. M. Diedenhofen and A. Klamt, "COSMO-RS as a tool for property prediction of IL mixtures – a review," *Fluid Phase Equilibria*, vol. 294, pp. 31–38, 2010.

140. C. Dai, G. Yu and Z. Lei, "Chapter 7 – Predictive molecular thermodynamic models for ionic liquids," in *Theoretical and Computational Approaches to Predicting Ionic Liquid Properties*, A. Joseph and S. Mathew, Eds. Elsevier, 2021, pp. 209–241.

141. L. F. Zubeir *et al.*, "Thermophysical properties of imidazolium tricyanomethanide ionic liquids: Experiments and molecular simulation," *Physical Chemistry Chemical Physics*, vol. 18, no. 33, pp. 23121–23138, 2016, doi:10.1039/C6CP01943A.

142. M. A. A. Rocha *et al.*, "Alkylimidazolium Based Ionic Liquids: Impact of Cation Symmetry on Their Nanoscale Structural Organization," *Journal of Physical Chemistry B*, vol. 117, no. 37, pp. 10889–10897, 2013, doi:10.1021/jp406374a.

143. M. Tariq, P. J. Carvalho, J. A. P. Coutinho, I. M. Marrucho, J. N. C. Lopes and L. P. N. Rebelo, "Viscosity of (C2–C14) 1-alkyl-3-methylimidazolium bis(trifluoromethylsulfonyl)amide ionic liquids in an extended temperature range," *Fluid Phase Equilibria*, vol. 301, no. 1, pp. 22–32, 2011, doi:10.1016/j.fluid.2010.10.018.

144. M. A. A. Rocha, F. M. S. Ribeiro, A. I. M. C. L. Ferreira, J. A. P. Coutinho and L. M. N. B. F. Santos, "Thermophysical properties of [CN−1C1im][PF6] ionic liquids," *Journal of Molecular Liquids*, vol. 188, pp. 196–202, 2013, doi:10.1016/j.molliq.2013.09.031.

145. D. Song and J. Chen, "Density and Viscosity Data for Mixtures of Ionic Liquids with a Common Anion," *Journal of Chemical & Engineering Data*, vol. 59, no. 2, pp. 257–262, 2014, doi:10.1021/je400332j.

146. D. Tomida, S. Kenmochi, T. Tsukada, K. Qiao, Q. Bao and C. Yokoyama, "Viscosity and Thermal Conductivity of 1-Hexyl-3-methylimidazolium Tetrafluoroborate and 1-Octyl-3-methylimidazolium Tetrafluoroborate at Pressures up to 20 MPa," *International Journal of Thermophysics*, vol. 33, no. 6, pp. 959–969, 2012, doi:10.1007/s10765-012-1233-x.

147. S. Bi, T. M. Koller, M. H. Rausch, P. Wasserscheid and A. P. Fröba, "Dynamic Viscosity of Tetracyanoborate- and Tricyanomethanide-Based Ionic Liquids by Dynamic Light Scattering," *Industrial & Engineering Chemistry Research*, vol. 54, no. 11, pp. 3071–3081, 2015, doi:10.1021/acs.iecr.5b00086.

148. C. M. S. S. Neves *et al.*, "Systematic Study of the Thermophysical Properties of Imidazolium-Based Ionic Liquids with Cyano-Functionalized Anions," *The Journal of Physical Chemistry B*, vol. 117, no. 35, pp. 10271–10283, 2013, doi:10.1021/jp405913b.

149. M. G. Freire *et al.*, "Thermophysical characterization of ionic liquids able to dissolve biomass," *Journal of Chemical & Engineering Data*, vol. 56, no. 12, pp. 4813–4822, 2011, doi:10.1021/je200790q.

150. R. Hayes, G. G. Warr and R. Atkin, "Structure and nanostructure in ionic liquids," *Chemical Reviews*, vol. 115, no. 13, pp. 6357–63426, 2015, doi:10.1021/cr500411q.

151. Y.-L. Wang *et al.*, "Microstructural and Dynamical Heterogeneities in Ionic Liquids," *Chemical Reviews*, vol. 120, no. 13, pp. 5798–5877, 2020, doi:10.1021/acs. chemrev.9b00693.

152. J. Habasaki and K. L. Ngai, "Multifractal nature of heterogeneous dynamics and structures in glass forming ionic liquids," *Journal of Non-Crystalline Solids*, vol. 357, no. 2, pp. 446–453, 2011, doi:10.1016/j.jnoncrysol.2010.06.047.

153. N. Vergadou, E. Androulaki, J.-R. Hill and I. G. Economou, "Molecular simulations of imidazolium-based tricyanomethanide ionic liquids using an optimized classical force field," *Physical Chemistry Chemical Physics*, vol. 18, no. 9, pp. 6850–6860, 2016, doi:10.1039/C5CP05892A.

154. S. M. Urahata and M. C. C. Ribeiro, "Unraveling Dynamical Heterogeneity in the Ionic Liquid 1-Butyl-3-methylimidazolium Chloride," *Journal of Physical Chemistry Letters*, vol. 1, no. 11, pp. 1738–1742, 2010, doi:10.1021/jz100411w.

155. T. Ishida and H. Shirota, "Dicationic versus monocationic ionic liquids: Distinctive ionic dynamics and dynamical heterogeneity," *Journal of Physical Chemistry B*, vol. 117, no. 4, pp. 1136–1150, 2013, doi:10.1021/jp3110425.

156. H. Jin, X. Li and M. Maroncelli, "Heterogeneous Solute Dynamics in Room Temperature Ionic Liquids," *The Journal of Physical Chemistry B*, vol. 111, no. 48, pp. 13473–13478, 2007, doi:10.1021/jp077226+.

157. N. Vergadou, "Molecular Simulation of Ionic Liquids: Complex Dynamics and Structure," *Springer Proceedings in Mathematics and Statistics*, vol. 219, pp. 297–312, 2017, doi:10.1007/978-3-319-68103-0_14.

158. E. Androulaki, N. Vergadou and I. G. Economou, "Analysis of the heterogeneous dynamics of imidazolium-based [Tf2N−] ionic liquids using molecular simulation," *Molecular Physics*, vol. 112, no. 20, pp. 2694–2706, 2014, doi:10.1080/00268976.2014. 906670.

159. M. G. Del Pópolo and G. A. Voth, "On the Structure and Dynamics of Ionic Liquids," *The Journal of Physical Chemistry B*, vol. 108, no. 5, pp. 1744–1752, 2004, doi:10.1021/ jp0364699.

160. H. Liu and E. Maginn, "A molecular dynamics investigation of the structural and dynamic properties of the ionic liquid 1-n-butyl-3-methylimidazolium bis(trifluoromethanesulfonyl) imide," *The Journal of Chemical Physics*, vol. 135, no. 12, 2011, Art no. 124507, doi:10.1063/1.3643124

161. J. Habasaki and K. L. Ngai, "Heterogeneous dynamics of ionic liquids from molecular dynamics simulations," *The Journal of Chemical Physics*, vol. 129, no. 19, 2008, Art no. 194501, doi:10.1063/1.3005372.

162. O. Borodin, "Polarizable Force Field Development and Molecular Dynamics Simulations of Ionic Liquids," *The Journal of Physical Chemistry B*, vol. 113, no. 33, pp. 11463–11478, 2009, doi:10.1021/jp905220k.

163. C. M. S. S. Neves *et al.*, "The impact of ionic liquid fluorinated moieties on their thermophysical properties and aqueous phase behaviour," *Physical Chemistry Chemical Physics*, vol. 16, no. 39, pp. 21340–21348, 2014, doi:10.1039/C4CP02008A.

164. S. Aparicio, M. Atilhan and F. Karadas, "Thermophysical Properties of Pure Ionic Liquids: Review of Present Situation," *Industrial & Engineering Chemistry Research*, vol. 49, no. 20, pp. 9580–9595, 2010, doi:10.1021/ie101441s.

165. F. Philippi and T. Welton, "Targeted modifications in ionic liquids – from understanding to design," *Physical Chemistry Chemical Physics*, vol. 23, no. 12, pp. 6993–7021, 2021, doi:10.1039/D1CP00216C.

166. T. Cremer *et al.*, "Towards a Molecular Understanding of Cation-Anion Interactions – Probing the Electronic Structure of Imidazolium Ionic Liquids by NMR Spectroscopy, X-ray Photoelectron Spectroscopy and Theoretical Calculations," *Chemistry – A European Journal*, vol. 16, pp. 9018–9033, 2010, doi:10.1002/chem.201001032.

167. B. L. Bhargava and S. Balasubramanian, "Refined potential model for atomistic simulations of ionic liquid [bmim][PF6]," *The Journal of Chemical Physics*, vol. 127, no. 11, 2007, Art no. 114510, doi:10.1063/1.2772268.

168. E. I. Izgorodina *et al.*, "Importance of dispersion forces for prediction of thermodynamic and transport properties of some common ionic liquids," *Physical Chemistry Chemical Physics*, vol. 16, no. 16, pp. 7209–7221, 2014, doi:10.1039/c3cp53035c.

169. E. I. Izgorodina, "Towards large-scale, fully ab initio calculations of ionic liquids," *Physical Chemistry Chemical Physics*, vol. 13, no. 10, pp. 4189–207, 2011, doi:10.1039/c0cp02315a.

170. E. I. Izgorodina, U. L. Bernard and D. R. MacFarlane, "Ion-Pair Binding Energies of Ionic Liquids: Can DFT Compete with Ab Initio-Based Methods?" *The Journal of Physical Chemistry A*, vol. 113, no. 25, pp. 7064–7072, 2009, doi:10.1021/jp8107649.

171. P. A. Hunt, B. Kirchner and T. Welton, "Characterising the electronic structure of ionic liquids: An examination of the 1-butyl-3-methylimidazolium chloride ion pair," *Chemistry*, vol. 12, no. 26, pp. 6762–6775, 2006, doi:10.1002/chem.200600103.

172. E. Androulaki, N. Vergadou, J. Ramos and I. G. Economou, "Structure, thermodynamic and transport properties of imidazolium-based bis(trifluoromethylsulfonyl)imide ionic liquids from molecular dynamics simulations," *Molecular Physics*, vol. 110, no. 11–12, pp. 1139–1152, 2012, doi:10.1080/00268976.2012.670280.

173. M. Bühl, A. Chaumont, R. Schurhammer and G. Wipff, "Ab Initio Molecular Dynamics of Liquid 1,3-Dimethylimidazolium Chloride," *Journal of Physical Chemistry B*, vol. 109, no. 39, pp. 18591–18599, 2005, doi:10.1021/jp0518299.

174. S. Tsuzuki, H. Tokuda, K. Hayamizu and M. Watanabe, "Magnitude and directionality of interaction in ion pairs of ionic liquids: Relationship with ionic conductivity," *The Journal of Physical Chemistry B*, vol. 109, no. 34, pp. 16474–16781, 2005, doi:10.1021/jp0533628.

175. W. Beichel *et al.*, "Charge-scaling effect in ionic liquids from the charge-density analysis of N,N'-dimethylimidazolium methylsulfate," *Angewandte Chemie*, vol. 53, no. 12, pp. 3143–3146, 2014, doi:10.1002/anie.201308760.

176. F. Dommert, K. Wendler, R. Berger, L. Delle Site and C. Holm, "Force fields for studying the structure and dynamics of ionic liquids: A critical review of recent developments," *ChemPhysChem*, vol. 13, no. 7, pp. 1625–1637, 2012, doi:10.1002/cphc.201100997.

177. K. Bica, M. Deetlefs, C. Schroder and K. R. Seddon, "Polarisabilities of alkylimidazolium ionic liquids," *Physical Chemistry Chemical Physics*, vol. 15, no. 8, pp. 2703–2711, 2013, doi:10.1039/c3cp43867h.

178. C. E. S. Bernardes *et al.*, "Additive polarizabilities in ionic liquids," *Physical Chemistry Chemical Physics*, vol. 18, pp. 1665–1670, 2016, doi:10.1039/c5cp06595j.

179. S. Koßmann, J. Thar, B. Kirchner, P. A. Hunt and T. Welton, "Cooperativity in ionic liquids," *The Journal Chemical Physics*, vol. 124, 2006, Art no. 174506.

180. K. Fumino, V. Fossog, P. Stange, D. Paschek, R. Hempelmann and R. Ludwig, "Controlling the Subtle Energy Balance in Protic Ionic Liquids: Dispersion Forces Compete with Hydrogen Bonds," *Angewandte Chemie*, vol. 54, no. 9, pp. 2792–2795, 2015, doi:10.1002/anie.201411509.

181. I. Skarmoutsos, D. Dellis, R. P. Matthews, T. Welton and P. A. Hunt, "Hydrogen Bonding in 1-Butyl- and 1-Ethyl-3-methylimidazolium Chloride Ionic Liquids," *The Journal of Physical Chemistry B*, vol. 116, no. 16, pp. 4921–4933, 2012, doi:10.1021/jp209485y.

182. K. Dong, S. J. Zhang and J. J. Wang, "Understanding the hydrogen bonds in ionic liquids and their roles in properties and reactions," *Chemical Communications*, vol. 52, no. 41, pp. 6744–6764, 2016, doi:10.1039/c5cc10120d.

183. P. A. Hunt, C. R. Ashworth and R. P. Matthews, "Hydrogen bonding in ionic liquids," *Chemical Society Reviews*, vol. 44, no. 5, pp. 1257–1288, 2015, doi:10.1039/c4cs00278d.

184. E. I. Izgorodina and D. R. MacFarlane, "Nature of hydrogen bonding in charged hydrogen-bonded complexes and imidazolium-based ionic liquids," *The Journal of Physical Chemistry B*, vol. 115, no. 49, pp. 14659–14667, 2011, doi:10.1021/jp208150b.

185. X. Song *et al.*, "Structural heterogeneity and unique distorted hydrogen bonding in primary ammonium nitrate ionic liquids studied by high-energy X-ray diffraction experiments and MD simulations," *The Journal of Physical Chemistry B*, vol. 116, no. 9, pp. 2801–2813, 2012, doi:10.1021/jp209561t.

186. B. Bhargava and S. Balasubramanian, "Probing anion–carbon dioxide interactions in room temperature ionic liquids: Gas phase cluster calculations," *Chemical Physics Letters*, vol. 444, pp. 242–246, 2007, doi:10.1016/j.cplett.2007.07.051.

187. M. I. Cabaço, M. Besnard, Y. Danten and J. A. P. Coutinho, "Solubility of CO_2 in 1-Butyl-3-methyl-imidazolium-trifluoro Acetate Ionic Liquid Studied by Raman Spectroscopy and DFT Investigations," *The Journal of Physical Chemistry B*, vol. 115, no. 13, pp. 3538–3550, 2011, doi:10.1021/jp111453a.

188. O. Hollóczki *et al.*, "Significant Cation Effects in Carbon Dioxide-Ionic Liquid Systems," *ChemPhysChem*, vol. 14, pp. 315–320, 2013, doi:10.1002/cphc.201200970.

189. D. S. Firaha and B. Kirchner, "CO_2 Absorption in the Protic Ionic Liquid Ethylammonium Nitrate," *Journal of Chemical & Engineering Data*, vol. 59, no. 10, pp. 3098–3104, 2014, doi:10.1021/je500166d.

190. X. Zhu, Y. Lu, C. Peng, J. Hu, H. Liu and Y. Hu, "Halogen bonding interactions between brominated ion pairs and CO_2 molecules: Implications for design of new and efficient ionic liquids for CO_2 absorption," *The Journal of Physical Chemistry B*, vol. 115, pp. 3949–3958, 2011, doi:10.1021/jp111194k.

191. Y. S. Sistla and V. Sridhar, "Molecular understanding of carbon dioxide interactions with ionic liquids," *Journal of Molecular Liquids*, vol. 325, 2021, Art no. 115162, doi:10.1016/J.Molliq.2020.115162.

192. E. I. Izgorodina, J. I. Hodgson, D. C. Weis, S. J. Pas and D. R. MacFarlane, "Physical Absorption of CO_2 in Protic and Aprotic Ionic Liquids: An Interaction Perspective," *The Journal of Physical Chemistry B*, vol. 119, no. 35, pp. 11748–11759, 2015, doi:10.1021/acs.jpcb.5b05115.

193. G. García, M. Atilhan and S. Aparicio, "Simultaneous CO_2 and SO_2 capture by using ionic liquids: A theoretical approach," *Physical Chemistry Chemical Physics*, vol. 19, no. 7, pp. 5411–5422, 2017, doi:10.1039/C6CP08151G.

194. B. Widom, "Some Topics in the Theory of Fluids," *The Journal of Chemical Physics*, vol. 39, no. 11, pp. 2808–2812, 1963, doi:10.1063/1.1734110.

195. C. H. Bennett, "Efficient estimation of free energy differences from Monte Carlo data," *J Comput Phys*, vol. 22, no. 2, pp. 245–268, 1976, doi:10.1016/0021-9991(76)90078-4.

196. J. G. Kirkwood, "Statistical Mechanics of Fluid Mixtures," *The Journal of Chemical Physics*, vol. 3, no. 5, pp. 300–313, 1935, doi:10.1063/1.1749657.

197. A. P. Lyubartsev, A. A. Martsinovski, S. V. Shevkunov and P. N. Vorontsov-Velyaminov, "New approach to Monte Carlo calculation of the free energy: Method of expanded ensembles," *The Journal of Chemical Physics*, vol. 96, no. 3, pp. 1776–1783, 1992, doi:10.1063/1.462133.

198. J. Deschamps, M. F. Costa Gomes and A. A. H. Pádua, "Molecular Simulation Study of Interactions of Carbon Dioxide and Water with Ionic Liquids," *ChemPhysChem*, vol. 5, no. 7, pp. 1049–1052, 2004, doi:10.1002/cphc.200400097.

199. J. K. Shah and E. J. Maginn, "A Monte Carlo simulation study of the ionic liquid 1-n-butyl-3-methylimidazolium hexafluorophosphate: Liquid structure, volumetric properties and infinite dilution solution thermodynamics of CO_2," *Fluid Phase Equilibria*, vol. 222, pp. 195–203, 2004, doi:10.1016/j.fluid.2004.06.027.

200. J. K. Shah and E. J. Maginn, "Monte Carlo simulations of gas solubility in the ionic liquid 1-n-butyl-3-methylimidazolium hexafluorophosphate," *The Journal of Physical Chemistry B*, vol. 109, no. 20, pp. 10395–10405, 2005, doi:10.1021/jp0442089.

201. D. Kerlé, R. Ludwig, A. Geiger and D. Paschek, "Temperature Dependence of the Solubility of Carbon Dioxide in Imidazolium-Based Ionic Liquids," *The Journal of Physical Chemistry B*, vol. 113, no. 38, pp. 12727–12735, 2009, doi:10.1021/jp9055285.

202. D. Kerlé, M. N. Jorabchi, R. Ludwig, S. Wohlrab and D. Paschek, "A simple guiding principle for the temperature dependence of the solubility of light gases in imidazolium-based ionic liquids derived from molecular simulations," *Physical Chemistry Chemical Physics*, vol. 19, no. 3, pp. 1770–1780, 2017, doi:10.1039/c6cp06792a.

203. M. N. Jorabchi, R. Ludwig and D. Paschek, "Quasi-Universal Solubility Behavior of Light Gases in Imidazolium-Based Ionic Liquids with Varying Anions: A Molecular Dynamics Simulation Study," *The Journal of Physical Chemistry B*, vol. 125, no. 6, pp. 1647–1659, 2021, doi:10.1021/acs.jpcb.0c10721.

204. A. F. Ghobadi, V. Taghikhani and J. R. Elliott, "Investigation on the Solubility of SO_2 and CO_2 in Imidazolium-Based Ionic Liquids Using NPT Monte Carlo Simulation," *Journal of Physical Chemistry B*, vol. 115, no. 46, pp. 13599–13607, 2011, doi:10.1021/jp2051239.

205. H. J. Liu, S. Dai and D. E. Jiang, "Molecular Dynamics Simulation of Anion Effect on Solubility, Diffusivity, and Permeability of Carbon Dioxide in Ionic Liquids," *Industrial & Engineering Chemistry Research*, vol. 53, no. 25, pp. 10485–10490, 2014, doi:10.1021/ie501501k.

206. N. Vergadou and I. G. Economou, *In Preparation*, 2021.

207. N. Vergadou, E. Androulaki and I. G. Economou, "Molecular simulation methods for CO_2 capture and gas separation with emphasis on ionic liquids," in *Process Systems and Materials for CO_2 Capture*. John Wiley & Sons, Ltd, 2017, pp. 79–111.

208. G. Romanos, "Membranes and microporous materials for environmental separations laboratory, national center for scientific research 'Demokritos'," Unpublished results.

209. I. Urukova, J. Vorholz and G. Maurer, "Solubility of CO_2, CO, and H_2 in the ionic liquid [bmim][PF6] from Monte Carlo simulations," *The Journal of Physical Chemistry B*, vol. 109, no. 24, pp. 12154–12159, 2005, doi:10.1021/jp050888j.

210. I. Urukova, J. Vorholz and G. Maurer, "Correction to 'Solubility of CO_2, CO, and H_2 in the Ionic Liquid [bmim][PF6] from Monte Carlo Simulations'," *The Journal of Physical Chemistry B*, vol. 110, no. 36, pp. 18072–18072, 2006, doi:10.1021/jp064806i.

211. J. Kumełan, A. P. S. Kamps, D. Tuma and G. Maurer, "Solubility of CO in the ionic liquid [bmim][PF6]," *Fluid Phase Equilibria*, vol. 228–229, pp. 207–211, 2005, doi:10.1016/j.fluid.2004.07.015.

212. W. Shi and E. J. Maginn, "Atomistic simulation of the absorption of carbon dioxide and water in the ionic liquid 1-n-hexyl-3-methylimidazolium bis(trifluoromethylsulfonyl) imide[hmim][Tf2N]," *Journal of Physical Chemistry B*, vol. 112, no. 7, pp. 2045–2055, 2008, doi:10.1021/jp077223x.

213. S. Duane, A. D. Kennedy, B. J. Pendleton and D. Roweth, "Hybrid Monte Carlo," *Physics Letters B*, vol. 195, no. 2, pp. 216–222, 1987, doi:10.1016/0370-2693(87)91197-X.

214. W. Shi and E. J. Maginn, "Molecular simulation and regular solution theory modeling of pure and mixed gas absorption in the ionic liquid 1-n-hexyl-3-methylimidazolium bis(trifluoromethylsulfonyl)amide ([hmim][Tf2N])," *The Journal of Physical Chemistry B*, vol. 112, no. 51, pp. 16710–16720, 2008, doi:10.1021/jp8075782.

215. W. Shi, D. C. Sorescu, D. R. Luebke, M. J. Keller and S. Wickramanayake, "Molecular simulations and experimental studies of solubility and diffusivity for pure and mixed gases of H_2, CO_2, and Ar absorbed in the ionic liquid 1-n-hexyl-3-methylimidazolium bis(trifluoromethylsulfonyl)amide ([hmim][Tf2N])," *The Journal of Physical Chemistry B*, vol. 114, no. 19, pp. 6531–6541, 2010, doi:10.1021/jp101897b.

216. R. Singh, E. Marin-Rimoldi and E. J. Maginn, "A Monte Carlo Simulation Study To Predict the Solubility of Carbon Dioxide, Hydrogen, and Their Mixture in the Ionic Liquids 1-Alkyl-3-methylimidazolium bis(trifluoromethanesulfonyl)amide ([Cnmim+] [Tf2N –], n = 4, 6)," *Industrial & Engineering Chemistry Research*, vol. 54, no. 16, pp. 4385–4395, 2015, doi:10.1021/ie503086z.

217. S. Budhathoki, J. K. Shah and E. J. Maginn, "Molecular Simulation Study of the Solubility, Diffusivity and Permselectivity of Pure and Binary Mixtures of CO_2 and CH_4 in the Ionic Liquid 1-n-Butyl-3-methylimidazolium bis(trifluoromethylsulfonyl) imide," *Industrial & Engineering Chemistry Research*, vol. 54, no. 35, pp. 8821–8828, 2015, doi:10.1021/acs.iecr.5b02500.

218. U. Kapoor and J. K. Shah, "Monte Carlo Simulations of Pure and Mixed Gas Solubilities of CO_2 and CH_4 in Nonideal Ionic Liquid–Ionic Liquid Mixtures," *Industrial & Engineering Chemistry Research*, vol. 58, no. 50, pp. 22569–22578, 2019, doi:10.1021/acs.iecr.9b03384.

219. X. Zhang, F. Huo, Z. Liu, W. Wang, W. Shi and E. J. Maginn, "Absorption of CO_2 in the ionic liquid 1-n-hexyl-3-methylimidazolium tris(pentafluoroethyl)trifluorophosphate ([hmim][FEP]): A molecular view by computer simulations," *The Journal of Physical Chemistry B*, vol. 113, no. 21, pp. 7591–7598, 2009, doi:10.1021/jp900403q.

220. K. Karanasiou, "Molecular Simulation of IL-Gas Mixtures," Diploma Thesis, School of Chemical Engineering, National Technical University of Athens and National Center for Scientific Research 'Demokritos', Athens, Greece, 2020.

221. K. Karanasiou, M. Panou and N. Vergadou, *In Preparation*, 2021.

222. E. Ricci, N. Vergadou, G. G. Vogiatzis, M. G. De Angelis and D. N. Theodorou, "Molecular Simulations and Mechanistic Analysis of the Effect of CO_2 Sorption on Thermodynamics, Structure, and Local Dynamics of Molten Atactic Polystyrene," *Macromolecules*, vol. 53, no. 10, pp. 3669–3689, 2020, doi:10.1021/acs.macromol.0c00323.

223. L. F. Zubeir, T. M. J. Nijssen, T. Spyriouni, J. Meuldijk, J.-R. Hill and M. C. Kroon, "Carbon Dioxide Solubilities and Diffusivities in 1-Alkyl-3-methylimidazolium Tricyanomethanide Ionic Liquids: An Experimental and Modeling Study," *Journal of Chemical & Engineering Data*, vol. 61, no. 12, pp. 4281–4295, 2016, doi:10.1021/acs.jced.6b00657.

224. A. Ayad, A. Negadi and F. Mutelet, "Carbon dioxide solubilities in tricyanomethanide-based ionic liquids: Measurements and PC-SAFT modeling," *Fluid Phase Equilibria*, vol. 469, pp. 48–55, 2018, doi:10.1016/j.fluid.2018.04.020.

225. F. Yan *et al.*, "Understanding the effect of side groups in ionic liquids on carbon-capture properties: A combined experimental and theoretical effort," *Physical Chemistry Chemical Physics*, vol. 15, no. 9, pp. 3264–3272, 2013, doi:10.1039/c3cp43923b.

226. J. X. Mao, J. A. Steckel, F. Yan, N. Dhumal, H. Kim and K. Damodaran, "Understanding the mechanism of CO_2 capture by 1,3 di-substituted imidazolium acetate based ionic liquids," *Physical Chemistry Chemical Physics*, vol. 18, no. 3, pp. 1911–1917, 2016, doi:10.1039/c5cp05713b.

227. C. Cadena, J. L. Anthony, J. K. Shah, T. I. Morrow, J. F. Brennecke and E. J. Maginn, "Why Is CO_2 So Soluble in Imidazolium-Based Ionic Liquids?" *Journal of the American Chemical Society*, vol. 126, no. 16, pp. 5300–5308, 2004, doi:10.1021/ja039615x.

228. F. Karadas *et al.*, "High pressure CO_2 absorption studies on imidazolium-based ionic liquids: Experimental and simulation approaches," *Fluid Phase Equilibria*, vol. 351, pp. 74–86, 2013, doi:10.1016/j.fluid.2012.10.022.

229. R. Babarao, S. Dai and D. E. Jiang, "Understanding the high solubility of CO_2 in an ionic liquid with the tetracyanoborate anion," *The Journal of Physical Chemistry B*, vol. 115, no. 32, pp. 9789–9794, 2011, doi:10.1021/jp205399r.

230. X. Zhang, X. Liu, X. Yao and S. Zhang, "Microscopic Structure, Interaction, and Properties of a Guanidinium-Based Ionic Liquid and Its Mixture with CO_2," *Industrial & Engineering Chemistry Research*, vol. 50, no. 13, pp. 8323–8332, 2011, doi:10.1021/ie1025002.

231. X. Huang, C. J. Margulis, Y. Li and B. J. Berne, "Why is the partial molar volume of CO_2 so small when dissolved in a room temperature ionic liquid? Structure and dynamics of CO_2 dissolved in [Bmim+] [PF6 –]," *Journal of American Chemical Society*, vol. 127, pp. 17842–17851, 2005, doi:10.1021/ja055315z.

232. J. Tong et al., "The dynamic behavior and intrinsic mechanism of CO_2 absorption by amino acid ionic liquids," *Physical Chemistry Chemical Physics*, vol. 23, no. 5, pp. 3246–3255, 2021, doi:10.1039/D0CP05735E.

233. M. P. Singh, Y. L. Verma, A. K. Gupta, R. K. Singh and S. Chandra, "Changes in dynamical behavior of ionic liquid in silica nano-pores," *Ionics*, vol. 20, no. 4, pp. 507–516, 2014, doi:10.1007/s11581-013-1008-9.

234. M. Klähn and A. Seduraman, "What Determines CO_2 Solubility in Ionic Liquids? A Molecular Simulation Study," *The Journal of Physical Chemistry B*, vol. 119, no. 31, pp. 10066–10078, 2015, doi:10.1021/acs.jpcb.5b03674.

235. T. C. Lourenço, M. F. C. Coelho, T. C. Ramalho, D. van der Spoel and L. T. Costa, "Insights on the Solubility of CO_2 in 1-Ethyl-3-methylimidazolium Bis(trifluoromethylsulfonyl) imide from the Microscopic Point of View," *Environmental Science & Technology*, vol. 47, no. 13, pp. 7421–7429, 2013, doi:10.1021/es4020986.

236. X. Liu, K. E. O'Harra, J. E. Bara and C. H. Turner, "Molecular insight into the anion effect and free volume effect of CO_2 solubility in multivalent ionic liquids," *Physical Chemistry Chemical Physics*, vol. 22, no. 36, pp. 20618–20633, 2020, doi:10.1039/D0CP03424J.

237. P. C. Rathi, R. F. Ludlow and M. L. Verdonk, "Practical High-Quality Electrostatic Potential Surfaces for Drug Discovery Using a Graph-Convolutional Deep Neural Network," *Journal of Medicinal Chemistry*, vol. 63, no. 16, pp. 8778–8790, 2020, doi:10.1021/acs.jmedchem.9b01129.

238. H. Liu, Z. Zhang, J. E. Bara and C. H. Turner, "Electrostatic Potential within the Free Volume Space of Imidazole-Based Solvents: Insights into Gas Absorption Selectivity," *The Journal of Physical Chemistry B*, vol. 118, no. 1, pp. 255–264, 2014, doi:10.1021/jp410143j.

239. G. Yu and X. Chen, "SO_2 capture by guanidinium-based ionic liquids: A theoretical study," *The Journal of Physical Chemistry B*, vol. 115, no. 13, pp. 3466–3477, 2011, doi:10.1021/jp107517t.

240. A. Mondal and S. Balasubramanian, "Understanding SO_2 Capture by Ionic Liquids," *The Journal of Physical Chemistry B*, vol. 120, no. 19, pp. 4457–4466, 2016, doi:10.1021/acs.jpcb.6b02553.

241. L. J. Siqueira, R. A. Ando, F. F. Bazito, R. M. Torresi, P. S. Santos and M. C. Ribeiro, "Shielding of ionic interactions by sulfur dioxide in an ionic liquid," *The Journal of Physical Chemistry B*, vol. 112, no. 20, pp. 6430–6435, 2008, doi:10.1021/jp800665y.

242. B. R. Prasad and S. Senapati, "Explaining the differential solubility of flue gas components in ionic liquids from first-principle calculations," *The Journal of Physical Chemistry B*, vol. 113, no. 14, pp. 4739–4743, 2009, doi:10.1021/jp805249h.

243. D. S. Firaha, M. Kavalchuk and B. Kirchner, "SO_2 Solvation in the 1-Ethyl-3-Methylimidazolium Thiocyanate Ionic Liquid by Incorporation into the Extended Cation-Anion Network," *J Solution Chem*, vol. 44, no. 3–4, pp. 838–849, 2015, doi:10.1007/s10953-015-0321-5.

244. X. Zhang et al., "Insight into the Performance of Acid Gas in Ionic Liquids by Molecular Simulation," *Industrial & Engineering Chemistry Research*, vol. 58, no. 3, pp. 1443–1453, 2019, doi:10.1021/acs.iecr.8b04929.

245. G. Garcia, M. Atilhan and S. Aparicio, "A density functional theory insight towards the rational design of ionic liquids for SO_2 capture," *Physical Chemistry Chemical Physics*, vol. 17, no. 20, pp. 13559–13574, 2015, doi:10.1039/c5cp00076a.

246. O. Hollóczki et al., "Carbene Formation in Ionic Liquids: Spontaneous, Induced, or Prohibited?" *The Journal of Physical Chemistry B*, vol. 117, no. 19, pp. 5898–5907, 2013, doi:10.1021/jp4004399.

247. A. L. Li, Z. Q. Tian, T. Y. Yan, D. E. Jiang and S. Dai, "Anion-Functionalized Task-Specific Ionic Liquids: Molecular Origin of Change in Viscosity upon CO_2 Capture," *Journal of Physical Chemistry B*, vol. 118, no. 51, pp. 14880–14887, 2014, doi:10.1021/jp5100236.

248. N. A. Andreeva and V. V. Chaban, "Amino-functionalized ionic liquids as carbon dioxide scavengers. Ab initio thermodynamics for chemisorption," *The Journal of Chemical Thermodynamics*, vol. 103, pp. 1–6, 2016, doi:10.1016/j.jct.2016.07.045.

249. Q. R. Sheridan, W. F. Schneider and E. J. Maginn, "Role of Molecular Modeling in the Development of CO$_2$-Reactive Ionic Liquids," *Chemical Reviews*, vol. 118, no. 10, pp. 5242–5260, 2018, doi:10.1021/acs.chemrev.8b00017.

250. C. Pinilla, M. G. Del Pópolo, R. M. Lynden-Bell and J. Kohanoff, "Structure and Dynamics of a Confined Ionic Liquid. Topics of Relevance to Dye-Sensitized Solar Cells," *The Journal of Physical Chemistry B*, vol. 109, no. 38, pp. 17922–17927, 2005, doi:10.1021/jp052999o.

251. R. Atkin and G. G. Warr, "Structure in confined room-temperature ionic liquids," *J Phys Chem C*, vol. 111, no. 13, pp. 5162–5168, 2007, doi:10.1021/jp067420g.

252. N. Sieffert and G. Wipff, "Ordering of Imidazolium-Based Ionic Liquids at the α-Quartz(001) Surface: A Molecular Dynamics Study," *The Journal of Physical Chemistry C*, vol. 112, pp. 19590–19603, 2008, doi:10.1021/jp806882e.

253. R. Singh, J. Monk and F. R. Hung, "Heterogeneity in the Dynamics of the Ionic Liquid [BMIM+][PF6 –] Confined in a Slit Nanopore," *The Journal of Physical Chemistry C*, vol. 115, no. 33, pp. 16544–16554, 2011, doi:10.1021/jp2046118.

254. R. Singh, J. Monk and F. R. Hung, "A Computational Study of the Behavior of the Ionic Liquid [BMIM+][PF6–] Confined Inside Multiwalled Carbon Nanotubes," *The Journal of Physical Chemistry C*, vol. 114, no. 36, pp. 15478–15485, 2010, doi:10.1021/jp1058534.

255. G. Ori, F. Villemot, L. Viau, A. Vioux and B. Coasne, "Ionic liquid confined in silica nanopores: Molecular dynamics in the isobaric-isothermal ensemble," *Molecular Physics*, vol. 112, no. 9–10, pp. 1350–1361, 2014, doi:10.1080/00268976.2014.902138.

256. G. Kritikos, N. Vergadou and I. G. Economou, "Molecular Dynamics Simulation of Highly Confined Glassy Ionic Liquids," *The Journal of Physical Chemistry C*, vol. 120, no. 2, pp. 1013–1024, 2016, doi:10.1021/acs.jpcc.5b09947.

257. H. Zhou *et al.*, "Nanoscale perturbations of room temperature ionic liquid structure at charged and uncharged interfaces," *ACS Nano*, vol. 6, p. 9818, 2012.

258. C. D. Wick, T. M. Chang and L. X. Dang, "Molecular Mechanism of CO$_2$ and SO$_2$ Molecules Binding to the Air/Liquid Interface of 1-Butyl-3-methylimidazolium Tetrafluoroborate Ionic Liquid: A Molecular Dynamics Study with Polarizable Potential Models," *The Journal of Physical Chemistry B*, vol. 114, p. 14965, 2010, doi:10.1021/jp106768y.

259. R. M. Lynden-Bell, J. Kohanoff and M. G. Del Popolo, "Simulation of interfaces between room temperature ionic liquids and other liquids," *Faraday Discussions*, vol. 129, pp. 57–67, 2005, doi:10.1039/b405514d.

260. B. L. Bhargava and S. Balasubramanian, "Layering at an Ionic Liquid–Vapor Interface: A Molecular Dynamics Simulation Study of [bmim][PF6]," *Journal of the American Chemical Society*, vol. 128, no. 31, pp. 10073–10078, 2006, doi:10.1021/ja060035k.

261. T. Yan, S. Li, W. Jiang, X. Gao, B. Xiang and G. A. Voth, "Structure of the liquid-vacuum interface of room-temperature ionic liquids: A molecular dynamics study," *The Journal of Physical Chemistry B*, vol. 110, no. 4, pp. 1800–6, 2006, doi:10.1021/jp055890p.

262. J. D. Morganti, K. Hoher, M. C. C. Ribeiro, R. A. Ando and L. J. A. Siqueira, "Molecular Dynamics Simulations of Acidic Gases at Interface of Quaternary Ammonium Ionic Liquids," *The Journal of Physical Chemistry C*, vol. 118, no. 38, pp. 22012–22020, 2014, doi:10.1021/jp505853k.

263. S. Budhathoki, J. K. Shah and E. J. Maginn, "Molecular Simulation Study of the Performance of Supported Ionic Liquid Phase Materials for the Separation of Carbon Dioxide from Methane and Hydrogen," *Industrial & Engineering Chemistry Research*, vol. 56, no. 23, pp. 6775–6784, 2017, doi:10.1021/acs.iecr.7b00763.

264. F. P. Kinik, C. Altintas, V. Balci, B. Koyuturk, A. Uzun and S. Keskin, "[BMIM][PF6] Incorporation Doubles CO_2 Selectivity of ZIF-8: Elucidation of Interactions and Their Consequences on Performance," *ACS Applied Materials & Interfaces*, vol. 8, no. 45, pp. 30992–31005, 2016, doi:10.1021/acsami.6b11087.

265. A. Thomas and M. Prakash, "The Role of Binary Mixtures of Ionic Liquids in ZIF-8 for Selective Gas Storage and Separation: A Perspective from Computational Approaches," *The Journal of Physical Chemistry C*, vol. 124, no. 48, pp. 26203–26213, 2020, doi:10.1021/acs.jpcc.0c07090.

266. K. Kumar and A. Kumar, "Enhanced CO_2 Adsorption and Separation in Ionic-Liquid-Impregnated Mesoporous Silica MCM-41: A Molecular Simulation Study," *The Journal of Physical Chemistry C*, vol. 122, no. 15, pp. 8216–8227, 2018, doi:10.1021/acs.jpcc.7b11529.

267. Y. Chen, Z. Hu, K. M. Gupta and J. Jiang, "Ionic Liquid/Metal–Organic Framework Composite for CO_2 Capture: A Computational Investigation," *The Journal of Physical Chemistry C*, vol. 115, no. 44, pp. 21736–21742, 2011, doi:10.1021/jp208361p.

268. A. M. O. Mohamed, S. Moncho, P. Krokidas, K. Kakosimos, E. N. Brothers and I. G. Economou, "Computational investigation of the performance of ZIF-8 with encapsulated ionic liquids towards CO_2 capture," *Molecular Physics*, vol. 117, no. 23–24, pp. 3791–3805, 2019, doi:10.1080/00268976.2019.1666170.

269. K. M. Gupta, Y. Chen, Z. Hu and J. Jiang, "Metal-organic framework supported ionic liquid membranes for CO_2 capture: Anion effects," *Physical Chemistry Chemical Physics*, vol. 14, pp. 5785–5794, 2012, doi:10.1039/C2CP23972H.

270. A. M. O. Mohamed, P. Krokidas and I. G. Economou, "Encapsulation of [bmim+] [Tf2N−] in different ZIF-8 metal analogues and evaluation of their CO_2 selectivity over CH_4 and N_2 using molecular simulation," *Molecular Systems Design & Engineering*, vol. 5, no. 7, pp. 1230–1238, 2020, doi:10.1039/D0ME00021C.

271. T. Yu *et al.*, "Mechanisms behind high CO_2/CH_4 selectivity using ZIF-8 metal organic frameworks with encapsulated ionic liquids: A computational study," *Chemical Engineering Journal*, vol. 419, p. 129638, 2021, doi:10.1016/j.cej.2021.129638.

272. A. Varnek, N. Kireeva, I. V. Tetko, I. I. Baskin and V. P. Solovév, "Exhaustive QSPR Studies of a Large Diverse Set of Ionic Liquids: How Accurately Can We Predict Melting Points?" *Journal of Chemical Information and Modeling*, vol. 47, pp. 1111–1122, 2007, doi:10.1021/ci600493x.

273. C. Nieto-Draghi *et al.*, "A General Guidebook for the Theoretical Prediction of Physicochemical Properties of Chemicals for Regulatory Purposes," *Chemical Reviews*, vol. 115, no. 24, pp. 13093–13164, 2015, doi:10.1021/acs.chemrev.5b00215.

274. Q. Shang, F. Yan, S. Xia, Q. Wang and P. Ma, "Predicting the surface tensions of ionic liquids by the quantitative structure property relationship method using a topological index," *Chem Eng Sci*, vol. 101, pp. 266–270, 2013, doi:10.1016/j.ces.2013.05.053.

275. K. Paduszyński, "In Silico Calculation of Infinite Dilution Activity Coefficients of Molecular Solutes in Ionic Liquids: Critical Review of Current Methods and New Models Based on Three Machine Learning Algorithms," *Journal of Chemical Information and Modeling*, vol. 56, no. 8, pp. 1420–1437, 2016, doi:10.1021/acs.jcim.6b00166.

276. A. Maiti, "Theoretical screening of ionic liquid solvents for carbon capture," *ChemSusChem*, vol. 2, no. 7, pp. 628–31, 2009, doi:10.1002/cssc.200900086.

277. M. Gonzalez-Miquel *et al.*, "COSMO-RS Studies: Structure-Property Relationships for CO_2 Capture by Reversible Ionic Liquids," *Industrial & Engineering Chemistry Research*, vol. 51, no. 49, pp. 16066–16073, 2012, doi:10.1021/ie302449c.

278. M. Gonzalez-Miquel, J. Palomar, S. Omar and F. Rodriguez, "CO_2/N_2 Selectivity Prediction in Supported Ionic Liquid Membranes (SILMs) by COSMO-RS," *Industrial & Engineering Chemistry Research*, vol. 50, pp. 5739–5748, 2011, doi:10.1021/ie102450x.

279. K. Paduszyński, "An overview of the performance of the COSMO-RS approach in predicting the activity coefficients of molecular solutes in ionic liquids and derived properties at infinite dilution," *Physical Chemistry Chemical Physics*, vol. 19, no. 19, pp. 11835–11850, 2017, doi:10.1039/C7CP00226B.

280. K. Z. Sumon and A. Henni, "Ionic liquids for CO_2 capture using COSMO-RS: Effect of structure, properties and molecular interactions on solubility and selectivity," *Fluid Phase Equilibria*, vol. 310, no. 1–2, pp. 39–55, 2011, doi:10.1016/j.fluid.2011.06.038.

281. X. C. Zhang, Z. P. Liu and W. C. Wang, "Screening of ionic liquids to capture CO_2 by COSMO-RS and experiments," *AIChE Journal*, vol. 54, no. 10, pp. 2717–2728, 2008, doi:10.1002/aic.11573.

282. M. B. Shiflett and E. J. Maginn, "The solubility of gases in ionic liquids," *AIChE Journal*, vol. 63, no. 11, pp. 4722–4737, 2017, doi:10.1002/aic.15957.

283. B. R. Mellein, A. M. Scurto and M. B. Shiflett, "Gas solubility in ionic liquids," *Current Opinion in Green and Sustainable Chemistry*, vol. 28, p. 100425, 2021, doi:10.1016/j.cogsc.2020.100425.

284. K. Huang and H.-L. Peng, "Solubilities of Carbon Dioxide in 1-Ethyl-3-methylimidazolium Thiocyanate, 1-Ethyl-3-methylimidazolium Dicyanamide, and 1-Ethyl-3-methylimidazolium Tricyanomethanide at (298.2 to 373.2) K and (0 to 300.0) kPa," *Journal of Chemical & Engineering Data*, vol. 62, no. 12, pp. 4108–4116, 2017, doi:10.1021/acs.jced.7b00476.

285. Z. Baramaki, Z. Arab Aboosadi and N. Esfandiari, "Thermodynamic modeling of ternary systems containing imidazolium-based ionic liquids and acid gases using SRK, Peng-Robinson, CPA and PC-SAFT equations of state," *Petroleum Science and Technology*, vol. 37, no. 24, pp. 2420–2428, 2019, doi:10.1080/10916466.2019.1610774.

286. F. Zareiekordshouli, A. Lashanizadehgan and P. Darvishi, "Experimental and theoretical study of CO_2 solubility under high pressure conditions in the ionic liquid 1-ethyl-3-methylimidazolium acetate," *The Journal of Supercritical Fluids*, vol. 133, pp. 195–210, 2018, doi:10.1016/j.supflu.2017.10.008.

287. F. Shaahmadi, B. Hashemi Shahraki and A. Farhadi, "The CO_2/CH_4 gas mixture solubility in ionic liquids [Bmim][Ac], [Bmim][BF4] and their binary mixtures," *The Journal of Chemical Thermodynamics*, vol. 141, p. 105922, 2020, doi:10.1016/j.jct.2019.105922.

288. A. Afsharpour and A. Kheiri, "The solubility of acid gases in the ionic liquid [C8mim] [PF6]," *Petroleum Science and Technology*, vol. 36, no. 3, pp. 232–238, 2018, doi:10.1080/10916466.2017.1416630.

289. A. Kheiri and A. Afsharpour, "The CPA EoS application to model CO_2 and H_2S simultaneous solubility in ionic liquid [C2mim][PF6]," *Petroleum Science and Technology*, vol. 36, no. 13, pp. 944–950, 2018, doi:10.1080/10916466.2018.1454953.

290. H. Soltani Panah, "Modeling H_2S and CO_2 solubility in ionic liquids using the CPA equation of state through a new approach," *Fluid Phase Equilibria*, vol. 437, pp. 155–165, 2017, doi:10.1016/j.fluid.2017.01.023.

291. J. S. Andreu and L. F. Vega, "Modeling the Solubility Behavior of CO_2, H_2, and Xe in [Cn-mim][Tf2N] Ionic Liquids," *The Journal of Physical Chemistry B*, vol. 112, no. 48, pp. 15398–15406, 2008, doi:10.1021/jp807484g.

292. F. Llovell, O. Vilaseca and L. F. Vega, "Thermodynamic Modeling of Imidazolium-Based Ionic Liquids with the [PF6]– Anion for Separation Purposes," *Separation Science and Technology*, vol. 47, no. 2, pp. 399–410, 2011, doi:10.1080/01496395.2011.635625.

293. I. I. I. Alkhatib *et al.*, "Screening of Ionic Liquids and Deep Eutectic Solvents for Physical CO_2 Absorption by Soft-SAFT Using Key Performance Indicators," *Journal of Chemical & Engineering Data*, vol. 65, no. 12, pp. 5844–5861, 2020, doi:10.1021/acs.jced.0c00750.

294. Y. Sun, A. Schemann, C. Held, X. Lu, G. Shen and X. Ji, "Modeling Thermodynamic Derivative Properties and Gas Solubility of Ionic Liquids with ePC-SAFT," *Industrial & Engineering Chemistry Research*, vol. 58, no. 19, pp. 8401–8417, 2019, doi:10.1021/acs.iecr.9b00254.

295. Y. Sun, A. Laaksonen, X. Lu and X. Ji, "How to Detect Possible Pitfalls in ePC-SAFT Modeling. 2. Extension to Binary Mixtures of 96 Ionic Liquids with CO_2, H_2S, CO, O_2, CH_4, N_2, and H_2," *Industrial & Engineering Chemistry Research*, vol. 59, no. 49, pp. 21579–21591, 2020, doi:10.1021/acs.iecr.0c04485.

296. E. K. Karakatsani, I. G. Economou, M. C. Kroon, M. D. Bermejo, C. J. Peters and G.-J. Witkamp, "Equation of state modeling of the phase equilibria of ionic liquid mixtures at low and high pressure," *Physical Chemistry Chemical Physics*, vol. 10, no. 40, pp. 6160–6168, 2008, doi:10.1039/b806584p.

297. E. K. Karakatsani, T. Spyriouni and I. G. Economou, "Extended statistical associating fluid theory (SAFT) equations of state for dipolar fluids," *AIChE Journal*, vol. 51, no. 8, pp. 2328–2342, 2005, doi:10.1002/aic.10473.

298. X. Ji and H. Adidharma, "Prediction of molar volume and partial molar volume for CO_2/ionic liquid systems with heterosegmented statistical associating fluid theory," *Fluid Phase Equilibria*, vol. 315, pp. 53–63, 2012, doi:10.1016/j.fluid.2011.11.014.

299. W. Ge *et al.*, "Analytical Multi-Scale Method for Multi-Phase Complex Systems in Process Engineering – Bridging Reductionism and Holism," *Chem Eng Sci*, vol. 62, no. 13, pp. 3346–3377, 2007, doi:10.1016/j.ces.2007.02.049.

300. X. Wang, H. Dong, X. Zhang, L. Yu, S. Zhang and Y. Xu, "Numerical Simulation of Single Bubble Motion in Ionic Liquids," *Chem Eng Sci*, vol. 65, no. 22, pp. 6036–6047, 2010, doi:10.1016/j.ces.2010.08.030.

301. X. Liu, Y. Jiang, C. Liu, W. Wang and J. Li, "Hydrodynamic Modeling of Gas-Solid Bubbling Fluidization Based on Energy-Minimization Multiscale (EMMS) Theory," *Industrial & Engineering Chemistry Research*, vol. 53, no. 7, pp. 2800–2810, 2014, doi:10.1021/ie4029335.

302. J. Li, "Approaching Virtual Process Engineering with Exploring Mesoscience," *Chemical Engineering Journal*, vol. 278, pp. 541–555, 2015, doi:10.1016/j.cej.2014.10.005.

303. B. Xu *et al.*, "Fixation of CO_2 into Cyclic Carbonates Catalyzed by Ionic Liquids: A Multiscale Approach," *Green Chemistry*, vol. 17, pp. 108–122, 2015, doi:10.1039/C4GC01754D.

304. K. Sumida *et al.*, "Carbon dioxide capture in metal-organic frameworks," *Chemical Reviews*, vol. 112, no. 2, pp. 724–81, 2012, doi:10.1021/cr2003272.

305. G. T. Rochelle, "Amine Scrubbing for CO_2 Capture," *Science*, vol. 325, no. 594, pp. 1652–1654, 2009, doi:10.1126/science.1176731.

306. M. E. Boot-Handford *et al.*, "Carbon capture and storage update," *Energy & Environmental Science*, vol. 7, no. 1, pp. 130–189, 2014, doi:10.1039/c3ee42350f.

307. A. Dibenedetto, M. Aresta, C. Fragale and M. Narracci, "Reaction of silylalkylmono- and silylalkyldi-amines with carbon dioxide: Evidence of formation of inter- and intra-molecular ammonium carbamates and their conversion into organic carbamates of industrial interest under carbon dioxide catalysis," *Green Chemistry*, vol. 4, no. 5, pp. 439–443, 2002, doi:10.1039/B205319P.

308. L. I. Eide and D. W. Bailey, "Precombustion Decarbonisation Processes," *Oil & Gas Science and Technology – Revue d'IFP Energies Nouvelles*, vol. 60, no. 3, pp. 475–484, 2005. [Online]. doi:10.2516/ogst:2005029.

309. H. Liu and Y. Shao, "Predictions of the impurities in the CO_2 stream of an oxy-coal combustion plant," *Applied Energy*, vol. 87, no. 10, pp. 3162–3170, 2010, doi:10.1016/j.apenergy.2010.04.014.

310. A. B. Rao and E. S. Rubin, "A Technical, Economic, and Environmental Assessment of Amine-Based CO_2 Capture Technology for Power Plant Greenhouse Gas Control," *Environmental Science & Technology*, vol. 36, no. 20, pp. 4467–4475, 2002, doi:10.1021/es0158861.

311. D. Aaron and C. Tsouris, "Separation of CO_2 from Flue Gas: A Review," *Separation Science and Technology*, vol. 40, no. 1–3, pp. 321–348, 2005, doi:10.1081/SS-200042244.

312. S. Choi, J. H. Drese and C. W. Jones, "Adsorbent Materials for Carbon Dioxide Capture from Large Anthropogenic Point Sources," *ChemSusChem*, vol. 2, no. 9, pp. 796–854, 2009, doi:10.1002/cssc.200900036.

313. S. A. Basnayake, J. Su, X. Zou and K. J. Balkus, "Carbonate-Based Zeolitic Imidazolate Framework for Highly Selective CO_2 Capture," *Inorganic Chemistry*, vol. 54, no. 4, pp. 1816–1821, 2015, doi:10.1021/ic5027174.

314. L.-S. Fan, L. Zeng, W. Wang and S. Luo, "Chemical looping processes for CO_2 capture and carbonaceous fuel conversion – prospect and opportunity," *Energy & Environmental Science*, vol. 5, no. 6, pp. 7254–7280, 2012, doi:10.1039/C2EE03198A.

315. H. Thakkar, S. Eastman, A. Al-Mamoori, A. Hajari, A. A. Rownaghi and F. Rezaei, "Formulation of Aminosilica Adsorbents into 3D-Printed Monoliths and Evaluation of Their CO_2 Capture Performance," *ACS Applied Materials & Interfaces*, vol. 9, no. 8, pp. 7489–7498, 2017, doi:10.1021/acsami.6b16732.

316. D. M. D'Alessandro, B. Smit and J. R. Long, "Carbon Dioxide Capture: Prospects for New Materials," *Angewandte Chemie International Edition*, vol. 49, no. 35, pp. 6058–6082, 2010, doi:10.1002/anie.201000431.

317. A. D. Ebner and J. A. Ritter, "State-of-the-Art Adsorption and Membrane Separation Processes for Carbon Dioxide Production from Carbon Dioxide Emitting Industries," *Separation Science and Technology*, vol. 44, no. 6, pp. 1273–1421, 2009, doi:10.1080/01496390902733314.

318. C. A. Scholes, G. Q. Chen, H. T. Lu and S. E. Kentish, "Crosslinked PEG and PEBAX Membranes for Concurrent Permeation of Water and Carbon Dioxide," *Membranes*, vol. 6, no. 1, 2016, doi:10.3390/membranes6010001.

319. E. Esposito *et al.*, "Pebax®/PAN hollow fiber membranes for CO_2/CH_4 separation," *Chemical Engineering and Processing – Process Intensification*, vol. 94, pp. 53–61, 2015, doi:10.1016/j.cep.2015.03.016.

320. B. Zornoza, C. Tellez, J. Coronas, J. Gascon and F. Kapteijn, "Metal organic framework based mixed matrix membranes: An increasingly important field of research with a large application potential," *Microporous and Mesoporous Materials*, vol. 166, pp. 67–78, 2013, doi:10.1016/j.micromeso.2012.03.012.

321. C. A. Scholes and S. Kanehashi, "Polymer of Intrinsic Microporosity (PIM-1) Membranes Treated with Supercritical CO_2," *Membranes*, vol. 9, no. 3, 2019, doi:10.3390/membranes9030041.

322. G. Bengtson, S. Neumann and V. Filiz, "Membranes of Polymers of Intrinsic Microporosity (PIM-1) Modified by Poly(ethylene glycol)," *Membranes*, vol. 7, no. 2, 2017, doi:10.3390/membranes7020028.

323. S. Rafiq, L. Deng and M.-B. Hägg, "Role of Facilitated Transport Membranes and Composite Membranes for Efficient CO_2 Capture – A Review," *ChemBioEng Reviews*, vol. 3, no. 2, pp. 68–85, 2016, doi:10.1002/cben.201500013.

324. M. Zalewski, T. Krawczyk, A. Siewniak and A. Sobolewski, "Carbon dioxide capture using water-imidazolium ionic liquids-amines ternary systems," *International Journal of Greenhouse Gas Control*, vol. 105, 2021, Art no. 103210, doi:10.1016/j.ijggc.2020.103210.

325. F. L. Bernard *et al.*, "Anticorrosion Protection by Amine–Ionic Liquid Mixtures: Experiments and Simulations," *Journal of Chemical & Engineering Data*, vol. 61, no. 5, pp. 1803–1810, 2016, doi:10.1021/acs.jced.5b00996.

326. J. Yang, X. Yu, J. Yan and S.-T. Tu, "CO_2 Capture Using Amine Solution Mixed with Ionic Liquid," *Industrial & Engineering Chemistry Research*, vol. 53, no. 7, pp. 2790–2799, 2014, doi:10.1021/ie4040658.

327. X. L. Papatryfon *et al.*, "CO$_2$ Capture Efficiency, Corrosion Properties and Ecotoxicity Evaluation of Amine Solutions Involving Newly Synthesized Ionic Liquids," *Industrial & Engineering Chemistry Research*, vol. 53, no. 30, pp. 12083–12102, 2014, doi:10.1021/ie501897d.

328. J. J. Wang, Y. Tian, Y. Zhao and K. L. Zhuo, "A volumetric and viscosity study for the mixtures of 1-n-butyl-3-methylimidazolium tetrafluoroborate ionic liquid with acetonitrile, dichloromethane, 2-butanone and N, N-dimethylformamide," *Green Chemistry*, vol. 5, pp. 618–622, 2003, doi:10.1039/B303735E.

329. Q. Zhou, L. S. Wang and H. P. Chen, "Densities and Viscosities of 1-Butyl-3-Methylimidazolium Tetrafluoroborate + H$_2$O Binary Mixtures from (303.15 to 353.15) K," *Journal of Chemical & Engineering Data*, vol. 51, no. 3, pp. 905–908, 2006, doi:10.1021/je050387r.

330. O. Ciocirlan, O. Croitoru and O. Iulian, "Densities and Viscosities for Binary Mixtures of 1-Butyl-3-Methylimidazolium Tetrafluoroborate Ionic Liquid with Molecular Solvents," *Journal of Chemical & Engineering Data*, vol. 56, no. 4, pp. 1526–1534, 2011, doi:10.1021/je101206u.

331. A. Finotello, J. E. Bara, S. Narayan, D. Camper and R. D. Noble, "Ideal gas solubilities and solubility selectivities in a binary mixture of room-temperature ionic liquids," *The Journal of Physical Chemistry B*, vol. 112, no. 8, pp. 2335–2339, 2008, doi:10.1021/jp075572l.

332. M. B. Shiflett and A. Yokozeki, "Phase Behavior of Carbon Dioxide in Ionic Liquids: [emim][Acetate], [emim][Trifluoroacetate], and [emim][Acetate] + [emim] [Trifluoroacetate] Mixtures," *Journal of Chemical & Engineering Data*, vol. 54, no. 1, pp. 108–114, 2008, doi:10.1021/je800701j.

333. E. Kühne, S. Santarossa, E. Perez, G. J. Witkamp and C. J. Peres, "New approach in the design of reactions and separations using an ionic liquid and carbon dioxide as solvents: Phase equilibria in two selected ternary systems," *The Journal of Supercritical Fluids*, vol. 46, no. 2, pp. 93–98, 2008, doi:10.1016/j.supflu.2008.04.016.

334. M. Ramdin *et al.*, "Solubility of CO$_2$ and CH$_4$ in Ionic Liquids: Ideal CO$_2$/CH$_4$ Selectivity," *Industrial & Engineering Chemistry Research*, vol. 53, no. 40, pp. 15427–15435, 2014, doi:10.1021/ie4042017.

335. A. Yokozeki and M. B. Shiflett, "Hydrogen Purification Using Room-Temperature Ionic Liquids," *Applied Energy*, vol. 84, no. 3, pp. 351–361, 2007, doi:10.1016/j.apenergy. 2006.06.002.

336. M. Solinas, A. Pfaltz, P. G. Cozzi and W. Leitner, "Enantioselective hydrogenation of imines in ionic liquid/carbon dioxide media," *Journal of the American Chemical Society*, vol. 126, no. 49, pp. 16142–7, 2004, doi:10.1021/ja046129g.

337. V. A. Toussaint, E. Kühne, A. Shariati and C. J. Peters, "Solubility Measurements of Hydrogen in 1-Butyl-3-methylimidazolium Tetrafluoroborate and the Effect of Carbon Dioxide and a Selected Catalyst on the Hydrogen Solubility in the Ionic Liquid," *The Journal of Chemical Thermodynamics*, vol. 59, pp. 239–242, 2013, doi:10.1016/j.jct.2012.12.013.

338. D. G. Hert, J. L. Anderson, S. N. V. K. Aki and J. F. Brennecke, "Enhancement of Oxygen and Methane Solubility in 1-Hexyl-3-methylimidazolium Bis(trifluoromethylsulfonyl) imide Using Carbon Dioxide," *Chemical Communications*, pp. 2603–2605, 2005, doi:10.1039/B419157A.

339. M. B. Shiflett, A. M. Niehaus and A. Yokozeki, "Separation of CO$_2$ and H$_2$S Using Room-TemperatureIonic Liquid [bmim][MeSO4]," *Journal of Chemical & Engineering Data*, vol. 55, no. 11, pp. 4785–4793, 2010, doi:10.1021/je1004005.

340. M. B. Shiflett and A. Yokozeki, "Separation of CO$_2$ and H$_2$S using room-temperature ionic liquid [bmim][PF6]," *Fluid Phase Equilibria*, vol. 294, no. 1–2, pp. 105–113, 2010, doi:10.1016/j.fluid.2010.01.013.

341. S. N. V. K. Aki, A. M. Scurto and J. F. Brennecke, "Ternary Phase Behavior of Ionic Liquid (IL)–Organic–CO$_2$ Systems," *Industrial & Engineering Chemistry Research*, vol. 45, no. 16, pp. 5574–5585, 2006, doi:10.1021/ie0511783.

342. S. N. V. K. Aki, B. R. Mellein, E. M. Saurer and J. F. Brennecke, "High-Pressure Phase Behavior of Carbon Dioxide with Imidazolium-Based Ionic Liquids," *The Journal of Physical Chemistry B*, vol. 108, no. 52, pp. 20355–20365, 2004, doi:10.1021/jp046895+.

343. D. Fu, X. Sun, J. Pu and S. Zhao, "Effect of Water Content on the Solubility of CO_2 in the Ionic Liquid [bmim][PF6]," *Journal of Chemical & Engineering Data*, vol. 51, no. 2, pp. 371–375, 2006, doi:10.1021/je0502501.

344. R. E. Baltus, B. H. Culbertson, S. Dai, H. Luo and D. W. DePaoli, "Low-Pressure Solubility of Carbon Dioxide in Room-Temperature Ionic Liquids Measured with a Quartz Crystal Microbalance," *The Journal of Physical Chemistry B*, vol. 108, no. 2, pp. 721–727, 2004, doi:10.1021/jp036051a.

345. G. E. Romanos *et al.*, "Enhanced CO_2 capture in binary mixtures of 1-alkyl-3-methylimidazolium tricyanomethanide ionic liquids with water," *The Journal of Physical Chemistry B*, vol. 117, no. 40, pp. 12234–12251, 2013, doi:10.1021/jp407364e.

346. M. Hasib-ur-Rahman, M. Siaj and F. Larachi, "Ionic liquids for CO_2 capture – Development and progress," *Chem Eng Process*, vol. 49, no. 4, pp. 313–322, 2010, doi:10.1016/j.cep.2010.03.008.

347. M. B. Shiflett, D. J. Kasprzak, C. P. Junk and A. Yokozeki, "Phase behavior of (carbon dioxide plus [bmim][Ac]) mixtures," *J Chem Thermodyn*, vol. 40, no. 1, pp. 25–31, 2008, doi:10.1016/j.jct.2007.06.003.

348. I. S. Molchan *et al.*, "Microscopic study of the corrosion behaviour of mild steel in ionic liquids for CO_2 capture applications," *RSC Advances*, vol. 5, no. 44, pp. 35181–35194, 2015, doi:10.1039/C5RA01097G.

349. M. J. Schneider, M. Haumann and P. Wasserscheid, "Asymmetric hydrogenation of methyl pyruvate in the continuous gas phase using Supported Ionic Liquid Phase (SILP) catalysis," *Journal of Molecular Catalysis A: Chemical*, vol. 376, pp. 103–110, 2013, doi:10.1016/j.molcata.2013.04.022.

350. L. A. Banu, D. Wang and R. E. Baltus, "Effect of Ionic Liquid Confinement on Gas Separation Characteristics," *Energy & Fuels*, vol. 27, no. 8, pp. 4161–4166, 2013, doi:10.1021/ef302038e.

351. M. S. Raja Shahrom, A. R. Nordin and C. D. Wilfred, "The improvement of activated carbon as CO_2 adsorbent with supported amine functionalized ionic liquids," *Journal of Environmental Chemical Engineering*, vol. 7, no. 5, 2019, Art no. 103319, doi:10.1016/j.jece.2019.103319.

352. F. P. Kinik, A. Uzun and S. Keskin, "Ionic Liquid/Metal-Organic Framework Composites: From Synthesis to Applications," *ChemSusChem*, vol. 10, no. 14, pp. 2842–2863, 2017, doi:10.1002/cssc.201700716.

353. S. Xiong *et al.*, "Ionic Liquids-Based Membranes for Carbon Dioxide Separation," *Israel Journal of Chemistry*, vol. 59, no. 9, pp. 824–831, 2019, doi:10.1002/ijch.201900062.

354. K. Friess *et al.*, "A Review on Ionic Liquid Gas Separation Membranes," *Membranes*, vol. 11, no. 2, 2021, Art no. 97, doi:10.3390/membranes11020097.

355. L. M. Robeson, "The upper bound revisited," *Journal of Membrane Science*, vol. 320, no. 1, pp. 390–400, 2008, doi:10.1016/j.memsci.2008.04.030.

356. G. Huang *et al.*, "Pebax/ionic liquid modified graphene oxide mixed matrix membranes for enhanced CO_2 capture," *Journal of Membrane Science*, vol. 565, pp. 370–379, 2018, doi:10.1016/j.memsci.2018.08.026.

357. F. Shi, Q. Zhang, D. Li and Y. Deng, "Silica-gel-confined ionic liquids: A new attempt for the development of supported nanoliquid catalysis," *Chemistry – A European Journal*, vol. 11, no. 18, pp. 5279–5288, 2005, doi:10.1002/chem.200500107.

358. G. E. Romanos *et al.*, "CO_2 Capture by Novel Supported Ionic Liquid Phase Systems Consisting of Silica Nanoparticles Encapsulating Amine-Functionalized Ionic Liquids," *The Journal of Physical Chemistry C*, vol. 118, no. 42, pp. 24437–24451, 2014, doi:10.1021/jp5062946.

359. A. Erto *et al.*, "Carbon-supported ionic liquids as innovative adsorbents for CO_2 separation from synthetic flue-gas," *J Colloid Interf Sci*, vol. 448, pp. 41–50, 2015, doi:10.1016/j.jcis.2015.01.089.

360. J. Lemus, J. Palomar, M. A. Gilarranz and J. J. Rodriguez, "Characterization of Supported Ionic Liquid Phase (SILP) materials prepared from different supports," *Adsorption*, vol. 17, no. 3, pp. 561–571, 2011, doi:10.1007/s10450-011-9327-5.

361. J. Palomar, J. Lemus, N. Alonso-Morales, J. Bedia, M. A. Gilarranz and J. J. Rodriguez, "Encapsulated ionic liquids (ENILs): From continuous to discrete liquid phase," *Chemical Communications*, vol. 48, no. 80, pp. 10046–10048, 2012, doi:10.1039/c2cc35291e.

362. W. Shi and D. C. Sorescu, "Molecular Simulations of CO_2 and H_2 Sorption into Ionic Liquid 1-n-Hexyl-3-methylimidazolium Bis(trifluoromethylsulfonyl)amide ([Hmim] [Tf2N]) Confined in Carbon Nanotubes," *The Journal Physical Chemistry B*, vol. 114, pp. 15029–15041, 2010.

363. K. S. Han, X. Wang, S. Dai and E. W. Hagaman, "Distribution of 1-Butyl-3-methylimidazolium Bistrifluoromethylsulfonimide in Mesoporous Silica As a Function of Pore Filling," *The Journal of Physical Chemistry C*, vol. 117, no. 30, pp. 15754–15762, 2013, doi:10.1021/jp404990q.

364. W. Shi and D. R. Luebke, "Enhanced gas absorption in the ionic liquid 1-n-hexyl-3-methylimidazolium bis(trifluoromethylsulfonyl)amide ([hmim][Tf2N]) confined in silica slit pores: A molecular simulation study," *Langmuir*, vol. 29, no. 18, pp. 5563–5572, 2013, doi:10.1021/la400226g.

365. Q. Berrod *et al.*, "Enhanced ionic liquid mobility induced by confinement in 1D CNT membranes," *Nanoscale*, vol. 8, no. 15, pp. 7845–7848, 2016, doi:10.1039/c6nr01445c.

366. T. Ohba and V. V. Chaban, "A highly viscous imidazolium ionic liquid inside carbon nanotubes," *The Journal Physical Chemistry B*, vol. 118, no. 23, pp. 6234–6240, 2014, doi:10.1021/jp502798e.

367. V. V. Chaban and O. V. Prezhdo, "Nanoscale carbon greatly enhances mobility of a highly viscous ionic liquid," *ACS Nano*, vol. 8, no. 8, pp. 8190–8197, 2014, doi:10.1021/nn502475j.

368. K. Fujie and H. Kitagawa, "Ionic liquid transported into metal-organic frameworks," *Coordination Chemistry Reviews*, vol. 307, pp. 382–390, 2016.

369. W. X. Chen, H. R. Xu, G. L. Zhuang, L. S. Long, R. B. Huang and L. S. Zheng, "Temperature-dependent conductivity of emim+(Emim+= 1-ethyl-3-methyl imidazolium) confined in channels of a metal-organic framework," *Chemical Communications*, vol. 47, pp. 11933–11935, 2011.

370. O. C. Vangeli *et al.*, "Grafting of Imidazolium Based Ionic Liquid on the Pore Surface of Nanoporous Materials – Study of Physicochemical and Thermodynamic Properties," *Journal of Physical Chemistry B*, vol. 114, pp. 6480–6491, 2010.

371. Y. L. Wang and A. Laaksonen, "Interfacial structure and orientation of confined ionic liquids on charged quartz surfaces," *Physical Chemistry Chemical Physics*, vol. 16, no. 42, pp. 23329–23339, 2014, doi:10.1039/c4cp03077j.

372. O. C. Vangeli, G. E. Romanos, K. G. Beltsios, D. Fokas, C. P. Athanasekou and N. K. Kanellopoulos, "Development and characterization of chemically stabilized ionic liquid membranes – Part I: Nanoporous ceramic supports," *Journal of Membrane Science*, vol. 365, no. 1, pp. 366–377, 2010, doi:10.1016/j.memsci.2010.09.030.

373. A. V. Perdikaki *et al.*, "Ionic Liquid-Modified Porous Materials for Gas Separation and Heterogeneous Catalysis," *The Journal of Physical Chemistry C*, vol. 116, no. 31, pp. 16398–16411, 2012, doi:10.1021/jp300458s.

374. K. L. Stefanopoulos *et al.*, "Investigation of Confined Ionic Liquid in Nanostructured Materials by a Combination of SANS, Contrast-Matching SANS, and Nitrogen Adsorption," *Langmuir*, vol. 27, no. 13, pp. 7980–7985, 2011, doi:10.1021/la201261r.

375. G. E. Romanos *et al.*, "Investigation of Physically and Chemically Ionic Liquid Confinement in Nanoporous Materials by a Combination of SANS, Contrast-Matching SANS, XRD and Nitrogen Adsorption," *Journal of Physics: Conference Series*, vol. 340, 2012, Art no. 012087, doi:10.1088/1742-6596/340/1/012087.

376. S. Livi, J. Duchet-Rumeau, J.-F. Gérard and T. N. Pham, "Polymers and Ionic Liquids: A Successful Wedding," *Macromolecular Chemistry and Physics*, vol. 216, no. 4, pp. 359–368, 2015, doi:10.1002/macp.201400425.

377. L. C. Tomé and I. M. Marrucho, "Ionic liquid-based materials: A platform to design engineered CO_2 separation membranes," *Chemical Society Reviews*, vol. 45, no. 10, pp. 2785–2824, 2016, doi:10.1039/C5CS00510H.

378. Z. Dai, R. D. Noble, D. L. Gin, X. Zhang and L. Deng, "Combination of ionic liquids with membrane technology: A new approach for CO_2 separation," *Journal of Membrane Science*, vol. 497, pp. 1–20, 2016, doi:10.1016/j.memsci.2015.08.060.

379. T. D. Huan, R. Batra, J. Chapman, S. Krishnan, L. Chen and R. Ramprasad, "A universal strategy for the creation of machine learning-based atomistic force fields," *npj Comput. Mater.*, vol. 3, no. 1, 2017, Art no. 37, doi:10.1038/s41524-017-0042-y.

380. S. Chmiela, A. Tkatchenko, H. E. Sauceda, I. Poltavsky, K. T. Schutt and K. R. Muller, "Machine learning of accurate energy-conserving molecular force fields," *Science advances*, vol. 3, no. 5, 2017, Art no. e1603015, doi:10.1126/sciadv.1603015.

381. V. L. Deringer, M. A. Caro and G. Csányi, "Machine Learning Interatomic Potentials as Emerging Tools for Materials Science," *Advanced Materials*, vol. 31, no. 46, 2019, Art no. 1902765, doi:10.1002/adma.201902765.

382. R. Batra and S. Sankaranarayanan, "Machine learning for multi-fidelity scale bridging and dynamical simulations of materials," *Journal of Physics: Materials*, vol. 3, no. 3, 2020, Art no. 031002, doi:10.1088/2515-7639/ab8c2d.

383. J. Wang *et al.*, "Machine Learning of Coarse-Grained Molecular Dynamics Force Fields," *ACS Central Science*, vol. 5, no. 5, pp. 755–767, 2019, doi:10.1021/acscentsci.8b00913.

4 Gas Hydrates for CO$_2$ Capture

Nicolas von Solms

CONTENTS

4.1 INTRODUCTION: WHAT ARE GAS HYDRATES?

Gas hydrates, or clathrate hydrates, are crystalline, ice-like solid compounds of small gas molecules and water that form at low temperature and high pressure. Typical gas hydrate structures are shown in Figure 4.1 [1].Cage structures of hydrates consist of water molecules arranged around a stabilizing small molecule (e.g., carbon dioxide, nitrogen, methane, ethane, propane, and others), with an oxygen molecule at each vertex. The seminal book by Sloan and Koh provides a detailed description of the gas hydrates and their structures [2]. Three different crystalline structures form in gas hydrates, which vary according to the type and the relative number of different cages; these are shown in Figure 4.1. So, for example, a combination of two 5^{12} cages and six $5^{12}6^2$ cages forms a structure I (sI). Similarly, sixteen 5^{12} cages and eight $5^{12}6^2$ cages combine to form structure II (sII). Structure I is generally formed from pure components (i.e., only one type of guest molecule). The structure I type gas hydrates of

DOI: 10.1201/9781003162780-4

143

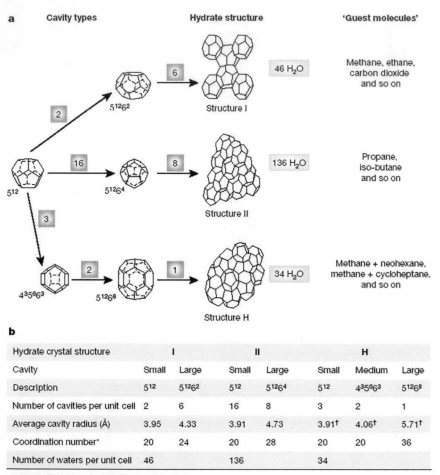

FIGURE 4.1 Gas hydrate structures [1]. Water molecules arrange themselves around small molecules like methane. The arrangement is stable enough to allow hydrates to exist as solids above the freezing point of water. The three crystal structures shown here (sI, sII, and sH) result in a combination of different numbers of small and large cages.

natural gas are found in nature (hydrate deposits in continental margins and permafrost). At least two hydrate-forming gases in a gas mixture combine to form structure II hydrates. Such mixtures are predominantly associated with oil and gas production. For example, natural gas essentially contains lower-end hydrocarbons, which can form structure II gas hydrate. Hence, hydrates formed from pure methane will be of structure I, whereas hydrates formed from a mixture containing 97% methane

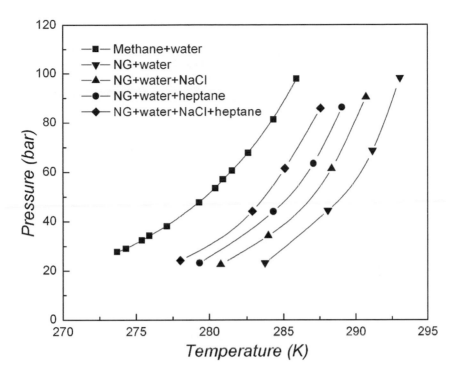

FIGURE 4.2 Three-phase coexistence curves of various hydrate-forming systems. Hydrates become more stable with increasing the pressure and decreasing the temperature [3]. The data points shown for the methane hydrate coexistence curve are from Bishnoi and Natarajan [4]. The natural gas hydrate coexistence curves were measured by Jensen *et al.* [3]. A 3.5 wt% NaCl solution was used for the experiments involving sodium chloride. For experiments involving heptane, the initial liquid phase contained an equal volume of heptane and water.

and 3% propane will be of structure II. At a given temperature, structure II hydrates form at significantly lower pressures than structure I. Hence, there is a greater risk of hydrate formation, and the scientific reason cannot be explained by the established theory. In addition, the hydrate formation temperature is higher than the freezing point of water at elevated pressures, typically in natural gas processing systems. For instance, the natural gas mixture used to produce the curves in Figure 4.2 [3] has a hydrate equilibrium temperature of around 20 °C at 100 bars. For methane hydrate at this pressure, the equilibrium temperature is around 12 °C. In simpler words, only 30 bars (and not 100 bars) is required to form hydrates at a temperature of 12 °C from the natural gas considered here. Operating conditions of certain INEOS-operated oil and gas platforms in the North Sea vary between temperatures of 5 °C and 20 °C and pressures of 150 to 200 bars. Hence, these oil and gas platforms operate well within the hydrate formation and stability zone.

4.2 THERMODYNAMICS OF GAS HYDRATES

Hydrate stability curves for various systems are shown in Figures 4.2 [3] and 4.3 [5]. These curves are also known as three-phase lines. The aqueous and hydrate phases can coexist on the equilibrium line for the hydrocarbon fluid phase. Field engineers find these curves in temperature-pressure space very useful. The curves allow them to establish at what conditions of temperature and pressure the hydrate formation is possible along the length of a pipeline.

It is logical to consider increasing the temperature of the gas mixture or heating (and insulating) the pipeline as a mitigation solution to hydrate formation. However, this is occasionally considered in real applications due to energy costs and impracticality in long pipelines. Pressure reduction is not practical since pressure is required for flow. Furthermore, where pressure drop occurs rapidly (such as across a valve), the pressure drop is accompanied by significant Joule-Thompson cooling, increasing the risk of hydrate formation. Hence, the addition of thermodynamic inhibitors to the gas mixtures is the widely used method for reducing hydrate formation. These thermodynamic inhibitors behave like an antifreeze. The most common chemicals used for this purpose are methanol, various glycols, or other alcohols. They are injected into the wellhead production stream and mixed with co-produced water. The hydrocarbons are thus in contact with an alcohol or glycol solution containing up to 50 wt% alcohol or glycol, which reduces the water activity. Hence, this shifts the equilibrium and the hydrate formation temperature drops, and it is more difficult to

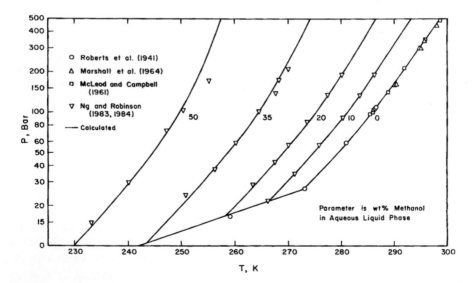

FIGURE 4.3 Inhibition of methane hydrate by methanol. The hydrate formation temperature is lowered by the addition of methanol at a certain fixed pressure. The weight percentages of the methanol/water mixtures are given for when in contact with the gas. Water is usually co-produced in oil and gas production. The 'water cut' (i.e., the ratio of water to oil or gas) increases as the field is consumed. Hence, more inhibitor (in this case methanol) is required to achieve the same degree of inhibition as the field is consumed [5].

form hydrates in such conditions. Figure 4.3 shows that at 150 bars, the equilibrium formation temperature of methane hydrate in water is around 290 K. Upon addition of enough methanol such that the produced water contains 50 wt%, the temperature for hydrate formation drops to near 250 K at the given pressure. This would be a sufficiently low temperature for most oil and gas operations. The amount of methanol varies according to the water present within the hydrocarbons. Sometimes, a large amount of methanol may be required that might be expensive. Since the ratio of water to hydrocarbons tends to increase during the lifetime of a field, the amount of methanol required for hydrated suppression will increase over time. Sloan [1] has estimated an expense of US$220 million annually.

Figure 4.2 can provide further insights on the gas hydrates. Natural gas hydrate has a substantially higher formation temperature than a pure methane hydrate. Natural gas is composed mostly of methane. However, other gases present in natural gas (like carbon dioxide, nitrogen, ethane, and propane) combine to form structure II hydrates. As mentioned earlier, these hydrate structures are more stable (or stable at lower pressures for a given temperature) than the structure I hydrates formed by the pure methane. The figure also shows that NaCl is a hydrate inhibitor (seawater hydrates form at slightly lower temperatures than those formed from pure water). Other chemicals such as tetrahydrofuran and cyclopentane act as hydrate promoters. The required hydrate formation pressure is lower for a specific temperature than for the system without the promoter. This is important for the application to CO_2 capture, where hydrate formation requires higher pressure and lower temperatures. This is evident from Figure 4.4, which shows how cyclopentane lowers the hydrate formation pressure for CO_2/N_2 mixture [6].

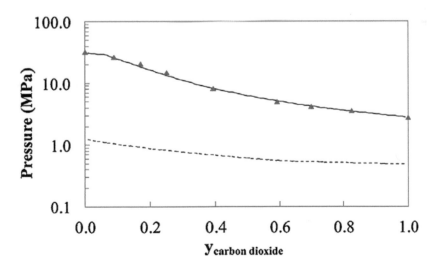

FIGURE 4.4 Mixed hydrate dissociation pressure as a function of carbon dioxide mole fraction in the vapor phase (binary vapor phase with nitrogen). The unpromoted system is the ternary system of CO_2, N_2, and H_2O. The promoted system is the quaternary system of CO_2, N_2, H_2O, and cyclopentane [6].

4.3 A PROCESS FOR CO₂ CAPTURE WITH GAS HYDRATES

Several works have recently appeared in the open literature using clathrate hydrates as the separation technology for CO_2 capture from flue gases [6–12]. Gas hydrates formed from flue gas are selective for CO_2. This property forms the basis of the separation technology for carbon capture. However, relatively high pressures are required to form the hydrates, meaning that competitive technologies will likely use so-called hydrate promoters such as tetrahydrofuran, cyclopentane, and their mixtures [7–9] to enable the formation of hydrates at lower pressures than original. Another option is to use a promoter, which forms semi-clathrate hydrates such as tetra-n-butylammonium bromide [12]. The promoter itself forms part of the lattice with the water molecules. Figure 4.5 shows the process layout for a hydrate-based post-combustion CO_2 separation system.

Figure 4.6 shows results for the mixture cyclopentane/tetrahydrofuran, demonstrating that the mixed promoter system lowers the hydrate formation pressure even further than using either promoter independently. The lines are model predictions using the Cubic-Plus-Association (CPA) equation of state combined with a hydrate model [9]. In the capture process considered here [8], CO_2 is physically adsorbed in solid gas hydrate crystals rather than chemically bonded to a solvent, such as in the amine process.

Nitrogen (N_2) and carbon dioxide (CO_2) are the two main constituents of flue gas from the power stations (alongside the water vapor). Both N_2 and CO_2 are known to form gas hydrates with water at low temperatures and high pressures. Pure CO_2 hydrates form at lower pressures than pure N_2 hydrates, hence a selectivity towards CO_2 in the mixed hydrate is expected. From a process point of view, the process is similar to the chemical absorption process, wherein the flue gas contacts directly with a liquid phase and the carbon dioxide is selectively removed. However, the process enjoys a benefit over the conventional amine process, as the solidified CO_2 hydrate particles can be separated from the remaining liquid phase. It allows operational ease, as only the CO_2 rich phase needs to

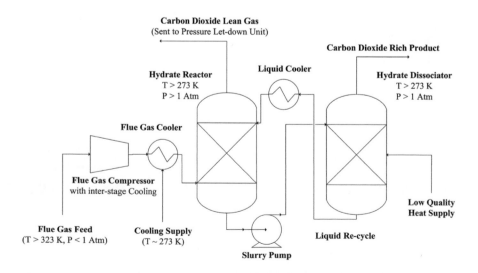

FIGURE 4.5 Simplified schematic of a suggested configuration for the gas clathrate hydrate–based post-combustion carbon dioxide capture technology [7].

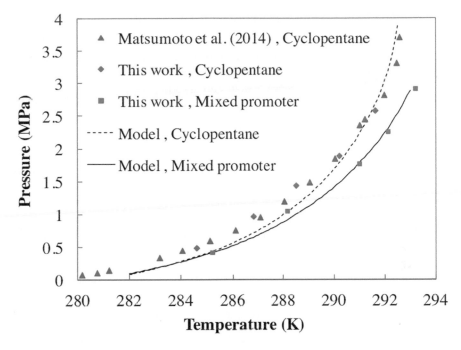

FIGURE 4.6 Comparison of experimental data and model results (using CPA) for incipient hydrate dissociation conditions in the ternary system of water + cyclopentane + carbon dioxide and the quaternary system of water + cyclopentane + tetrahydrofuran + carbon dioxide (mixed promoter) [9].

Source: Literature data from [13].

be treated in the CO_2 desorption process, where the hydrate is dissociated to release the trapped carbon dioxide. Hence, the process requires much less energy in the desorption step for carbon dioxide compared to the conventional amine process.

However, the process also suffers from a disadvantage, upon comparison to the chemical absorption process, as the feed flue gas must be compressed and cooled for the hydrates to form. Since compression is an expensive unit operation, low-pressure selective hydrate formation is needed to make this capture process competitive from an economic point of view. On the other hand, the advantages of the hydrate-based separation technology include low-temperature operation, where low-quality heat can be used in the release section of the process. A smaller amount of excess liquid is heated in the release part since the hydrate slurry may be concentrated before being heated. The captured gas is delivered at high pressure and low temperature, reducing costs for liquefaction of the final carbon dioxide product. To evaluate the energy efficiency of this process and to be able to calculate phase equilibrium in hydrate systems, accurate thermodynamic models are required over the appropriate range of temperatures and pressures. The phase equilibrium properties are crucial, since formation pressures for a given temperature and the distribution of the different species in each phase (aqueous, hydrate, or perhaps organic) are paramount. The model and phase equilibrium calculations are considered next.

4.4   MODELING PHASE EQUILIBRIUM IN HYDRATE SYSTEMS

The knowledge of the phase equilibrium is important for developing appropriate process design. For gas hydrate systems, the equation of state lies at the heart of most thermodynamic models, as fluid phases are always present and are applicable over a range of temperatures and pressures. Since hydrates are a solid phase, an additional model is needed to include the solid phase. There are a few variations of these, but most models are based on Langmuir's isotherm theory. This theory considers that hydrate guest molecules adsorb onto the hydrate cavities formed by water molecules. An equation of state model appropriate for gas hydrate systems is the CPA model. This is essentially the Soave-Redlich-Kwong (SRK) equation of state combined with a model that accurately accounts for hydrogen bonding. Since water is a precondition for hydrates, this is an important feature of the model. More detail can be found in the work of Herslund [8]. Even simple hydrates by their nature are already a quite complex system in thermodynamic terms. For example, the system shown in Figure 4.7 consists of three components and two phases. Here, the authors show how the pressure required for hydrate formation decreases as the content of CO_2 increases in the N_2/CO_2 mixture. Interestingly, the model also predicts a hydrate structure change (the discontinuity), which is not clear from the actual experimental data. A good hydrate model must be capable of calculating other thermodynamic properties not related to the hydrate themselves, specifically,

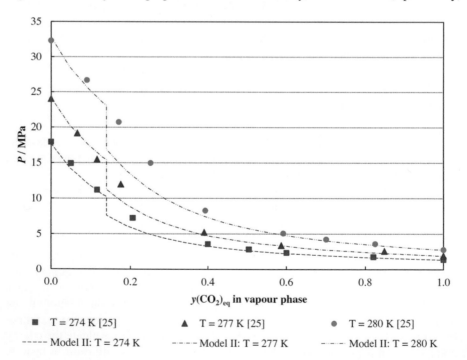

FIGURE 4.7   Dissociation pressures of mixed N_2/CO_2 hydrates as a function of equilibrium vapor phase compositions at three different temperatures. Interestingly, the model predicts a structure shift from structure II hydrate (nitrogen) to structure I (CO_2) at a CO_2 mole fraction of around 0.15. This trend is not very clear from the data [8].

here the solubility of CO_2 and N_2 in water. Secondary thermodynamic properties like the heat of hydrate formation, entropy changes, and so forth, are also of interest.

Another important feature of Figure 4.7 is that a pressure of at least around 7 MPa is required to form hydrate from flue gas, which is unlikely to be economical. Therefore, one could consider adding hydrate promoters to the water to lower the required formation pressure. This is explored in the next section.

4.5   HYDRATE FORMATION IN THE PRESENCE OF PROMOTERS

4.5.1   TETRAHYDROFURAN

Tetrahydrofuran (THF) is a five-sided cyclic ether that can form hydrates from its solutions with water at ambient temperature and atmospheric pressures. It has been shown that the addition of THF to water can lower the formation pressure of hydrates formed from flue gas and this THF solution. As an example, Figure 4.8 [7] shows the formation pressure of THF/CO_2 hydrates.

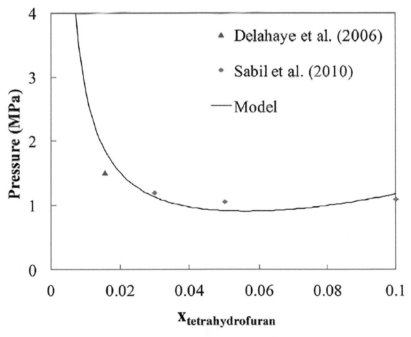

a)   Tetrahydrofuran + carbon dioxide

$$T = 286.1 \text{ K}$$

FIGURE 4.8   Dissociation pressures for the mixed sII hydrate of tetrahydrofuran and carbon dioxide as a function of the initial liquid phase mole fraction of tetrahydrofuran. Beyond about 5 mol%, adding more THF starts to inhibit the formation (pressure rises again) [7].

There is an optimum amount of THF to add (around 5 mol%), after which additional THF starts to raise the formation pressure again. This is because reducing the water activity due to the additional THF makes it difficult for hydrate to form. This optimum corresponds to a THF molecule surrounded by 17 water molecules in the hydrate cage. The THF/water system itself, even without hydrates, displays complex behavior. For instance, there is a liquid-liquid split at elevated pressures [10]. The formation of hydrogen bonds between the water and the THF molecules makes this whole system complex yet interesting and validates using the CPA model. An important point with any gas separation technology is the selectivity of the process. Figure 4.9 [7] shows how much CO_2 enters the hydrate phase as a function of the CO_2 content of the flue gas.

It is clear from the modeling results that while lowering the formation pressure, the selectivity towards CO_2 in the hydrate is decreased somewhat (i.e., the THF is in some sense parasitic). In either case, the behavior shown in Figure 4.9 would suggest that a multi-staged process is required.

4.5.2 CYCLOPENTANE

Cyclopentane (CP) is also interesting as a hydrate promoter since it forms hydrates at conditions like THF. The fluid phase behavior of this system is difficult to

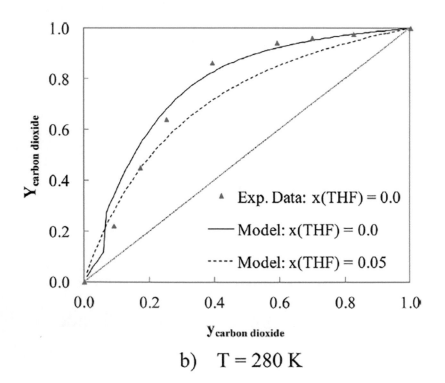

b)   T = 280 K

FIGURE 4.9   Carbon dioxide content in hydrate phase on water- and promoter-free basis as a function of initial vapor phase composition [7].

model since the mutual solubility of these components is very low. However, some solubility is a prerequisite for hydrate formation. The liquid-liquid equilibrium behavior of this binary system is shown in Figure 4.10 [6], where a CPA equation of state adequately models it. However, the minimum solubility is not captured. The mutual solubility of water and hydrocarbons is notoriously difficult to model. The system with hydrates is even more complex, since it requires dealing with three phases and four components. In terms of selectivity, cyclopentane is comparable to THF. As an additional note on promotion, Figure 4.6 shows that combining THF and cyclopentane results in an even lower formation pressure. The model was initially predicted and then subsequently verified experimentally by Herslund *et al.* [9].

4.5.3 PROPANE

Propane has also been used as a hydrate promoter similar to cyclopentane, although it is a gas at normal conditions [14–16]. The presence of propane reduces the induction time (effectively the time lag before noticeable hydrate growth occurs), which is advantageous. Still, the subsequent growth rate of hydrates was also reduced, which is undesirable. Like other promoters, the uptake of CO_2 was also somewhat reduced due to the competition for large cages with propane. The verdict was that propane was less efficient than THF in promoting the formation of CO_2 hydrates.

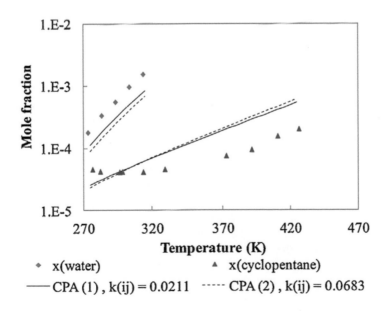

FIGURE 4.10 Liquid-liquid equilibrium in system water/cyclopentane. Note the mole fraction is on a log scale. These two components are insoluble in each other, but a degree of solubility is a prerequisite for hydrate formation [6].

4.5.4   TETRA-N-BUTYL AMMONIUM BROMIDE (TBAB)

Tetra-n-butyl ammonium bromide (TBAB) is a so-called semi-clathrate hydrate. Interestingly, the promoter does not enter the hydrate cage structure, which is somewhat opposite to THF and CP. Instead, TBAB forms part of the hydrate cage by forming hydrogen bonds with the water molecules forming the cage [12, 14]. This feature enables TBAB (and other semi-clathrate hydrates) to effectively expand the cage capacity of the water and is thus not a "parasitic" inhibitor such as those mentioned above (where the inhibitor competes with the CO_2 molecule for space in the hydrate cage). It was found that TBAB can lower the formation pressure by up to 90% compared to pure CO_2 hydrates [17]. Since the semi-clathrate hydrates operate differently from normal hydrate promoters, there is also potential for combining them with other promoters such as THF. Some other examples of these semi-clathrate hydrates have shown potential as additives to reduce hydrate formation from synthetic flue gases (N_2/CO_2 mixtures). This includes tetra n-butyl ammonium fluoride (TBAF), tetra n-butylphosphonium bromide (TBPB), and tetra n-butyl ammonium nitrate (TBANO_3).

4.5.5   KINETIC PROMOTERS

Up to now, all the considerations connected to CO_2 capture with gas hydrates have been of a thermodynamic nature: the three-phase lines (e.g. Figures 4.2, 4.3, and 4.6) and model calculations. However, the formation of hydrates is not an instantaneous process. It is a so-called stochastic process, where the formation of hydrate nuclei is preceded by an induction time, which is not deterministic. This means that if the temperature and pressure of operation are very close to the three-phase line, the rate of formation may be slow, perhaps even impractically slow for a CO_2 capture process. For example, consider Figure 4.6 for the mixed promoter system: At a temperature of 286 K and a pressure of 0.5 MPa, the system is inside the hydrate zone. This means that hydrates are the thermodynamically favored phase – at equilibrium. However, since the system is just barely inside the hydrate zone, the rate will likely be too slow to allow for a practical process under these conditions. Hence, the driving force for the forward equilibrium must be increased by venturing the system further into the hydrate zone. This can be achieved by increasing the operating pressure or reducing the temperature. To overcome this limitation (at least to some extent), the kinetic limitations of a hydrate-based process, kinetic promoters can be added to the process.

Typically, surfactants do not alter the thermodynamic equilibrium in the way the compounds mentioned above do; rather, they act as catalysts for the formation process. The surfactant's hydrophobic end forms a gas-surfactant complex that rapidly migrates through the water and attaches itself to the hydrophilic end, ensuring rapid hydrate formation and reduced induction time. Essentially, the reduction in water surface tension improves the hydrate formation kinetics. The surfactants investigated for hydrate-based CO_2 capture, Tween-80 (T-80), dodecyl-trimethyl-ammonium chloride (DTAC) and sodium dodecyl sulphate (SDS) have been the most widely investigated. It would make sense to combine a thermodynamic and a kinetic promoter, and several combinations have been tested [14], including SDS/THF and

DTAC/TBAB. The thermodynamic promoter CP has also been tried and produces an interesting effect: CP-containing light oils form emulsions with the water in the presence of surfactants. This in turn increases the CP/water interfacial area, increasing the hydrate formation rate.

4.5.6 PROCESS INNOVATIONS

The preceding discussions have mainly concerned the physicochemical aspects of hydrate formation. Most studies are conducted in laboratory equipment, which is frequently a stirred tank reactor or a variation [9]. However, one of the challenges of hydrate-based capture is the formation of agglomerates of hydrate crystals. These agglomerates reduce the water/gas interface, resist mass transfer for the gas, and reduce the rate of conversion of water to hydrate. One solution to this mass transfer issue is using a fixed bed crystallizer filled with porous silica gel [18]. The use of porous silica gel significantly increases the contact area between the water and the gas molecules, reducing induction time and increasing the uptake rate. The rapid reaction due to the greatly increased interfacial area eliminates the need for mechanical agitation or stirring, thus presenting at least one process advantage over a stirred tank reactor type of installation. The use of silica can be further divided into silica sand and silica gel. With silica gel, the pore size is significant. If the pores are too narrow, rapid formation of hydrates in the pores provides resistance to mass transfer, reducing the method's effectiveness. On the other hand, using larger pores and particle sizes improve gas uptake, water conversion, and CO_2 recovery and contribute positively to the CO_2 separation factor [19–20] (i.e., the selectivity of the process for CO_2 compared to other gases in the mixture). For example, Figure 4.9 shows how the CO_2 mole fraction in hydrate depends on gas-phase composition for a N_2/CO_2 mixture. Good results have also been obtained with silica sand [20], which has the additional advantage of cheaper material.

Another process innovation that has been investigated is the use of a bubble column, where gas is bubbled up through a column of water (or promoter-containing liquid). Hydrates form on the bubble surface, thereby allowing the separation of the CO_2 from the gas mixture. The efficacy of this method is dependent on the bubble size, where best effects are achieved for smaller bubble sizes – around 50 μm [21]. A bubble size that is too large results in too much gas inside. Hence, the bubble is not accessible to the hydrate-forming gas once a resistant hydrate layer has been formed. The process is of limited interest as a potential hydrate-based carbon capture technique due to the very large size required for the bubble.

Another innovative method that seems promising is the temperature ramping technique. In this process, the temperature is cycled over a small range [21] (e.g., 276–279 K). At lower temperatures, CO_2 is more soluble in the liquid. Hence, more hydrate is formed during a subsequent temperature increase. The reasons for this are unclear but may be connected to the fact that hydrate formation kinetics are improved at slightly higher temperatures, provided there is sufficient CO_2 available for hydrate formation (i.e., the transfer of gas is not rate limiting). To some extent, this can be accomplished by cycling temperature within a narrow range. It was also found that the effects of temperature ramping were most evident at the start of hydrate formation and favored equipment at a larger scale. In [21], setups that differed in volume by a factor of 100 were employed.

4.6   ENGINEERING DESIGN OF HYDRATE-BASED CAPTURE PLANTS

Aside from a few very brief process sketches found in the literature (e.g., Figure 4.5 and [14]), most of the work in hydrate-based carbon capture studies is on the level of laboratory or possibly pilot-scale investigations of scientific research questions. However, a collaboration between Nexant, Simteche, and Los Alamos National Laboratory has produced detailed process design and flow sheeting studies of hydrate-based processes to separate CO_2 and H_2S from syngas [22–23]. The syngas is mainly a mix of H_2 (55 mol%) and CO_2 (40 mol%), with several impurities making up the remaining 5 mol%, notably around 0.6 mol% H_2S. Significantly, the syngas feed pressure is around 65–70 bars, which immediately suggests the possible advantage of a hydrate-based process, compared to capture from power plant flue gas, which is available at near atmospheric pressures. The various layouts considered should process around 8800 ton/day of syngas. The two most interesting options were a single-stage process using a promoter (tetrahydrofuran) and a two-stage process without the addition of a promoter see for example Figure 4.9. A single stage process is not adequate for the N_2/CO_2 mixture, perhaps not even a two-stage process. Moreover, the much higher formation pressure of pure H_2 hydrate – around 2000 bars at ambient temperature [24] – means that the hydrate-based separation will require fewer stages for the H_2/CO_2 mixture compared to a N_2/CO_2 mixture. A major design goal of the process is to recover 90% of the CO_2 in the syngas.

4.6.1   TETRAHYDROFURAN-PROMOTED SINGLE-HYDRATE REACTOR

In this process, the syngas is combined with a hydrate promoter (THF) in such a ratio to achieve a hydration number of 8.5 (as mentioned earlier, there is a maximum amount of promoter that should be added to reduce formation pressure, after which the promoter inhibits hydrate formation due to the reduced water activity). The mixture is then led to a single hydrate formation reactor that is cooled using ammonia refrigerant. The reactor operating conditions are 6 °C and 68 bars. The hydrate slurry formed in the reactor then proceeds to a separator stage, where the treated syngas is removed. The hydrate slurry containing the CO_2 and H_2S continues to a two-step flash process, recovering the gas. Both flash vessels operate at an elevated temperature of 21 °C as compared to the original hydrate formation process. Simultaneously, the first vessel's pressure is reduced to 43 bars and the second to 19 bars. The recovered gas is compressed to 150 bars and dried, while the promoter-containing water is recycled. Despite the relative simplicity of the process, it was unattractive compared with the baseline Selexol process [25], mainly due to the increased refrigeration loading and refrigeration compressor power. The Selexol process is based on the physical absorption (unlike the alkanolamine processes, which require reaction between the solvent and CO_2), which means that the heat required to release the gas from the rich solvent is less. However, it also means that Selexol is better suited to high-pressure separation (in the range 20–130 bars) since at lower pressures, the (physical) solubility of CO_2 is too low. Below 20 bars, the alkanolamines or other chemical solvents are generally preferred for CO_2 capture applications. The solvents used are a mixture of dimethyl ethers of polyethylene glycol. UOP LLC licenses the technology. A two-reactor-stage promoted process was more promising than the single-stage process but still not competitive with the Selexol process.

4.6.2 TWO-STEP REACTOR SYSTEM WITHOUT PROMOTER

Based on previous studies for single unpromoted reactor systems at 3000 psi (207 bars) and 1800 psi (124 bars), respectively (which were not competitive compared to the Selexol process), a two-step hydrate formation reactor process was studied. As CO_2 recovery increases from around 70% to 90% for one-step processes, CAPEX and OPEX significantly increase. The two reactors in the two-step process operate at an intermediate pressure of 2400 psi (165 bars). No separation equipment exists between the reactors, and the temperature in reactor 1 is only changed from 7 °C to around 1 °C in reactor 2. About two-thirds of the CO_2 is converted to hydrate in the first reactor. The process looks very similar to the single-stage processes, except that the hydrate formation reactor is essentially "split" into two, with additional cooling between the reactors. A detailed techno-economic study found this two-reactor process competitive with the Selexol process in terms of CAPEX and OPEX and the avoided capture cost per ton of CO_2 captured: $27 per ton opposed to $25 for the Selexol process.

4.7 PERSPECTIVES AND SUSTAINABILITY

As reported in Dashti et al. [14], in 1999 the US Department of Energy reported the cost of a ton of CO_2 captured using hydrate-based technology in an integrated gasification combined cycle to be US$8.75, compared with US$59 for a ton of CO_2 captured using conventional amine-based absorption and US$64 per ton for adsorption by zeolites. This surprisingly low value led to a flurry of research activities in the beginning of the new millennium. The potential for sustainability in this technology is considered high. This is because the capture solvent is essentially water with some stable additives, all of which are recycled, and the energy required for regeneration is low and required at low temperatures. However, there are drawbacks to the technology: the pressures required for hydrate formation are still too high compared with other competing technologies – especially at the temperature of interest (for example at 40 °C). Moreover, the rate of hydrate formation is slow if the absorption unit operates close to the equilibrium pressure, meaning that a large unit is required (high CAPEX). The selectivity of hydrates for CO_2 is good but probably not good enough to ensure the use of a single separation stage, especially if promoters are used, as discussed above. It is more likely that a staged process is more efficient, which will require a complex design since solids handling is needed. A possible way forward is further investigation of semi-clathrate hydrates as promoters [12].

REFERENCES

1. E. D. Sloan, "Fundamental principles and applications of natural gas hydrates," *Nature*, vol. 426, no. 6964, pp. 353–359, Nov. 2003, doi:10.1038/nature02135.

2. E. D. Sloan and C. A. Koh, *Clathrate Hydrates of Natural Gases – E. Dendy Sloan, J*, 3rd ed. CRC Press, 2007.

3. L. Jensen, K. Thomsen and N. Von Solms, "Inhibition of structure I and II gas hydrates using synthetic and biological kinetic inhibitors," *Energy and Fuels*, vol. 25, no. 1, pp. 17–23, Jan. 2011, doi:10.1021/ef100833n.

4. P.R. Bishnoi and V. Natarajan, "Formation and decomposition of gas hydrates," *Fluid Phase Equilib.*, vol. 117, no. 1–2, pp. 168–177, Mar. 1996, doi:10.1016/0378-3812(95)02950-8.

5. F. E. Anderson and J. M. Prausnitz, "Inhibition of gas hydrates by methanol," *AIChE J.*, vol. 32, no. 8, pp. 1321–1333, Aug. 1986, doi:10.1002/aic.690320810.

6. P. J. Herslund, K. Thomsen, J. Abildskov and N. Von Solms, "Modelling of cyclopentane promoted gas hydrate systems for carbon dioxide capture processes," *Fluid Phase Equilib.*, vol. 375, pp. 89–103, Aug. 2014, doi:10.1016/j.fluid.2014.04.039.

7. P. J. Herslund, K. Thomsen, J. Abildskov and N. Von Solms, "Modelling of tetrahydrofuran promoted gas hydrate systems for carbon dioxide capture processes," *Fluid Phase Equilib.*, vol. 375, pp. 45–65, Aug. 2014, doi:10.1016/j.fluid.2014.04.031.

8. P. J. Herslund, K. Thomsen, J. Abildskov and N. Von Solms, "Phase equilibrium modeling of gas hydrate systems for CO_2 capture," *J. Chem. Thermodyn.*, vol. 48, pp. 13–27, May 2012, doi:10.1016/j.jct.2011.12.039.

9. P. J. Herslund, N. Daraboina, K. Thomsen, J. Abildskov and N. von Solms, "Measuring and modelling of the combined thermodynamic promoting effect of tetrahydrofuran and cyclopentane on carbon dioxide hydrates," *Fluid Phase Equilib.*, vol. 381, pp. 20–27, Nov. 2014, doi:10.1016/j.fluid.2014.08.015.

10. P. J. Herslund, K. Thomsen, J. Abildskov and N. Von Solms, "Application of the cubic-plus-association (CPA) equation of state to model the fluid phase behaviour of binary mixtures of water and tetrahydrofuran," *Fluid Phase Equilib.*, vol. 356, pp. 209–222, Oct. 2013, doi:10.1016/j.fluid.2013.07.036.

11. P. J. Herslund *et al.*, "Thermodynamic promotion of carbon dioxide-clathrate hydrate formation by tetrahydrofuran, cyclopentane and their mixtures," *Int. J. Greenh. Gas Control*, vol. 17, pp. 397–410, Sep. 2013, doi:10.1016/j.ijggc.2013.05.022.

12. F. Tzirakis, P. Stringari, N. von Solms, C. Coquelet and G. Kontogeorgis, "Hydrate equilibrium data for the $CO_2 + N_2$ system with the use of tetra-n-butylammonium bromide (TBAB), cyclopentane (CP) and their mixture," *Fluid Phase Equilib.*, vol. 408, pp. 240–247, Jan. 2016, doi:10.1016/j.fluid.2015.09.021.

13. Y. Matsumoto, T. Makino, T. Sugahara and K. Ohgaki, "Phase equilibrium relations for binary mixed hydrate systems composed of carbon dioxide and cyclopentane derivatives," *Fluid Phase Equilib.*, vol. 362, pp. 379–382, Jan. 2014, doi:10.1016/j.fluid.2013.10.057.

14. H. Dashti, L. Zhehao Yew and X. Lou, "Recent advances in gas hydrate-based CO_2 capture," *Journal of Natural Gas Science and Engineering*, vol. 23. Elsevier B.V., pp. 195–207, Mar. 1, 2015, doi:10.1016/j.jngse.2015.01.033.

15. R. Kumar, P. Englezos, I. Moudrakovski and J. A. Ripmeester, "Structure and composition of CO_2/H_2 and $CO_2/H_2/C_3H_8$ hydrate in relation to simultaneous CO_2 capture and H_2 production," *AIChE Journal*, vol. 55, no. 6, pp. 1584–1594, Jun. 2009, doi:10.1002/aic.11844.

16. P. Linga, R. Kumar and P. Englezos, "The clathrate hydrate process for post and pre-combustion capture of carbon dioxide," *J. Hazard. Mater.*, vol. 149, no. 3, pp. 625–629, Nov. 2007, doi:10.1016/j.jhazmat.2007.06.086.

17. X. Sen Li, Z. M. Xia, Z. Y. Chen, K. F. Yan, G. Li and H. J. Wu, "Equilibrium hydrate formation conditions for the mixtures of CO_2 + H_2 + tetrabutyl ammonium bromide," *J. Chem. Eng. Data*, vol. 55, no. 6, pp. 2180–2184, Jun. 2010, doi:10.1021/je900758t.

18. Y. Seo and S. P. Kang, "Enhancing CO_2 separation for pre-combustion capture with hydrate formation in silica gel pore structure," *Chem. Eng. J.*, vol. 161, no. 1–2, pp. 308–312, Jul. 2010, doi:10.1016/j.cej.2010.04.032.

19. A. Adeyemo, R. Kumar, P. Linga, J. Ripmeester and P. Englezos, "Capture of carbon dioxide from flue or fuel gas mixtures by clathrate crystallization in a silica gel column," *Int. J. Greenh. Gas Control*, vol. 4, no. 3, pp. 478–485, May 2010, doi:10.1016/j.ijggc.2009.11.011.

20. P. Babu, R. Kumar and P. Linga, "Pre-combustion capture of carbon dioxide in a fixed bed reactor using the clathrate hydrate process," *Energy*, vol. 50, no. 1, pp. 364–373, Feb. 2013, doi:10.1016/j.energy.2012.10.046.

21. C. G. Xu, X. Sen Li, Q. N. Lv, Z. Y. Chen and J. Cai, "Hydrate-based CO_2 (carbon dioxide) capture from IGCC (integrated gasification combined cycle) synthesis gas using bubble method with a set of visual equipment," *Energy*, vol. 44, no. 1, pp. 358–366, Aug. 2012, doi:10.1016/j.energy.2012.06.021.

22. C. G. Xu, J. Cai, X. Sen Li, Q. N. Lv, Z. Y. Chen and H. W. Deng, "Integrated process study on hydrate-based carbon dioxide separation from integrated gasification combined cycle (IGCC) synthesis gas in scaled-up equipment," *Energy and Fuels*, vol. 26, no. 10, pp. 6442–6448, Oct. 2012, doi:10.1021/ef3011993.

23. Nexant Simteche, "2006 engineering analysis tasks of contract mod 017 – final report," 2006. [Online]. Available: www.osti.gov/servlets/purl/915435 (accessed Jul. 1, 2021).

24. G. S. Smirnov and V. V. Stegailov, "Toward determination of the new hydrogen hydrate clathrate structures," *J. Phys. Chem. Lett.*, vol. 4, no. 21, pp. 3560–3564, Nov. 2013, doi:10.1021/jz401669d.

25. A. R. Kohl and R. Nielsen, *Gas Purification*, 5th ed. Gulf Professional Publishing, 1997.

5 Sustainable Metal-Organic Framework Technologies for CO_2 Capture

Hira Naveed, Haleefa Shaheen, Ranjana Kumari, Reshma Lakra, Asim Laeeq Khan, and Subhankar Basu

CONTENTS

5.1 INTRODUCTION

CO_2 capture through adsorption is an effective technology and makes use of the high capacity and affinity of adsorbents such as zeolites, activated carbon, aluminophosphates, silica gel, metal-oxide sieves, carbon nanotubes, pillared clays, porous organic materials, and polymeric and inorganic resins [1–4]. Gas separation methods include membrane separation, cryogenic distillation, adsorption, and absorption (chemical and physical) techniques [5–8]. By examining the various technologies, adsorptive separation at high pressure receives more importance because of its low energy use, uncomplicated maintenance, and ease in tuning for different applications, simple operation, and environmentally benign characteristics [9]. This process simultaneously involves two mechanisms, adsorption and desorption. In adsorption, a certain gas is separated from a gaseous mixture in a packed bed or a fixed bed reactor filled with appropriate adsorbents [10]. Once the adsorbents are exhausted, they are regenerated by desorption. Many cyclic adsorption-desorption methods are

DOI: 10.1201/9781003162780-5

employed in gas separation methods, like inert purge cycles, cycles of displacement, temperature swing adsorption (TSA) [11], pressure swing adsorption (PSA), electrical swing adsorption (ESA), and vacuum swing adsorption (VSA) [12]. PSA and TSA are the most preferred methods for CO_2 capture. TSA involves the heating of adsorbent for desorption of adsorbed gas, and it may take anywhere between an hour to over a day. In contrast, the PSA occurs by cycling the pressure of the adsorbent. The PSA has an advantage over TSA due to its rapid adsorbent recovery. The VSA process is relatively new; it is attractive to researchers as it operates near ambient (negative gauge, as the name implies) pressure [13].

The separation in an adsorption-desorption process occurs by selective adsorption of gas, and it happens when the functional groups on the adsorbent surface have a high affinity for the diffused gas molecule. Activated carbon and zeolites have been widely studied, but they have a low CO_2 adsorption capacity and only perform well at high temperature for a gaseous mixture with substantial moisture [14]. Activated carbon is widely used on a commercial scale and has a significantly high CO_2 adsorption capacity than zeolites. Still, the interaction of CO_2 molecules with activated carbon does not have sufficient bond strength, and therefore the adsorbents are temperature dependent. Hence, the CO_2 adsorption efficiency of the activated carbon decreases with an increase in temperature and shows poor gas selectivity (e.g., in CO_2/N_2 mixture) [15].

Metal-organic frameworks (MOFs) are reported as an alternative to conventional adsorbents due to their increased porosity, surface areas, and stability under different operating environments. Figure 5.1 shows the research trends on CO_2 capture using MOFs over the years. MOFs are considered coordination polymers; their formation is achieved by the metal ions or clusters' organic ligands as they become attached by the coordinate bonding (Figure 5.2). They have several advantages over other porous materials (e.g., well-developed 3D structures, high surface area, controlled porous

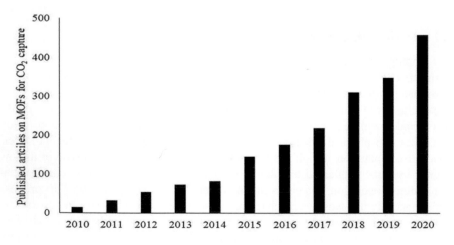

FIGURE 5.1   Research on the development of MOFs for CO_2 captures.

Source: **www.sciencedirect.com.**

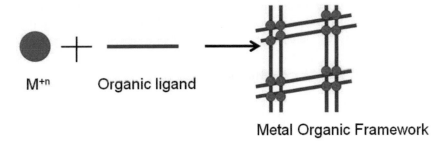

FIGURE 5.2 Schematic process of MOF formation.

TABLE 5.1
MOF Manufacturers and Their Applications for CO$_2$ Capture [19–24]

Suppliers	Applications
BASF	For gas storage
MOF Technologies	For CO$_2$ capture
Immaterial Labs Ltd	For gas separation and storage
Unisieve	For gas purification
Framergy	For CO$_2$ capture
GCE	For CO$_2$ capture

structures, and surface properties) [16]. The combination of all the suitable properties indicates that MOFs are an ideal adsorbent. Over the last few decades, MOFs have been used in drug delivery, gas storage, gas separations, catalysis, sensors, purifications, and so forth [17]. They have excellent surface properties and surface area, which are scalable for industrial applications [18]. Table.5.1 shows MOF manufacturers and their applications in CO$_2$ capture, storage, and purification.

5.2 MECHANISMS OF CO$_2$ ADSORPTION BY MOFs

Over the years, MOFs have been reported to have high selectivity while separating CO$_2$ from gas mixtures [25–28]. It is an essential criterion for selecting a MOF for CO$_2$ adsorption along with the kinetic diameter of comparable CO$_2$ molecules (3.3 Å) and the pore structure of the selected MOF. This mechanism of separation is called the molecular sieving effect (shape exclusion and size). Further, MOFs with polar (–OH, –N=N–, –NH$_2$, and –N= C (R)–) pores show increased carbon dioxide (CO$_2$) uptake because of quadrupole moments of the molecules of CO$_2$. As per the separation requirements, when a MOF structure is tuned, the CO$_2$ adsorption capacity of the adsorbents is significantly enhanced, and the selectivity improves. MOFs are mainly categorized into two broad types: rigid MOFs and dynamic/flexible MOFs. The rigid MOFs show complex frameworks that result in permanent pores like zeolites. In contrast, dynamic MOFs show soft frameworks, in which the structures change with the process parameters/external factors (e.g., temperature, guest molecules and their pressure) [29].

5.2.1 Rigid MOFs and CO_2 Adsorption

Several rigid metal-organic frameworks are reported for the adsorption of selective gases. The mechanism of separation in the rigid MOFs is the same as the zeolites. Selective separation in rigid MOFs is achieved via the molecular sieving effect. The adsorption on MOFs occurs due to the different interaction strengths between the gas molecules and the MOFs. The factors that are responsible for selective separation in the rigid MOFs are pore size and shape exclusion, surface-adsorbate interactions and size-shape exclusion, and MOF surface-gas interactions (Figure 5.3) [28].

Various MOFs are suitable and used for the adsorption of different gases [29–30]. The molecules having a compatible kinetic diameter with the pore diameters of the adsorbent can pass through the pores and be adsorbed (Figure 5.3). The kinetic diameter of the gas molecules CO_2, N_2, and CH_4 are 3.3 Å, 3.8 Å, and 3.64 Å, respectively. Based on the molecular sieving, the Mn-MOFs are one of the most evaluated rigid MOFs for the selective separation of CO_2 from the CO_2-CH_4 mixture. These types of MOFs consist of a 3D framework and a 1D channel. Wide cages exist in such MOFs, which connect through narrow pores. Based on the experimental work, CO_2 adsorbed preferably over CH_4 on rigid MOFs at 195 K. Smaller molecules (i.e., H_2) adsorb

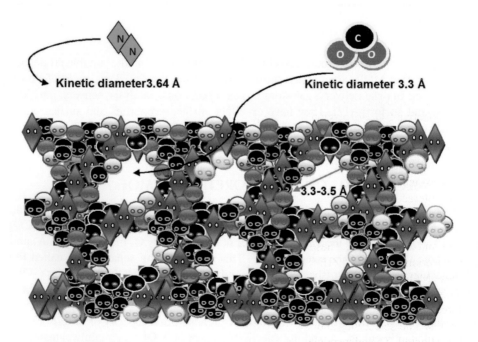

FIGURE 5.3 A graphical illustration of CO_2 adsorption in a rigid MOF.

preferentially over N_2 and Ar at 78 K [29]. The adsorption capacities for N_2 and CH_4 remained insignificant in both cases. The selective CO_2 adsorption is attained due to the narrow pores present in the channels. Based on the size and shape, MIL-96 [30] and $Zn_2(cnc)_2(dpt)$ [31] were studied for the selective uptake of CO_2 from the CO_2-CH_4 mixture. These MOFs are used for natural gas purification and the separation of land-fill gases. A rigid MOF, PCN-26, has a three-dimensional structure. This MOF was reported to be suitable for selective CO_2 adsorption because of its size separation effect. The pore diameter of this MOF is 3.68Å. The CO_2 adsorption capacity of activated PCN-26 was high, at 298 K and 1 bar for the N_2-CH_4 mixture [32].

The properties of gas molecules and the interaction of these molecules with the MOFs surface is significant in the selective gas adsorption process. Adsorption of gas molecules on rigid MOFs relies on the thermodynamic or kinetic equilibrium effects at a mentioned equilibrium time. The polarity, quadrupole moment, H-bonding, and pore surface properties are responsible for the selective adsorption [33]. A pillared-layer MOF $Zn_2(ndc)_2(dpni)$ is usually prepared by microwave heating. The investigation showed that the adsorbing capacity of CO_2 was higher than CH_4. It was attributed to its significant quadrupole moment and kinetic effect [34]. A 3D MOF with a porous surface and a 1D channel was synthesized for the effective adsorption of carbon dioxide and methane [35]. This MOF contained coordinate unsaturated Mn II sites. At ambient temperature, the adsorption capability of this 3D MOF was higher for CO_2 than for CH_4. Some other rigid MOFs (e.g., ZIFs frameworks) also have high selectivity towards carbon dioxide (CO_2) compared to carbon monoxide [36]. Large cages are present in the ZIFs, which connects through narrow pores. ZIF-68, ZIF-69, and ZIF-70 belong to those classes of ZIFs that show large cages with diameters 7.2 Å, 10.2 Å, and 15.9 Å, respectively, and they are interlinked via pores on the surface with diameters 4.4 Å, 7.5 Å, and 13.1 Å, respectively. At a temperature of 273 K, all these ZIFs showed high selectivity and adsorption capacity for CO_2 over CO due to molecular sieving and high CO_2 quadrupole moment [36].

The pore size influences the interaction of gaseous molecules with MOF, defining how gas molecules are adsorbed selectively by some MOFs. For example, $Er_2(pda)_3$ for selective CO_2 adsorption over N_2 uses this effect [37]. This MOF has a 3D structure with a 1D channel that consists of unsaturated Er III sites. Because of two different mechanisms, the CO_2 uptake capacity of rigid MOFs was observed to be higher than N_2. The pore sizes of 3.4Å diameter and the presence of the π-electrons, polar groups and unsaturated Er III sites contribute to this behavior. This electrical field can interact with the quadrupole moment of the CO_2 molecules and thus increase the potential of CO_2 adsorption. Some other examples of rigid MOFs employed for the selective adsorption of CO_2 are (based on the aforementioned effects) are ZIF-100 and ZIF-95 [38]. These MOFs have large cavities (3.65 Å). The CO_2 selective adsorption capacity of ZIF-95 and ZIF-100 compared to CH_4, N_2, and CO shows that these MOFs have a high selective adsorption capacity for CO_2 over other gases [38]. This is because of the presence of suitable pore size and its strong surface-adsorbate interactive effect due to the high quadrupole moment of CO_2 molecules and N_2 in MOFs. The selectivity and adsorption capacity of UTSA-16 for CO_2 has also been investigated. The adsorption capacity of CO_2 for UTSA-16 was 160 m^3m^{-3}. UTSA-16 showed high CO_2 selectivity and adsorption capacity because of the secured binding sites to the molecule of CO_2 and the optimum size of the pores suitable for the kinetic diameter of the CO_2 [39]. The results from these selected studies reveal that owing to the tenability of their shapes, pore size, and surface properties, rigid MOFs are thus suitable for gas separation and CO_2 sequestration.

5.2.2 FLEXIBLE/DYNAMIC MOFs AND CO_2 ADSORPTION

The behavior of the flexible MOFs makes them different, and the gas separation mechanism becomes more complicated than the rigid ones. During adsorption/desorption processes, these distinctive MOFs entrain a flexible framework and exhibit active behaviors. Flexible MOFs have this remarkable feature of the structural transformability to create the desired properties through external catalysts (i.e., temperature, pressure) [40–41] and molecular uniqueness [42]. The said behavior is attributed to the stretch activity of organic ligands, which includes twisting, bending, rotating or tilting, and their secondary building unit's variation [43]. For example, based on the size of the guest molecules, flexible MOF pore sizes vary in the process of adsorption/desorption. Because of the readjustment of the framework during adsorption-desorption of gas molecules, the variation in flexible MOFs structure is accompanied by the hysteresis behavior [44–45]. The two major active behaviors of the concerned MOFs are the breathing phenomenon [46–47], and the gate opening and closing [48–50]. The process of breathing happens whenever sudden and unexpected compressing or expansion occurs for the MOF's unit cell, which ultimately causes the transitions in its structure because of the interactive and unique gas molecules. Second, the gate opening and closing happens through the outer catalyst in the close/nonporous MOFs, resulting in a transient, open-spongy structure [51].

A flexible MOF comprising Zn(ADC)(4,4′-Bpe)0.5, consisting of paddle-wheel shaped di-nuclear Zn_2 units, joined together by ADC di-anions and additionally shafted by 4,4′-Bpe, results in a 3D structure of intertwined molecules [44]. The sieving effect of molecules reduces the compatibility nature of the size of the MOF pores (3.4 × 3.4 Å) and CO_2 molecules (3.3 Å). It results in an increased uptake of CO_2 over CH_4 (3.8 Å). For Ni_2 (cyclam)$_2$(mtb) at 1 bar and 195 K, selective adsorption of CO_2 over CH_4 is reported [52]. High CO_2 adsorption over N_2 and CH_4 is well documented due to carbon dioxide's similar size to the adsorbent MOF. However, at a lower temperature (77 K), the studied gases did not adsorb on the MOF [45]. The size/shape exclusion depends on the selective gas's adsorbing behavior noted in flexible PCN-5 and the acquisition of a twofold, intertwined structure. PCN-5 MOF selectively separates CO_2 from CO_2 and CH_4 mixture at 195 K and 1 bar. This is due to the comparable pore size of the adsorbent MOF and kinetic diameter of the separating gas molecules (CH_4 and CO_2); the CO_2 uptake was 210 mg g^{-1} (4.8 mmol g^{-1}) and the CH_4 uptake was 30 mg g^{-1} (1.9 mmol g^{-1}) [53].

For the selective adsorption performance of flexible MOFs, the surface properties of the MOFs are of interest. The hydrated and dehydrated forms of the MIL-53 series demonstrate such properties. MIL-53 series MOFs possess a 3D structure and consist of a 1D channel, with a formula of $MO_4(OH)_2$ being paired, because of the 1,4-benzene dicarboxylic acid. The breathing characteristics of MIL-53 series MOFs occur through the hydration and dehydration process [42, 54–55]. During the hydration process of MIL-53 series MOF, the hydrogen bonding (HB) occurs between the oxygen atoms of the water molecules and the carboxylate group of the organic ligands that results in a change in the pore size and consequently becomes constricted [55].

CO_2 adsorption on MIL-53s was less at low pressure (<5 bars) because of the presence of the water molecules that ended up with narrow pores [42]. Therefore, due to rising gas pressure, the MOF's framework modulates itself into a larger pore size. The extension of MOF's framework results in an increased CO_2 selective adsorption to about 7.2 mmol g^{-1} at 18 bars. Additionally, the increment observed in pore volume ranged from 1012.8 $Å^3$ in hydrated to 1522.5 $Å^3$ for the hydrated and CO_2. In dehydrated MIL-53, adsorption behavior contrasts with hydrated behavior [42]. The dehydrated MIL-53 can be represented by the two steps for the structural transition in the case of selective CO_2 adsorption. In the first step, the CO_2 uptake capacity rises dramatically (3 mmol g^{-1}) at low pressure (<5 bars); the second step begins at approximately 8 bars, wherein the CO_2 uptake was roughly 7.7 mmol g^{-1}. It refers to an alteration in structure from a large to narrow pore size, owing to the selective adsorption of CO_2. It is further accredited to the high gas pressure resulting in the structural transition again to the enlarged porous structure of MIL-53, because of the steric restriction of the water molecules [56]. Dehydrated MIL-53 reported CH_4 uptake of 4.6 mmol g^{-1} at 20 bars. Contrarily, selective CH_4 adsorption was nearly zero (0.2 mmol g^{-1}). This difference in CH_4 adsorption was due to the repulsion effect of polarity between the molecules of H_2O in the MIL-53 MOF and nonpolar CH_4 molecules. Consequently, this current work confirms a high capability of such MOFs towards the application for separation and gas storage and particularly for selective CO_2 adsorption over methane molecules [56]. Another study was reported, [Ni(bpe)$_2$(N(CN)$_2$)](N(CN)$_2$) [57], with an intertwined 3D framework and interlinked with 1D channels. Increasing the temperature to 195 K, a notable amount of selective CO_2 adsorption was achieved into the mentioned framework. This phenomenal interaction was due to the swelling of Nickel's π-electron clouds and polar groups in the host framework. The gate-opening phenomenon is an exciting function of a few flexible MOFs seen during the mechanism of adsorption. The mechanism suggests that when the host MOF structure is related to the guest molecules, the pores start to enlarge due to interaction between them. Such enlargement and expansion are directly related to the adsorbate gas molecules and interactions among the functional groups of the MOF.

ZIF-20 is one of the examples of the flexible MOF, the structural rearrangement happens during CO_2 adsorption from the CO_2-CH_4 mixture [58]. The concerned MOF also has a three-dimensional framework along with larger cages connected through narrow openings. The magnitude of CO_2 adsorption is five times greater than CH_4 at 273 K and 1 bar. It is due to the powerful interlinkage of the MOF porous surface and the CO_2 molecules. The pore size window is 2.8 Å (calculated for the crystallized structure), less than the kinetic diameters of CH_4 and CO_2 (i.e., 3.8 Å and 3.24 Å, respectively). As mentioned above, the expansion of the small windows due to the window-widening process makes the large cages approachable; hence, the gas molecules diffuse via the framework. Likewise, the quadrupolar moment of CO_2 imposes an uncommon impact on the flexibility in a few MOFs framework such as ZIF-20.

Cu(dhbc)$_2$(4,4'bpy) [50] in the hydrated structure has a 2D-shaped sheet accompanied with a 1D channel. This flexible MOF's framework consists of rigidly interconnected sheets, and the pores as provided by the 'dhbc' benzene are arranged in a vertical shape into the concerned sheets and intertwined. The area of the cross section for the framework is enclosed with the H$_2$O molecules (3.6 × 4.2 Å). The consolidation of the 3D structure occurs through some interlinked networks between the neighboring ligands of dhbc via π-π stacking. The adsorption isotherm of CO$_2$, CH$_4$, O$_2$, and N$_2$ starts from 0.4, 9, 35, and 50 bars, respectively, confirming an instant improvement in the specified pressure for the mentioned gas molecules. Each gas molecule has its gate-opening and gate-closing pressures, depending upon the intermolecular forces of interaction between the gas molecule and MOF. Another MOF, Cu (pyrdc)(bpp) [57], shows the same dynamic attachment with Cu (dhbc)2(4,4'-bpy) [58], having a 2D pillared-bilayer structure and pathways consisting of a nonporous framework. This MOF shows a porous dynamic behavior during the adsorption/desorption process with the bond formation through the inclusion and breaking stimulation by the guest removal. In this study, no O$_2$ and N$_2$ adsorption is reported at 77K. In contrast, CO$_2$ (3.3 Å), with kinetic diameter compared to N$_2$ (3.6 Å) and O$_2$ (3.4 Å), diffuses through and adsorbs in the structure. This behavior indicates that the specific opening-gate pressure of individual gas results in selective CO$_2$ adsorption in the framework comparable to O$_2$ and N$_2$.

Although the concerned mechanism of ambiguous rearrangement is associated with the MOFs, this type of adsorbent creates an interest to search for novel materials for application in the adsorption and separation processes. The most exclusive factor with the flexible MOF is the strong and weak intermolecular interactions in their framework that differentiate them from the rigid one or other adsorbents. The MOF's structural integrity and porosity are resultant of strong bonds, whereas the flexible and active behaviors are attributed to the weak interactions.

5.3 CO$_2$ ADSORPTION CAPACITY

An important criterion in the adsorption technology is the capacity of the adsorbate (i.e., surface area to volume ratio). MOFs are characterized by high to ultra-high surface area, which facilitates high adsorption capacity of CO$_2$ compared to other adsorbents (e.g., activated carbons, zeolites). In addition, the adsorbent-adsorbate interaction, the temperature of adsorption, and the pressure are significant factors in the uptake capacity of CO$_2$ [59]. Millward and Yaghi [60] first studied eight MOFs with different geometries and characteristics (CuBTC, MOF-505, MOF-2, IRMOF-11, IRMOF-1, IRMOFs-3, MOF-177, and IRMOF-6), all at a pressure of 35 bars and a temperature of 298 K. MOF-177 had a surface area of 4508 m^2 g^{-1} and exhibited a maximum of 60 wt% CO$_2$ adsorption. It improved with the development of MOF-210 (71 wt%) at 50 bars and 298 K. The outstanding CO$_2$ uptake capacity is interlinked with the surface area of MOFs (BET area of surface 6240 m^2 g^{-1} and Langmuir surface area 10,400 m^2 g^{-1}). Furukawa et al. [61] studied five MOFs with similar metal sites; they showed that by enlarging the ligands of the structure, the MOFs' capacity of adsorption improved. By increasing the ligand of MOF-177 from BTB, 1,3,5 benzenetribenzoate, to BBC, 4,4',4"-(benzene-1,3,5-triyl-tris (benzene-4,1-iyl)) tribenzoate (MOF-200), CO$_2$ uptake increased from 60 wt% to 70 wt % [62]. Table 5.2

TABLE 5.2
The CO_2 Adsorption Capacities of Different MOFs

MOFs	Surface area (m²/g)	Capacity (wt%) low/high	Pressure (bar) low/high	Temperature (K)	Selectivity	Reference
MOF-177	4508	60	35	298	–	[61]
MOF-2	345	12.3	35	298	–	[61]
MOF-505	1547	31	35	298	–	[60]
IRMOF-6	2516	46.2	35	298	–	[60]
IRMOF-11	2096	39.3	35	298	–	[60]
IRMOF-3	2160	45.1	35	298	–	[60]
HKUST-1	1781	32	35	298	–	[60]
Zn-MOF-74	816	31.4	35	298	–	[60]
UiO(bpdc)	2646	8/79.7	1/20	298	–	[63]
DGC-MIL-101	4198	14.5/59.8	1/40	298	–	[64]
HTS-MIL-101	3482	52.8	1/40	298	–	[64]
DMOF-A	760	10.6/17.1	20	298	–	[65]
DMOF-DM	1120	25.4	20	298	23	[65]
DMOF-C12	1180	8.8/26.4	20	298	17	[65]
Mg-MOF-74	1416	30.1	1	298	–	[66]
Ni-MOF-74	1252	19.4	1	298	–	[66]
Mg/DOBDC	1415.1	25	1	298	–	[67]
Co/DOBDC	1089.3	21.6	1	298	–	[67]
Ni/DOBDC	1017.5	20.5	1	298	–	[67]
MIL-100(Cr)	1528.7	9.5	1	298	–	[66]
ZIF-7	312	9.1	1	298	–	[66]
CPM-5	2187	8.8	1	298	16.1	[68]
ZIF-7-R	5	8.7	1	303	–	[69]
MIL-101 (Cr)	2549	24.2	30	303	–	[64]
DMOF-TM	1050	13.3	1/20	298	28	[65]

presents some of the most notable MOFs and their CO_2 capacity at varying operating conditions. The recent method of improving CO_2 adsorption is producing MOF composites. CO_2 composite MOFs are reported to have an increased surface area and carbon dioxide adsorption capacity in contrast to the parent MOFs. A list of such MOFs is mentioned in the review work of Ghanbari *et al.* [33] as aminated Cu-BTC-graphite oxide, GO@MOF-505, GrO@MIL-101, GRO@CuBTC, GO@HKUST-1, MWCNT@MIL-101, and ZIF-8/CNT.

5.4   MOF TECHNOLOGY APPLICATIONS FOR CO_2 CAPTURE

The commonly reported CO_2 capture technologies are post-combustion, pre-combustion, oxyfuel combustion, and direct-fuel combustion (Figure 5.4). Post-combustion

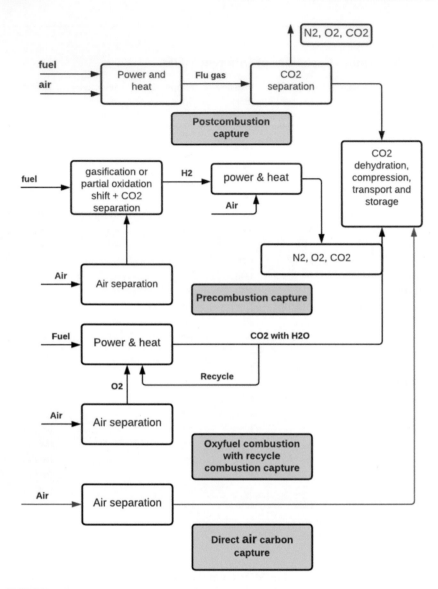

FIGURE 5.4 Process flow chart of CO₂-capturing technologies.

CO_2 capture is a traditional process where CO_2 is captured after the combustion and the flue gas is released into the atmosphere, utilized, or stored irrevocably. The process involves enormous volumes of flue gas to treat and requires extensive pretreatment involving the removal of corrosive gases such as H_2S. The pretreatment step also requires cooling since the combustion of fuels results in high-temperature flue gases. Some of the energy is saved by the flue gas heat recovery heat exchanger.

The high volumes of flue gas and low partial pressure of CO_2 in the mixture bring technological challenges to the process requiring a large size of equipment. The flue gas from the existing power plants accounts for 60% of the global CO_2 emissions and consists primarily of CO_2 (3%–16%) and N_2 (73%–77%) at a pressure of 1 bar [70].

The process conditions for the post-combustion CO_2 capture process favors the use of MOFs. Mg-MOF-74 shows unique properties of high CO_2 adsorption with the low-pressure drying conditions (0.15 bar and 5.28 mmol/g at 40 °C) [71]. The high uptake of Mg-MOF-74 is due to the presence of open metal sites. The open metal sites bring high affinity with CO_2 molecules. One of the bottlenecks in applying this MOF is the drastic drop in performance due to water vapors in the flue gas stream. With the water vapor (5%–7%) in the gas, it is necessary to either incorporate an expensive dehydration step or prepare MOFs with high selectivity towards CO_2 in the presence of water. Martínez et al. evaluated the post-combustion CO_2 separation performance of three commonly and commercially available MOFs: ZIF-8, HKUST-1 and MIL53 [72]. The MOFs were functionalized with tetra-ethylene-pent-amine (TEPA) in a bid to improve their affinity with CO_2. The results showed that the amine-functionalized MOFs showed very high adsorption capacities for the carbon dioxide, even with a high amount of moisture. This study opened new avenues in the application of MOFs for post-combustion CO_2 capture. Maurya and Singh [73] carried out a comparative study involving three MOFs, with a single-wall carbon nanotube (SWCNT) and a covalent organic framework (COFs), for similar flue gas treatment conditions. The results of this study concluded that the CO_2 capacity of adsorption can be arranged in descending order (SWCNT > InOF-1 > COF-300 > UiO-66 > COF-108 > ZIF-8) for post-combustion carbon capture. The results showed that the studied MOF inflicted trans-to-cis transition of the material resulting in an aggressive CO_2 uptake. The selected studies point to an enormous potential of MOF application in the post-combustion CO_2 capture technology for power production plants.

The CO_2 capture from pre-combustion requires extensive pretreatment of fuel before combustion. The pretreatment steps depend on fuel/industrial gas and may include a variety of constituent removal depending on the type of the process (reforming of natural gas, biomass/coal gasification, etc.). The discussion here does not include CO_2 removal from raw natural gas, as this has been covered in detail in the previous sections, but it entails discussion related to product gases like hydrogen, carbon monoxide, and others. The advantage of pre-combustion carbon capture is that these processes are relatively less expensive than the post-combustion and have a better capacity for regeneration, partially attributed to the absence of inert N_2. Syngas comprising H_2, CO, and CO_2 mixture needs separation of H_2 for its further use as a carrier fuel. The differences in kinetic diameters of H_2 (2.89 Å) and CO_2 (3.30 Å) separate two gas molecules based on their distinctive sizes using favorable molecular sieve membranes. The major challenge associated with the pre-combustion technology is the low temperature requirement for optimum performance, so some cooling mechanism is needed. The application of MOFs in pre-combustion capture of CO_2 is due to high affinity and adsorption capacities for CO_2. This results in the production of purified H_2. Since pre-combustion processes run at moderate to low pressures, MOFs outperform the competing zeolite and silica type of porous particles owing to their better adsorption under such conditions. Several studies are reported on the application of MOFs for pre-combustion CO_2 removal [74]. The $Zn_2(bIm)_4$-type layered MOF (synthesized with an aperture of 2.9 Å due to its unique cage-sized clusters) has a molecular sieve aperture between that of CO_2 and H_2, leading to the easy passage of H_2 while restricting CO_2. The molecular sieving approach resulted in a very high selectivity. In another study, Nandi et al. [75] prepared highly selective ultra-microporous MOFs based on Ni-(4-pyridylcarboxylate) with a CO_2 uptake capacity of 3.95 mmol g^{-1}. The application of MOF in PSA exhibited adsorption-desorption carbon cycling and a self-diffusivity of carbon was 3×10^{-9} m^2/s, which is considerably higher than the zeolites. Herm et al. [74] evaluated the performances of several MOFs with flexible structure, high surface area, or open cation metal sites for application in pre-combustion CO_2 capture using pressure swing adsorption. The study concluded that by varying CO_2 concentration in the feed, MOFs with a high content of major metal cation sites (e.g., Cu-BTTri, Mg_2 (dobdc)) have better performance than the conventional adsorbents. These selected studies reflect that MOFs provide a promising pathway for pre-combustion carbon removal while offering high productivity and good energy efficiency.

The oxyfuel combustion process allows ease in capturing the new CO_2 emissions from coal and other carbon-rich fuels. The flue gas from the oxyfuel combustion process contains mostly CO_2, once the moisture and minor impurities are removed. The major disadvantage of this method is the operation at a very high temperature (since the inert nitrogen does not lower the exhaust gas temperatures). Moreover, the energy-intensive process for cryogenic O_2/air separation and the use of costly materials also overburdens the economics [76]. The oxygen's separation from the air is possible through microporous MOFs, which can potentially minimize the energy cost otherwise used for the cryogenic separation of oxygen from the air. In contrast to N_2, O_2 accepts electrons from the open coordination sites which are abundantly present in electron-rich MOFs. The redox-active metal sites interact with O_2 from the air through reversible electron transfer [77]. MOFs with high reduction ability are reported to have a strong affinity and selectivity towards O_2. In $Cr_3(BTC)_2$, the electron transformation from Cr and O_2 produced Cr-superoxide adsorption of 11 wt% at 298 K and 2 mbar and selectivity of O_2/N_2 of 19.3 [78]. Similarly, Fe_2(dobdc) at 298 K and 1 bar resulted in a 10 wt% O_2 adsorption capacity in comparison to a 1.3 wt% N_2 uptake capacity under similar conditions.

In recent years, nonconventional technologies like direct air carbon capture have received attention for the mitigation of the global CO_2 concentration in the atmosphere, where CO_2 is captured directly from the air [79]. Direct air carbon capture (DACC), which is also quoted as a 'negative carbon technology', includes CO_2 capture from the air using liquid sorbent or solids under atmospheric conditions. It generates a concentrated CO_2 stream for sequestration or reuse [80]. DACC claims to reduce CO_2 emissions from both point sources and varied sources. MOFs tested for DACC applications include Mg-MOF-74, SIFSIX-3-Zn, SIFSIX-3-Cu, en-Mg_2(dobpdc) and mmen-Mg_2(dobpdc)). The strong bonds between the CO_2 and the amine groups present in en-Mg_2(dobpdc) and mmen-Mg_2(dobpdc) resulted in a maximum CO_2 uptake of 2.83 and 2 mmol g^{-1}, respectively, at 298 K and 1 bar [81]. A comparative study of the CO_2 capture processes is mentioned in Table 5.3.

TABLE 5.3
Comparative Study of CO_2 Capture Technologies

Process	Advantages	Disadvantages
Pre-combustion carbon capture	Mostly used in industrial processes Require much less energy as compared to other carbon capture methods Less water used as compared to post-combustion capture	High equipment cost Extensive support system required Loss of energy is high in comparison with post-combustion Applicable to new plants only
Post-combustion carbon capture	Applicable for both old and new coal-powered plants Most common technology for carbon capture Process is well developed, hence research on new sorbents and capture equipment is available	High water requirement Low CO_2 partial pressure at ambient pressure The amine technologies used are not efficient, causing an 11% efficiency reduction
Oxyfuel combustion	High-efficiency steam cycles are produced Low level of pollutant emission Does not require on-site chemical operation	Expensive technology Not yet proven on large scale Risk of CO_2 leakage High energy requirement

5.5 MOF-BASED MEMBRANES FOR CO_2 SEPARATION

MOFs are used as filler material in the polymer matrix to fabricate hybrid film-like structures called mixed matrix membranes (MMMs) (Figure 5.5). The appearance of MOFs in the polymer matrix improves the gas separation performance (selectivity and permeability) of the membrane. MOF-based MMMs include an organic linker that binds with the organic polymer matrix. It results in defect-free membranes and allows the exploitation of high flux and molecular sieving abilities of MOFs and the polymer matrix's high processability. In one of the earliest studies on MOFs-based MMMs, Basu *et al.* [82] incorporated the CuBTC MOF in the Matrimid® and polysulfone matrix blends. Asymmetric MMMs reported increased permeability and CO_2/N_2 selectivity than the unfilled membranes, thereby proving the potential of MOFs for post-combustion CO_2 capture. The enhanced selectivity indicates a good interfacial contact and high steric hindrance and CO_2 sorption of MOF particles. In a subsequent study, Basu *et al.* [83] fabricated both dense and asymmetric MMMs using three different MOFs (CuBTC, MIL-53 (Al), and ZIF-8) and compared their results for post-combustion CO_2 capture. Optimized loading of MOF particles was achieved for all three types, leading to high CO_2 permeability and selectivity. A three-layered structure of MOF-based thin film nanocomposite MMMs were evaluated for CO_2 capture [84]. The top selective layer comprised MOF embedded in PIM-1. The intermediate PDMS gutter layer was followed by a polymeric substrate. The layered MMMs made use of a MOF particle blend (NH_2-UiO-66 and MOF-74-Ni), resulting

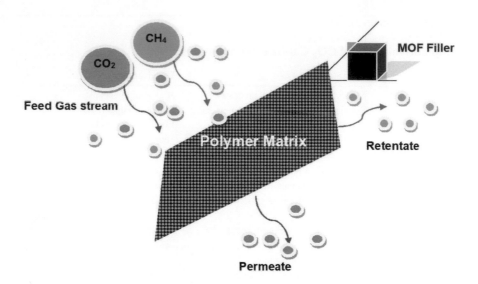

FIGURE 5.5 CO_2 separation from the gas mixture using MOF-based mixed matrix membrane (MMM).

in an increased CO_2 permeation of 4660–7460 gas permeation units (GPU) and CO_2/ N_2 selectivity of 26–33, which was a substantial increase from pristine PIM-1 with 4320 GPU CO_2 permeance and CO_2/N_2 selectivity of 19. The successful utilization of UiO66 type MOF made in several subsequent studies employing post-functionalization and modulation approaches in a bid to push to the limits of high CO_2 permeability and selectivity. Tahir *et al.* [85] imparted sulfonic acid functionalization on UiO66 by grafting a silane coupling agent (mercaptopropyl tri-methoxy-silane). The modified MOF and polysulfone-based MMMs demonstrated that increased sulfonic group concentration results in an improved selectivity owing to their high affinity towards CO_2 gas molecules. The membranes also exhibited increased CO_2 permeability and selectivity with single and binary gas mixtures. Anjum *et al.* [86] used a modulation approach, in combination with amine-functionalized linkers, to apply MOFs to enhance the MOF selectivity. The amine functionalization in the pores of the MOFs resulted in an increased separation performance with a CO_2 selectivity of up to 47.7 and permeability of up to 19.4 Barrer compared to the bare Matrimid membrane (i.e., 50% increase in the CO_2 selectivity and 540% increase in the permeability).

   Yasmeen *et al.* [87] used an ionic liquid modified ZIF-67 type of MOF based MMMs with polysulfone. The results showed that the ionic liquid had a $-NH_2$ group and acetate group combines with ZIF-67 nanoporous structure. The resulting large surface area improved the CO_2/N_2 selectivity up to 90% compared to the bare membranes. The study opened a new avenue for applying a three-component system comprising ionic liquids, MOFs, and polymers for post-combustion CO_2 capture. The potential of novel MOF-801 into a polyether-block-amide (PEBA) polymer was also reported [88]. The results showed defect-free MMMs and highly selective and permeable membranes due to the preferred uptake of CO_2 over N_2. Ishaq *et al.* [89]

showed high-performance MMMs comprising a bio metal-organic framework (Bio-MOF-1) and polysulfone. The use of nano-bars fillers, which were rectangular in shapes of lengths ~0.5–4.0 μm and widths ~0.05–0.20 μm, reported good dispersion and interaction in the polymer matrix, even at increased filler loadings. The MMMs showed a CO_2 permeability of 16.57 Barrer and ideal selectivity of 45.6 for CO_2/N_2, indicating an improvement of 168% and 58% for CO_2 permeability and ideal selectivity, respectively, compared to the bare membrane. The presence of adeninate amino and pyrimidine Lewis's bases was the prominent influencer in enhancing the separation performance due to the adsorption of the CO_2 molecules.

NOTT-300 MOF with high selectivity and permeability was introduced in a Pebax®1657 matrix and studied for CO_2 separation from CO_2/N_2 gas mixture under different operating conditions [90]. NOTT-300 was selected due to its exceptional capture abilities for CO_2 and high stability in harsh conditions. It is a competent membrane material due to its green synthesis and high CO_2 uptake capacity. The addition of NOTT-300 resulted in an improved performance with increased CO_2 permeability and CO_2/N_2 selectivity. The permeability of CO_2 increased to 380% and selectivity to 26%, when the filler loading was increased to 40%. Altintas and Keskin, [91] carried out an extensive computational study to evaluate the MOF library for CO_2 adsorption. Monte Carlo and molecular dynamics simulations were run to identify the most promising MOFs for application in MMMs. The objective of this study was to achieve MMMs that could surpass the so-called target region of Robeson's upper bound. This study provides the researchers with a new pathway in material design and development for MMMs fabrication. Prasetya and Ladewig [92] used a light-responsive MOF comprising Zn metal and 1,4-diazabicyclo octane (DABCO) and 2-phenyldiazenyl terephthalic acid linkers. MMMs with PIM-1 matrix were demonstrated. Benzaqui et al. [93] incorporated a microporous MIL-96-(Al) of CO_2/N_2 for the sake of the CO_2 capture in the post-combustion. The fabricated MMMs showed fault-free structure and facilitated CO_2 transport through the highly porous network of the MIL-96 MOF. Results were much superior to that of pure polymeric membranes.

These selected studies employing MOFs in MMMs prove their potential as a promising alternative to conventional fillers. Their inclusion increased CO_2 permeability and CO_2/N_2 selectivity and allows for the development of defect-free MMMs, high CO_2 uptake capacity, and molecular sieving of the MOF cages. Considering the availability of hundreds of possible MOF structures, computation simulation studies are essential to screen and select the most optimum MOFs. Despite their high performance, challenges still exist to explore the usefulness of long-term stability under harsh conditions.

5.6 CO_2 CONVERSION USING MOF AS A CATALYST

CO_2 is converted to valuable chemicals by a carbon conversion process, employing a photo-catalyst. The various products of photochemical reactions are shown in Table 5.4. This can be further studied in the review article of Elhenawy et al. [94]. The studies report the formation of CO, CH_3OH, CH_4, HCOOH, and C_2H_4

TABLE 5.4
MOF-Based Catalysts for CO_2 Conversion to Chemicals

Photo-catalytic conversion		Electrochemical and electrocatalytic conversion	
MOF	Product	MOF	Product
MOF4	CO	Zn-BTC	CH_4
Zn_2GeO_4/ZIF-8	CH_3OH	M-PMOF	CO
NH_2-MIL-125 (Ti)	$HCOO^-$	Re-SURMOF	CO
$Cu_3(BTC)_2$ @TiO_2	CH_4	ZIF-8	CO
Cu-porphyrin MOF	CH_3OH	ZIF-CNT-FA-p	CO
Pt-NH_2-MIL-125 (Ti)	HCOO	$Al_2(OH)_2$TCPP-Co	CO
Au-NH_2-MIL-125 (Ti)			
NH_2-UiO-66 (Zr)	HCOO	CR-MOF	HCOOH
NH_2-UiO-66 (Zr/Ti)			
Ui-66-CrCAT	HCOOH	Ru (III)-doped	CH_3OH, C_2H_5OH
Ui-66-GaCAT		HKUST1	
Co-ZIF-9	CO	Ag_2O/layer ZIF	CO
Co-MOF-74			
Mn-MOF-74			
Zn-ZI-8			
CPO-27-Mg/TiO_2	CO	C-AFC@ZIF-8	CO
TiO_2			
CPO-27-Mg			
Co-ZIF-9/TiO_2	CO	ZIF-8 derived Fe-N-C	CO
Zn/PMO	CH_4		
PCN-22	HCOO		
2Cu/ZI-8N_2	CH_3OH		
Ag@Co-ZIF-9	CO		
Ni MOLs	CO		
TiO_2/Cu_2O/$Cu_3(BTC)_2$	CO		
CdS/UiO-bppy/Co	CO		
NH_2-rGO/Al-MOF	HCOO		
Zn-MOF nanosheets [CO_2 (OH) L] $(ClO_4)_3$	CO		

Source: Adapted from [97].

and other products. MOFs act as catalysts in converting carbon dioxide to fuels, by carrying out cyclo-addition, photo-reduction, and hydrogenation of carbon. The photocatalytic reaction is carried out in the presence of UV or visible light. The photo-oxidation of CO_2 on the surface of the MOF involves many steps. The MOF catalysts initially adsorb photon, resulting in the separation of the electron-hole pair. Second, the excited electron moves from a high molecular orbit to a low molecular orbit [95]. Finally, the CO_2 molecules adsorbed on the MOF's catalytic center accepts the electron. This results in different chemical products (e.g., CO, CH_4, HCOOH).

The process of CO_2 photo-reduction on MOFs includes a series of reactions in which an electron is transferred, and CO_2 is reduced to HCOOH, HCHO and so forth at different potentials [94]. The MOFs studied as catalysts in the reaction and its corresponding products includes MOF4: CO; Zn_2GeO_4/ZIF-8: CH_3OH; NH_2-MIL-125 (Ti): $HCOO^-$; $Cu_3(BTC)_2$ @TiO_2: CH_4; and Cu-porphyrin MOF: CH_3OH.

Zn-Ni bimetal MOFs have been used to synthesize Ni-Ni-doped porous interconnected carbon (NiNPIC) catalyst which was used to increase the CO_2 reduction reaction and electro-catalytic activity. The interconnected pores and surface area of catalyst resulted in a channel for highly accessible Ni-N sites, which allowed mass diffusion and led to rapid electron transfer, high electrolyte/gas transport, low interface resistance, and resulted in carbon dioxide reduction. The result shows that the synthesized catalyst is efficient in converting CO_2 to CO with moderate potential [96]. Electrochemical reduction technology performs as a catalyst in converting CO_2 to methanol using alternative energy sources, where electro-catalyst plays a significant role. Dong et al. reported a stable three-dimensional porphyrin-based MOF of PCN-222 (Fe) for heterogeneous catalysis. The study reported increased catalytic performance for the electrochemical conversion of CO_2 to CO and 91% FECO in a saturated aqueous solution of 0.5 M $KHCO_3$ and carbon dioxide (CO_2) [97]. MOF-based catalysts for electrolytic CO_2 reduction have also been reported. The ZIF-CNT-FA-p shows the highest value for CO, and ZIF-8 shows the lowest value.

REFERENCES

1. M. Nisar, F. L. Bernard, E. Duarte, V. V. Chaban and S. Einloft, "New polysulfone microcapsules containing metal oxides and ([BMIM][NTf2]) ionic liquid for CO_2 capture," *Journal of Environmental Chemical Engineering*, vol. 9, no. 1, p. 104781, 2021.
2. M. Irani, A. T. Jacobson, K. A. M. Gasem and M. Fan, "Modified carbon nanotubes/tetraethylenepentamine for CO_2 capture," *Fuel*, vol. 206, pp. 10–18, 2017.
3. L. Kong, S. Han, T. Zhang, L. He and L. Zhou, "Developing hierarchical porous organic polymers with tunable nitrogen base sites via theoretical calculation-directed monomers selection for efficient capture and catalytic utilization of CO_2," *Chemical Engineering Journal*, vol. 420, p. 127621, 2021.
4. E. Davarpanah et al., "CO_2 capture on natural zeolite clinoptilolite: Effect of temperature and role of the adsorption sites," *Journal of Environmental Management*, vol. 275, p. 111229, 2020.
5. K. Sumida et al., "Carbon Dioxide Capture in Metal-Organic Frameworks," *Chemical Reviews*, vol. 112, no. 2, pp. 724–781, 2012.
6. X. Ge and S. Ma, "CO_2 Capture and Separation of Metal-Organic Frameworks," *Materials for Carbon Capture*, doi:10.1002/9781119091219.ch2 pp. 5–27, 2020.
7. Y. Wang, L. Zhao, A. Otto, M. Robinius and D. Stolten, "A review of post-combustion CO_2 capture technologies from coal-fired power plants," *Energy Procedia*, vol. 114, pp. 650–665, 2017.
8. H. Li, K. Wang, Y. Sun, C. T. Lollar, J. Li and H.-C. Zhou, "Recent advances in gas storage and separation using metal-organic frameworks," *Materials Today*, vol. 21, no. 2, pp. 108–121, 2018.
9. A. Sattari, A. Ramazani, H. Aghahosseini and M. K. Aroua, "The application of polymer containing materials in CO_2 capturing via absorption and adsorption methods," *Journal of CO_2 Utilization*, vol. 48, p. 101526, 2021.

10. S. Trevisan, R. Guédez and B. Laumert, "Thermo-economic optimization of an air driven supercritical CO_2 Brayton power cycle for concentrating solar power plant with packed bed thermal energy storage," *Solar Energy*, vol. 211, pp. 1373–1391, 2020.

11. W.-Y. Gao, C.-Y. Tsai, L. Wojtas, T. Thiounn, C.-C. Lin and S. Ma, "Interpenetrating Metal- Metalloporphyrin Framework for Selective CO_2 Uptake and Chemical Transformation of CO_2," *Inorganic Chemistry*, vol. 55, no. 15, pp. 7291–7294, 2016.

12. L. Riboldi and O. Bolland, "Overview on pressure swing adsorption (PSA) as CO_2 capture technology: State-of-the-art, limits and potentials," *Energy Procedia*, vol. 114, pp. 2390–2400, 2017.

13. S. G. Subraveti, S. Roussanaly, R. Anantharaman, L. Riboldi and A. Rajendran, "Techno-economic assessment of optimised vacuum swing adsorption for post-combustion CO_2 capture from steam-methane reformer flue gas," *Separation and Purification Technology*, vol. 256, p. 117832, 2021.

14. S. Lawson, C. Griffin, K. Rapp, A. A. Rownaghi and F. Rezaei, "Amine-Functionalized MIL-101 Monoliths for CO_2 Removal from Enclosed Environments," *Energy & Fuels*, vol. 33, no. 3, pp. 2399–2407, 2019.

15. J. Zhang, P. A. Webley and P. Xiao, "Effect of process parameters on power requirements of vacuum swing adsorption technology for CO_2 capture from flue gas," *Energy Conversion and Management*, vol. 49, no. 2, pp. 346–356, 2008.

16. L. A. Darunte, A. D. Oetomo, K. S. Walton, D. S. Sholl and C. W. Jones, "Direct Air Capture of CO_2 Using Amine Functionalized MIL-101(Cr)," *ACS Sustainable Chemistry & Engineering*, vol. 4, no. 10, pp. 5761–5768, 2016.

17. R. J. Kuppler *et al.*, "Potential applications of metal-organic frameworks," *Coordination Chemistry Reviews*, vol. 253, no. 23, pp. 3042–3066, 2009.

18. A. J. Howarth *et al.*, "Chemical, thermal and mechanical stabilities of metal-organic frameworks," *Nature Reviews Materials*, vol. 1, no. 3, p. 15018, 2016.

19. BASF. (2013, 22–06–2021). *BASF to showcase metal organic frameworks (MOFs) for energy storage at NGV Americas Conference*. Available: www.nanowerk.com/news2/newsid=33225.php

20. MOF Technologies. (2021, 22–06–2021). *MOFs for Carbon Dioxide*. Available: www.moftechnologies.com

21. Immaterial Labs. (2021, 20–05–2021). *Materials*. Available: https://immaterial.com.

22. Unisieve, "Sustainable separation – cleaner gasses," 2020. [Online]. Available: www.unisieve.com/technology

23. U. Ryu *et al.*, "Recent advances in process engineering and upcoming applications of metal-organic frameworks," *Coordination Chemistry Reviews*, vol. 426, p. 213544, 2021.

24. M. Burgess. (2020, 22–05–2021). *CARMOF project investigating MOFs for increased carbon capture efficiency*. Available: www.gasworld.com/carmof-project-investigating-mofs-for-carbon-capture/2019864.article

25. B. Bhattacharya and D. Ghoshal, "Selective carbon dioxide adsorption by mixed-ligand porous coordination polymers," *CrystEngComm*, 10.1039/C5CE01246E vol. 17, no. 44, pp. 8388–8413, 2015.

26. T. Fukushima *et al.*, "Solid Solutions of Soft Porous Coordination Polymers: Fine-Tuning of Gas Adsorption Properties," *Angewandte Chemie International Edition*, doi:10.1002/anie.201000989 vol. 49, no. 28, pp. 4820–4824, 2010.

27. J.-R. Li *et al.*, "Carbon dioxide capture-related gas adsorption and separation in metal-organic frameworks," *Coordination Chemistry Reviews*, vol. 255, no. 15, pp. 1791–1823, 2011.

28. J.-R. Li, R. J. Kuppler and H.-C. Zhou, "Selective gas adsorption and separation in metal-organic frameworks," *Chemical Society Reviews*, 10.1039/B802426J vol. 38, no. 5, pp. 1477–1504, 2009.

29. D. N. Dybtsev, H. Chun, S. H. Yoon, D. Kim and K. Kim, "Microporous Manganese Formate: A Simple Metal-Organic Porous Material with High Framework Stability and Highly Selective Gas Sorption Properties," *Journal of the American Chemical Society*, vol. 126, no. 1, pp. 32–33, 2004.

30. T. Loiseau *et al.*, "MIL-96, a Porous Aluminum Trimesate 3D Structure Constructed from a Hexagonal Network of 18-Membered Rings and μ3-Oxo-Centered Trinuclear Units," *Journal of the American Chemical Society*, vol. 128, no. 31, pp. 10223–10230, 2006.

31. M. Xue *et al.*, "Robust Metal-Organic Framework Enforced by Triple-Framework Interpenetration Exhibiting High H$_2$ Storage Density," *Inorganic Chemistry*, vol. 47, no. 15, pp. 6825–6828, 2008.

32. W. Zhuang, D. Yuan, D. Liu, C. Zhong, J.-R. Li and H.-C. Zhou, "Robust Metal-Organic Framework with an Octatopic Ligand for Gas Adsorption and Separation: Combined Characterization by Experiments and Molecular Simulation," *Chemistry of Materials*, vol. 24, no. 1, pp. 18–25, 2012.

33. T. Ghanbari, F. Abnisa and W. M. A. Wan Daud, "A review on production of metal organic frameworks (MOF) for CO$_2$ adsorption," *Science of the Total Environment*, vol. 707, p. 135090, 2020.

34. Y.-S. Bae *et al.*, "Separation of CO$_2$ from CH4 Using Mixed-Ligand Metal-Organic Frameworks," *Langmuir*, vol. 24, no. 16, pp. 8592–8598, 2008.

35. H. R. Moon, N. Kobayashi and M. P. Suh, "Porous Metal-Organic Framework with Coordinatively Unsaturated MnII Sites: Sorption Properties for Various Gases," *Inorganic Chemistry*, vol. 45, no. 21, pp. 8672–8676, 2006.

36. R. Banerjee *et al.*, "High-Throughput Synthesis of Zeolitic Imidazolate Frameworks and Application to CO$_2$ Capture," *Science*, vol. 319, no. 5865, p. 939, 2008.

37. L. Pan *et al.*, "Porous Lanthanide Organic Frameworks: Synthesis, Characterization, and Unprecedented Gas Adsorption Properties," *Journal of the American Chemical Society*, vol. 125, no. 10, pp. 3062–3067, 2003.

38. B. Wang, A. P. Côté, H. Furukawa, M. O'Keeffe and O. M. Yaghi, "Colossal cages in zeolitic imidazolate frameworks as selective carbon dioxide reservoirs," *Nature*, vol. 453, no. 7192, pp. 207–211, 2008.

39. S. Xiang *et al.*, "Microporous metal-organic framework with potential for carbon dioxide capture at ambient conditions," *Nature Communications*, vol. 3, no. 1, p. 954, 2012.

40. S. Henke, A. Schneemann and R. A. Fischer, "Massive Anisotropic Thermal Expansion and Thermo-Responsive Breathing in Metal-Organic Frameworks Modulated by Linker Functionalization," *Advanced Functional Materials*, doi:10.1002/adfm.201301256 vol. 23, no. 48, pp. 5990–5996, 2013.

41. T. D. Keene *et al.*, "Solvent-modified dynamic porosity in chiral 3D kagome frameworks," *Dalton Transactions*, 10.1039/C3DT00096F vol. 42, no. 22, pp. 7871–7879, 2013.

42. M. Alhamami, H. Doan and C.-H. Cheng, "A Review on Breathing Behaviors of Metal-Organic-Frameworks (MOFs) for Gas Adsorption," *Materials*, vol. 7, no. 4, 2014.

43. J. Seo, C. Bonneau, R. Matsuda, M. Takata and S. Kitagawa, "Soft Secondary Building Unit: Dynamic Bond Rearrangement on Multinuclear Core of Porous Coordination Polymers in Gas Media," *Journal of the American Chemical Society*, vol. 133, no. 23, pp. 9005–9013, 2011.

44. B. Chen, S. Ma, E. J. Hurtado, E. B. Lobkovsky and H.-C. Zhou, "A Triply Interpenetrated Microporous Metal-Organic Framework for Selective Sorption of Gas Molecules," *Inorganic Chemistry*, vol. 46, no. 21, pp. 8490–8492, 2007.

45. B. Chen, S. Ma, F. Zapata, F. R. Fronczek, E. B. Lobkovsky and H.-C. Zhou, "Rationally Designed Micropores within a Metal-Organic Framework for Selective Sorption of Gas Molecules," *Inorganic Chemistry*, vol. 46, no. 4, pp. 1233–1236, 2007.

46. G. Férey and C. Serre, "Large breathing effects in three-dimensional porous hybrid matter: Facts, analyses, rules and consequences," *Chemical Society Reviews*, 10.1039/B804302G vol. 38, no. 5, pp. 1380–1399, 2009.

47. C. Serre *et al.*, "An Explanation for the Very Large Breathing Effect of a Metal-Organic Framework during CO_2 Adsorption," *Advanced Materials*, doi:10.1002/adma.200602645 vol. 19, no. 17, pp. 2246–2251, 2007.

48. D. Fairen-Jimenez, S. A. Moggach, M. T. Wharmby, P. A. Wright, S. Parsons and T. Düren, "Opening the Gate: Framework Flexibility in ZIF-8 Explored by Experiments and Simulations," *Journal of the American Chemical Society*, vol. 133, no. 23, pp. 8900–8902, 2011.

49. C. Gücüyener, J. van den Bergh, J. Gascon and F. Kapteijn, "Ethane/Ethene Separation Turned on Its Head: Selective Ethane Adsorption on the Metal-Organic Framework ZIF-7 through a Gate-Opening Mechanism," *Journal of the American Chemical Society*, vol. 132, no. 50, pp. 17704–17706, 2010.

50. R. Kitaura, K. Seki, G. Akiyama and S. Kitagawa, "Porous Coordination-Polymer Crystals with Gated Channels Specific for Supercritical Gases," *Angewandte Chemie International Edition*, doi:10.1002/anie.200390130 vol. 42, no. 4, pp. 428–431, 2003.

51. F.-X. Coudert, C. Mellot-Draznieks, A. H. Fuchs and A. Boutin, "Prediction of Breathing and Gate-Opening Transitions upon Binary Mixture Adsorption in Metal-Organic Frameworks," *Journal of the American Chemical Society*, vol. 131, no. 32, pp. 11329–11331, 2009.

52. Y. E. Cheon and M. P. Suh, "Multifunctional Fourfold Interpenetrating Diamondoid Network: Gas Separation and Fabrication of Palladium Nanoparticles," *Chemistry – A European Journal*, doi:10.1002/chem.200701813 vol. 14, no. 13, pp. 3961–3967, 2008.

53. S. Ma, X.-S. Wang, E. S. Manis, C. D. Collier and H.-C. Zhou, "Metal-Organic Framework Based on a Trinickel Secondary Building Unit Exhibiting Gas-Sorption Hysteresis," *Inorganic Chemistry*, vol. 46, no. 9, pp. 3432–3434, 2007.

54. P. L. Llewellyn, S. Bourrelly, C. Serre, Y. Filinchuk and G. Férey, "How Hydration Drastically Improves Adsorption Selectivity for CO_2 over CH_4 in the Flexible Chromium Terephthalate MIL-53," *Angewandte Chemie International Edition*, doi:10.1002/anie.200602278 vol. 45, no. 46, pp. 7751–7754, 2006.

55. S. Bourrelly, P. L. Llewellyn, C. Serre, F. Millange, T. Loiseau and G. Férey, "Different Adsorption Behaviors of Methane and Carbon Dioxide in the Isotypic Nanoporous Metal Terephthalates MIL-53 and MIL-47," *Journal of the American Chemical Society*, vol. 127, no. 39, pp. 13519–13521, 2005.

56. M. P. Suh, H. J. Park, T. K. Prasad and D.-W. Lim, "Hydrogen Storage in Metal-Organic Frameworks," *Chemical Reviews*, vol. 112, no. 2, pp. 782–835, 2012.

57. T. K. Maji, R. Matsuda and S. Kitagawa, "A flexible interpenetrating coordination framework with a bimodal porous functionality," *Nature Materials*, vol. 6, no. 2, pp. 142–148, 2007.

58. H. Hayashi, A. P. Côté, H. Furukawa, M. O'Keeffe and O. M. Yaghi, "Zeolite A imidazolate frameworks," *Nature Materials*, vol. 6, no. 7, pp. 501–506, 2007.

59. B. Liu, R. Zhao, K. Yue, J. Shi, Y. Yu and Y. Wang, "New amine-functionalized cobalt cluster-based frameworks with open metal sites and suitable pore sizes: Multipoint interactions enhanced CO_2 sorption," *Dalton Transactions*, 10.1039/C3DT51258D vol. 42, no. 38, pp. 13990–13996, 2013.

60. A. R. Millward and O. M. Yaghi, "Metal-Organic Frameworks with Exceptionally High Capacity for Storage of Carbon Dioxide at Room Temperature," *Journal of the American Chemical Society*, vol. 127, no. 51, pp. 17998–17999, 2005.

61. H. Furukawa, K. E. Cordova, M. O'Keeffe and O. M. Yaghi, "The Chemistry and Applications of Metal-Organic Frameworks," *Science*, vol. 341, no. 6149, p. 1230444, 2013.

62. H. Furukawa *et al.*, "Ultrahigh Porosity in Metal-Organic Frameworks," *Science*, vol. 329, no. 5990, p. 424, 2010.
63. B. Li *et al.*, "Enhanced Binding Affinity, Remarkable Selectivity, and High Capacity of CO_2 by Dual Functionalization of a *rht*-Type Metal-Organic Framework," *Angewandte Chemie International Edition*, doi:10.1002/anie.201105966 vol. 51, no. 6, pp. 1412–1415, 2012.
64. J. Kim, Y.-R. Lee and W.-S. Ahn, "Dry-gel conversion synthesis of Cr-MIL-101 aided by grinding: High surface area and high yield synthesis with minimum purification," *Chemical Communications*, 10.1039/C3CC44559C vol. 49, no. 69, pp. 7647–7649, 2013.
65. N. C. Burtch, H. Jasuja, D. Dubbeldam and K. S. Walton, "Molecular-Level Insight into Unusual Low Pressure CO_2 Affinity in Pillared Metal-Organic Frameworks," *Journal of the American Chemical Society*, vol. 135, no. 19, pp. 7172–7180, 2013.
66. S. Shruti *et al.*, "Curcumin release from cerium, gallium and zinc containing mesoporous bioactive glasses," *Microporous and Mesoporous Materials*, vol. 180, pp. 92–101, 2013.
67. L. Li, J. Yang, J. Li, Y. Chen and J. Li, "Separation of CO_2/CH_4 and CH_4/N_2 mixtures by M/DOBDC: A detailed dynamic comparison with MIL-100(Cr) and activated carbon," *Microporous and Mesoporous Materials*, vol. 198, pp. 236–246, 2014.
68. R. Sabouni, H. Kazemian and S. Rohani, "Carbon dioxide adsorption in microwave-synthesized metal organic framework CPM-5: Equilibrium and kinetics study," *Microporous and Mesoporous Materials*, vol. 175, pp. 85–91, 2013.
69. W. Cai *et al.*, "Thermal Structural Transitions and Carbon Dioxide Adsorption Properties of Zeolitic Imidazolate Framework 7 (ZIF-7)," *Journal of the American Chemical Society*, vol. 136, no. 22, pp. 7961–7971, 2014.
70. Z. Hu, M. Khurana, Y. H. Seah, M. Zhang, Z. Guo and D. Zhao, "Ionized Zr-MOFs for highly efficient post-combustion CO_2 capture," *Chemical Engineering Science*, vol. 124, pp. 61–69, 2015.
71. S. R. Caskey, A. G. Wong-Foy and A. J. Matzger, "Dramatic Tuning of Carbon Dioxide Uptake via Metal Substitution in a Coordination Polymer with Cylindrical Pores," *Journal of the American Chemical Society*, vol. 130, no. 33, pp. 10870–10871, 2008.
72. F. Martínez, R. Sanz, G. Orcajo, D. Briones and V. Yángüez, "Amino-impregnated MOF materials for CO_2 capture at post-combustion conditions," *Chemical Engineering Science*, vol. 142, pp. 55–61, 2016.
73. M. Maurya and J. K. Singh, "Effect of Ionic Liquid Impregnation in Highly Water-Stable Metal-Organic Frameworks, Covalent Organic Frameworks, and Carbon-Based Adsorbents for Post-combustion Flue Gas Treatment," *Energy & Fuels*, vol. 33, no. 4, pp. 3421–3428, 2019.
74. Z. R. Herm, J. A. Swisher, B. Smit, R. Krishna and J. R. Long, "Metal-Organic Frameworks as Adsorbents for Hydrogen Purification and Precombustion Carbon Dioxide Capture," *Journal of the American Chemical Society*, vol. 133, no. 15, pp. 5664–5667, 2011.
75. S. Nandi *et al.*, "A single-ligand ultra-microporous MOF for precombustion CO_2 capture and hydrogen purification," *Science Advances*, vol. 1, no. 11, p. e1500421, 2015.
76. Y. Peng *et al.*, "Metal-organic framework nanosheets as building blocks for molecular sieving membranes," *Science*, vol. 346, no. 6215, p. 1356, 2014.
77. A. Kather and G. Scheffknecht, "The oxycoal process with cryogenic oxygen supply," *Naturwissenschaften*, vol. 96, no. 9, pp. 993–1010, 2009/09/01 2009.
78. L. J. Murray *et al.*, "Highly-Selective and Reversible O_2 Binding in Cr3(1,3,5-benzenetricarboxylate)2," *Journal of the American Chemical Society*, vol. 132, no. 23, pp. 7856–7857, 2010.
79. K. S. Lackner, "The thermodynamics of direct air capture of carbon dioxide," *Energy*, vol. 50, pp. 38–46, 2013.

80. E. S. Sanz-Pérez, C. R. Murdock, S. A. Didas and C. W. Jones, "Direct Capture of CO_2 from Ambient Air," *Chemical Reviews*, vol. 116, no. 19, pp. 11840–11876, 2016.

81. A. Kumar *et al.*, "Direct Air Capture of CO_2 by Physisorbent Materials," *Angewandte Chemie International Edition*, doi:10.1002/anie.201506952 vol. 54, no. 48, pp. 14372–14377, 2015.

82. S. Basu, A. Cano Odena and I. F. J. Vankelecom, "Asymmetric Matrimid®/[Cu3(BTC)2] mixed-matrix membranes for gas separations," *Journal of Membrane Science*, vol. 362, no. 1, pp. 478–487, 2010.

83. S. Basu, A. Cano-Odena and I. F. J. Vankelecom, "Asymmetric Matrimid®/[Cu3(BTC)2] mixed-matrix membranes for gas separations," *Journal of Membrane Science*, vol. 362, no. 1, pp. 478–487, 2011.

84. M. Liu, M. D. Nothling, P. A. Webley, J. Jin, Q. Fu and G. G. Qiao, "High-throughput CO_2 capture using PIM-1@MOF based thin film composite membranes," *Chemical Engineering Journal*, vol. 396, p. 125328, 2020.

85. Z. Tahir *et al.*, "SO3H functionalized UiO-66 nanocrystals in polysulfone based mixed matrix membranes: Synthesis and application for efficient CO_2 capture," *Separation and Purification Technology*, vol. 224, pp. 524–533, 2019.

86. M. W. Anjum, F. Vermoortele, A. L. Khan, B. Bueken, D. E. D. Vos and I. F. Vankelecom, "Modulated UiO-66-Based Mixed-Matrix Membranes for CO_2 Separation," *ACS Applied Material Interfaces*, vol. 7, no. 45, pp. 25193–25201, 2015.

87. I. Yasmeen *et al.*, "Synergistic effects of highly selective ionic liquid confined in nano-cages: Exploiting the three component mixed matrix membranes for CO_2 capture," *Chemical Engineering Research and Design*, vol. 155, pp. 123–132, 2020.

88. J. Sun, Q. Li, G. Chen, J. Duan, G. Liu and W. Jin, "MOF-801 incorporated PEBA mixed-matrix composite membranes for CO_2 capture," *Separation and Purification Technology*, vol. 217, pp. 229–239, 2019.

89. S. Ishaq, R. Tamime, M. R. Bilad and A. L. Khan, "Mixed matrix membranes comprising of polysulfone and microporous Bio-MOF-1: Preparation and gas separation properties," *Separation and Purification Technology*, vol. 210, pp. 442–451, 2019.

90. N. Habib *et al.*, "Development of highly permeable and selective mixed matrix membranes based on Pebax®1657 and NOTT-300 for CO_2 capture," *Separation and Purification Technology*, vol. 234, p. 116101, 2020.

91. C. Altintas and S. Keskin, "Molecular Simulations of MOF Membranes and Performance Predictions of MOF/Polymer Mixed Matrix Membranes for CO_2/CH_4 Separations," *ACS Sustainable Chemistry & Engineering*, vol. 7, no. 2, pp. 2739–2750, 2019.

92. N. Prasetya and B. P. Ladewig, "New Azo-DMOF-1 MOF as a Photoresponsive Low-Energy CO_2 Adsorbent and Its Exceptional CO_2/N_2 Separation Performance in Mixed Matrix Membranes," *ACS Applied Materials & Interfaces*, vol. 10, no. 40, pp. 34291–34301, 2018.

93. M. Benzaqui *et al.*, "Revisiting the Aluminum Trimesate-Based MOF (MIL-96): From Structure Determination to the Processing of Mixed Matrix Membranes for CO_2 Capture," *Chemistry of Materials*, vol. 29, no. 24, pp. 10326–10338, 2017.

94. S. E. Elhenawy, M. Khraisheh, F. AlMomani and G. Walker, "Metal-Organic Frameworks as a Platform for CO_2 Capture and Chemical Processes: Adsorption, Membrane Separation, Catalytic-Conversion, and Electrochemical Reduction of CO_2," *Catalysts*, vol. 10, no. 11, 2020.

95. S. Omar, M. Shkir, M. Ajmal Khan, Z. Ahmad and S. AlFaify, "A comprehensive study on molecular geometry, optical, HOMO-LUMO, and nonlinear properties of 1,3-diphenyl-2-propen-1-ones chalcone and its derivatives for optoelectronic applications: A computational approach," *Optik*, vol. 204, p. 164172, 2020.

96. Z. Ma *et al.*, "Ultrasonic assisted synthesis of Zn-Ni bi-metal MOFs for interconnected Ni-N-C materials with enhanced electrochemical reduction of CO$_2$," *Journal of CO$_2$ Utilization*, vol. 32, pp. 251–258, 2019.

97. B.-X. Dong *et al.*, "Electrochemical Reduction of CO$_2$ to CO by a Heterogeneous Catalyst of Fe–Porphyrin-Based Metal-Organic Framework," *ACS Applied Energy Materials*, vol. 1, no. 9, pp. 4662–4669, 2018.

6 Novel CO_2 Separation Membranes

Asif Jamil, Muhammad Latif, Alamin Idris Abdulgadir, Danial Qadir, and Hafiz Abdul Mannan

CONTENTS

6.1 IMPORTANCE OF CO_2 REMOVAL

Greenhouse gases have been a dire concern to environmentalists, industrialists, and academics alike. Greenhouse gases such as carbon dioxide, methane, and nitrous oxide have been of particular interest because of their severe threats to human and natural life on planet Earth [1]. However, studies have shown that among these gases, the management of carbon dioxide has gained prime importance for its enormous emissions volume to disrupt the natural environment and habitat of all sorts of life on the planet. A few of the significant adverse effects are acid rain, global warming, and disturbed ecosystems [2–3].

DOI: 10.1201/9781003162780-6

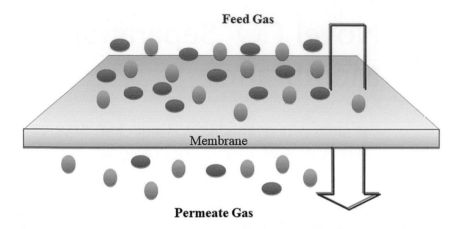

FIGURE 6.1   Schematic illustration of gas permeation across the membrane.

A quarter of CO_2 emissions comes from the burning of fossil fuels during the power/electricity generation. Concerted efforts are underway to effectively cut down on these emissions [4]. Moreover, low CO_2 content in natural gas is equally desirable to limit corrosion, improve the calorific value of the gas, and reduce the diameter of transmission pipelines [2]. Hence, removal of carbon dioxide is an imminent exigency. However, most conventional carbon dioxide removal technologies are riddled with drawbacks, namely, high capital cost, limited CO_2 loading capability, and operational complications [4]. Membrane offers low capital cost, high capacity, low area footprint, high energy efficiency, and simple process operation, filling the gaps left by other techniques.

6.2   INTRODUCTION TO MEMBRANE TECHNOLOGY

A membrane is a selective barrier for a particular gas permeates. The driving force for the separation of penetrant gases is either concentration or pressure gradient across the membrane's surface. This gradient is created by application of pressure or enrichment of the gas (by increasing concentration) in the upstream side of the membrane. A schematic illustration of gas permeation through a simple flat sheet membrane is shown in Figure 6.1.

In an ideal scenario, a membrane permits the passage of one or more gas molecules in a selective manner from a multicomponent mixture. Nowadays, membrane technology finds application in many aspects of separation engineering, ranging from solid-liquid, ion-liquid, reverse osmosis, liquid-liquid, forward osmosis, and liquid-gas to gas-gas.

6.3 PARAMETERS FOR DETERMINING THE MEMBRANE'S PERFORMANCE

Gas permeability and selectivity are the two primary factors in determining the performance of a membranes. Gas permeability can be calculated from the pressure and the thickness normalized flux, and it is usually reported as in barrer given in equation 6.1.

$$P_A = (J_A \times l)/\Delta P_A \tag{6.1}$$

where J, Δp, and l denotes the flux, differential partial pressure, and the thickness of the membrane, respectively.

On the other hand, for asymmetric membranes, thickness cannot be evaluated easily; an alternate term, 'pressure normalized flux' (i.e., P/l), is used instead. This is called permeance and reported in GPU. Hence, Eq. 6.1 takes the form of Eq. 6.2.

$$P_A /l = J_A/\Delta P_A \tag{6.2}$$

For the solution diffusion mechanism, the permeation (P_A) of gas across the membrane is defined as the product of the coefficients of gas diffusion (D_A) and sorption (S_A). The ideal selectivity (α) for a gas mixture containing two components, A and B, can be represented by Eq. 6.3.

$$\text{Selectivity}(\alpha_{AB}) = P_A/P_B = D_A.S_A/D_B.S_B \tag{6.3}$$

6.4 CLASSIFICATION OF MEMBRANES

With a large scope and numerous applications, it is difficult to classify membranes in a tight-knitted manner. Many classification schemes are reported. Some of the chosen few are detailed below:

Origin: Membranes are widely available in nature or can be developed synthetically. The cell, the basic component of life, has multiple membranes that perform distinctive functions. Synthetic membranes could further be divided into categories like 'organic' (e.g., polymers, macromolecules) and 'inorganic' (e.g., ceramics, metallics). Furthermore, mixed matrix membranes combined the synergistic advantages of both organic and inorganic phases [5].

Structure and morphology: The structure and morphology of the membranes could be of a symmetric or asymmetric type. The symmetry of the membranes defines associated characteristics and applications. The membrane structure could be classified as thin, dense, porous, homogenous, or heterogeneous. Transportation across the membranes could be active, passive, or reactive, depending upon the interactions with the incident gas molecules.

Geometry and configuration: Membranes based on geometry or configuration are flat sheet, tubular, and hollow fibers. The structural geometry plays a vital role in terms of gas molecule permeation across membranes. For instance, hollow fiber membranes provide a higher surface-to-volume ratio for gas molecules, thus higher separation ability than flat sheet and tubular type membranes.

Operational and separation processes: The separation mechanism is classified into different types such as sieve, solubility-diffusivity, and charge interactions. In the sieve mechanism, separation is governed by the difference in the size of the pores and gas molecules. The solubility-diffusivity mechanism explains the solubility and diffusivity of the incident molecules through the membrane, whereas the difference in charges of the incident species drives or controls the separation for electrochemical mechanism. The interactions among the incident molecules and membranes surface are responsible for the separation process.

Applications: Membranes find applications in liquid or gas phase separation systems. Moreover, they are vaguely categorized based on their end-use applications such as water treatment, pharmaceutical, petrochemical, food, and oil and gas industries [5].

6.5   GAS TRANSPORT MECHANISMS IN MEMBRANES

Various transport mechanisms have been devised for gas separations (represented in Figure 6.2), namely, solution-diffusion, Knudsen diffusion, surface diffusion, capillary condensation, and molecular sieving diffusion. Most mechanisms occur in porous membranes except for the solution diffusion mechanism [6].

Molecular-sieving membranes contain pores smaller than 7 Å, which is sufficient to allow smaller molecules to pass and exclude larger ones. The separation mechanism pertains to the molecular size and orientation of the penetrant gas molecules [7].

In *Knudsen-diffusion*, the collision of penetrant gas molecules with the pore walls is dominant rather than the mutual collision of penetrant molecules. The pore size

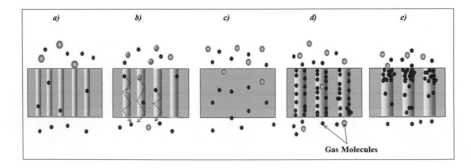

FIGURE 6.2   Schematic representation of gas transport mechanisms: (a) molecular diffusion, (b) Knudsen diffusion, (c) solution diffusion, (d) surface diffusion, and (e) capillary condensation [6].

is sufficiently large enough to accommodate the colliding gas molecules within its volume. The Knudsen-diffusion membranes exhibit low selectivity, thus restricting their industrial applications [8].

In the *surface-diffusion mechanism*, the penetrant gas molecules are adsorbed on the wall surface of the pores. Also, gas molecules travel across the membrane under the influence of the concentration gradient [9].

Capillary condensation occurs when the gas molecules interact with the pore walls. The pore size causes the condensation of gas molecules, resulting in subsequent diffusion of the gas molecules across the membrane [10].

Most porous membranes exhibit attractive permeation performance and appreciable chemical and thermal stability; however, these are expensive to fabricate, difficult to process, and mechanically weak [11]. Nonporous, dense membranes are commercially accepted where a solution diffusion mechanism controls the movement of gas molecules. In the mechanism mentioned above, molecules of incident gas sorb at the membrane surface (i.e., the feed side). This molecule travels across the membrane through diffusion and finally emerges on the permeate side due to desorption from the membrane.

6.6 MEMBRANE TECHNOLOGY FOR GAS SEPARATION

The potential of membrane technology was well known long before the first commercial-scale gas separation membrane was introduced. Thomas Graham was the first to report the use of polymeric membranes for gas separation. Later, he developed Graham's law by experimenting with gas diffusion through a tube with a sealed end [12]. Loeb and Sourirajan's novel phase inversion method laid the foundation stone for the commercial application of gas membranes. The time line of membrane development is shown in Figure 6.3.

The development of membrane for CO$_2$ separation started in the early 1990s and has seen rapid commercialization, as currently more than 200 plants are operating worldwide. Cellulose acetate and polyimides are the dominating membrane materials at the commercial scale [13]. The largest natural gas processing facility using membrane technology was installed in Pakistan in 1995 using spiral wound modules, demonstrating the simplicity to scale up the membrane module [14].

6.6.1 ADVANTAGES OF MEMBRANE TECHNOLOGY

Various advantages have been documented for membrane technology applications in the last several decades. The membrane technology enables ease of operation and poses a small equipment footprint; hence, the basic membrane module may simply be scaled up or stacked in modular arrangement to meet the operational requirements. Additionally, the lack of phase transitions and chemical additives simplifies operation and manufacture. Other advantages include increased efficiency in the utilization of raw materials and the possibility of recycling by-products. Additionally, the

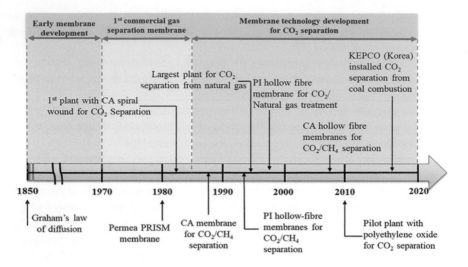

FIGURE 6.3 Time line of membrane development.

uninterrupted steady-state operation is achievable when gas separation membrane devices are used [15].

Finally, all the benefits outlined above contribute to cost-cutting and making the separation process more eco-friendly and more sustainable. Membrane methods are advantageous for handling a large quantity of impurities while maintaining a reasonable level of purity. Similarly, the supply stream must be sufficiently pressurized and devoid of contaminants that might damage the membranes.

6.6.2 Weaknesses of Membrane Technology

Despite the various developments and modifications of the membrane technology, inherent limitations such as membrane life, fouling, and plasticization still linger to reduce the separation performance. Fouling occurs when species adhere to the membrane's surface or become stuck in holes, blocking them. Sulfur-containing molecules, such as H_2S and SO_2, are among the species that foul gas separation membranes [16].

Gas separation membranes, which are typically polymeric, may also degrade due to compaction, as the pore diameters decrease because of pressurization. Since membrane degradation is irreversible (the loss being permanent), replacement of the membrane is the only option [17].

Numerous aspects are considered while designing an optimum membrane system. One of these issues is heat stress on the membrane's structural integrity, which might occur due to increasing temperature operation. As a result, consideration of thermal expansion throughout the system's design is critical for proper separation.

6.7 MEMBRANE MATERIALS FOR CO$_2$ SEPARATION MEMBRANES

The selection of a suitable membrane material is critical for satisfactory functioning of the membrane. This choice might be made based on the chemical or physical qualities of the substance. In each scenario, the critical factor to consider is the interaction between the gas molecules and the membrane material, as this contact is harmful to the separation process's overall efficiency [18].

6.7.1 INORGANIC MATERIALS FOR CO$_2$ SEPARATION

Inorganic membranes exhibit exceptional diffusivity-selectivity owing to their morphology (including the size and shape of the molecules). These membranes have sieve-like characteristics and hence perform better than polymeric membranes. Additionally, inorganic membranes have better thermal, chemical, and mechanical properties and membrane longevity [19].

Carbon molecular sieves and zeolite membranes are two kinds of inorganic membranes that exhibit superior separation properties due to their narrow pore size range (2–20 nm). Additionally, these membranes work better in high-pressure conditions (where CO$_2$ might cause swelling) and high temperatures [20–21].

Table 6.1 entails the CO$_2$ separation performance of various membranes. The zeolite membranes SAPO-34, MFI, and DDR have narrow pore size distribution of 0.40, 0.50, 0.36 nm, respectively, approaching the kinetic diameter of CO$_2$, and have shown better separation selectivity than silica carbon membranes [22]. Hasegawa *et al.* produced zeolite membranes of the FAU type using an α-alumina tube as a support layer. The outer surface of the support tube comprises a variable ratio of Si/Al

TABLE 6.1

Inorganic Membranes Used for CO$_2$/CH$_4$ and CO$_2$/N$_2$ Gas Pairs Reported in the Literature

Membrane	Pressure (bar)	Permeance CO$_2$	Selectivity CO$_2$/CH$_4$	Selectivity CO$_2$/N$_2$	Ref
SAPO-34 zeolite	2.2	1046 GPU	120	–	[24]
MFI	–	27 GPU	8	–	[25]
DDR zeolite	5	199 GPU	220	–	[22]
Carbonized (Polyamic acid)	1.6	6 GPU	70	–	[26]
Silica	1.5–4	750 GPU	326	–	[27]
Silica	–	8.1×10^{-8} mol.m^{-2} S^{-1} Pa^{-1}	–	8.9	[28]
DDR zeolite	5	293 GPU	500	–	[29]
Silicate	1	5755 GPU	2.4	–	[30]
Carbon	1	1.5×10^{-8} mol.m^{-2} S^{-1} Pa^{-1}	54	506	[31]

ions [23]. Later, an ion-exchange method was employed to replace the ions with Rb^+ and K^+ ions. The Na-zeolite containing Si/Al exhibited significant CO_2 selectivity, with 28 for the CO_2/CH_4 system. For the CO_2/N_2 system, the membrane showed a significantly higher selectivity of 78. It was established that the ion exchange effect was maximal for sodium-containing zeolite membranes that exchanged respective ions with potassium ions. Nevertheless, the selectivity of the potassium-zeolite membrane was found in the range of 25–40 along with the permeability in the range of 7.5–9.0 $\times 10^{-7}$ mol m^{-2} s^{-1} Pa^{-1}.

6.7.1.1 Limitations of Inorganic Membranes

Most of the porous membranes exhibit attractive permeation performance and superior thermal and chemical characteristics. However, inorganic membranes are expensive to fabricate, difficult to produce, and mechanically weak.

6.7.2 POLYMERIC MATERIALS FOR CO_2 SEPARATION

Polymeric materials are better than their counterparts since they possess higher chemical resistance, considerable thermal stability, and appreciable mechanical strength. To date, several organic polymers have been explored for membrane development, specifically for gas capturing applications [32–33]. These membranes can be glassy or rubbery, depending on the type of basic monomer. Due to the narrow intersegmental gaps, improved chain interaction, and crystalline structure of glassy polymers, they often have a low gas penetration and high intrinsic selectivity. Additionally, strong chain entanglements in glassy polymeric systems result in a transport of gas molecules dependent simply on their size and form.

On the contrary, rubbery polymers behave differently from glassy polymers. Due to their lack of crystallinity, an abundance of polar groups and a lower degree of crosslinking than glassy polymers, the rubbery polymers have high permeability but an imperfect selectivity. Such properties result in an unrestricted passage of gas molecules, resulting in a superior permeability than glassy polymers.

A long list of the polymers used in the manufacture of gas separation membranes can be compiled from the literature. Despite their potential for separation, polymeric membranes are confined to the Robeson upper bound performance curve. Polymer membranes are the most economically viable industrial gas separation membranes. The dense polymeric membranes for gas separation applications that are commercially available include polysulfone (PSF), cellulose acetate (CA), polyphenylene oxides (poly(2,6,-dimethyl-1,4-phenylene oxide), PPO, polyimides (PI) Matrimid, and polycarbonates (Tetrabromo Bisphenol A polycarbonate, TBBPC) as discussed in Table 6.2.

Commercial membranes made of cellulose acetate (CA) are used in enhanced oil recovery operations to extract carbon dioxide and hydrogen sulfide from natural gas and carbon dioxide from hydrocarbon mixtures. CA polymers contain semi-crystalline materials with varying degrees of substitution or acetylation of OH groups. The polymers are soluble in organic solvents due to the reduced hydrogen bonding. It was noticed that the degree of substitution or acetylation influences CO_2 permeability in

TABLE 6.2

Polymeric Membranes for CO_2/CH_4 and CO_2/N_2 Separations

Polymers	Permeability				Selectivity		Reference
	CO_2	N_2	CH_4	Units	CO_2/N_2	CO_2/CH_4	
PSF	5.6	0.25	0.25	Barrer	22.4	22.4	[35]
CA	4.8	0.15	0.15	Barrer	32	32	[34]
Matrimid	10	0.32	0.28	Barrer	31	36	[36]
PPO	61	4.1	4.3	Barrer	15	14	[37]
TBBPC	4.2	0.18	0.13	Barrer	23	32	[38]
PA	11	-	0.30	Barrer	-	36.3	[39]
6FDA-durene	458	-	28.44	Barrer	-	16.1	[40]
PDMS	3800	-	1198	Barrer	-	3.7	[41]
PC	11.7	0.58	0.66	GPU	20.18	17.66	[42]
PEI	0.74	-	-	GPU	-	43.1	[43]

CA polymers, as increasing the degree of acetylation from 1.75 to 2.85 enhanced the CO_2 permeability from 1.84 to 6.56 Barrer [34].

The increase in CO_2 permeability is due to a decrease in polymer packing density, resulting in an increased free volume. Thus, the substitution of polar hydroxyl group substitution with acetate groups reduced the polymer chain packing density. Despite being relatively inexpensive, CA membranes are prone to plasticization when subjected to CO_2 gas at high pressures, resulting in loss of CO_2/CH_4 selectivity.

Aromatic polyimides are attractive membrane materials known to possess good physical characteristics, high gas permeability, and high intrinsic selectivity. Matrimid is a polyimide type with good solubility in common organic solvents, a prerequisite for membrane fabrication [44]. The main limitation of Matrimid, aside from its high cost, is plasticization when exposed to condensable gas streams such as CO_2. For instance, Bos *et al.* [45] have reported a 45% decrease in selectivity of Matrimid membranes when the feed pressures of CO_2-CH_4 (55–45 mole%) mixed gas was increased from 5 to 50 bars.

Poly-2,6-dimethyl-1,4-phenylene oxide (PPO) is considered a high-performance thermoplastic, which offers appreciable mechanical strength along with thermo-oxidative stability. PPO exhibits relatively high CO_2 permeability and low selectivity, attributed to its conformational and chemical structure, precisely the ether bond's position. Gas diffusivity and permeability are substantial because of the availability of excessive free volume and flexibility caused by the phenyl rings around the ether linkages. In contrast, the absence of polar moieties in the main chain of polymers resulted in lower selectivity [46].

Limitations of polymeric membranes: In the gas separation process, despite the material flexibility, ease of processability, and good thermal and chemical stabilities, polymeric membranes face certain limitations, such as plasticization, permeability-selectivity trade-off, and physical aging.

Permeability-selectivity trade-off: Based on the structure-property data compiled from literature, there exists a trade-off (balance) relationship between permeability and selectivity.

Physical aging: Physical aging of polymeric membranes is another obstacle to industrial gas separation applications. Physical aging is considered as a phenomenon resulting from molecular level relaxation of polymer chains. It alters the physical properties of subsequent polymer on the macroscopic level, hence causing lower permeability.

Plasticization: Plasticization is a severe problem in gas separation membranes usually caused by a condensable CO_2 gas. Thus, when the CO_2 concentration in the polymer matrix increases, the polymer swells, increasing the free volume and chain mobility. The phenomenon leads to an increase in permeability and a loss in selectivity, particularly at high feed pressures.

6.7.3 MIXED MATRIX MEMBRANES (MMMs)

Mixed matrix membranes were found to remedy the drawbacks of pure polymeric and ceramic membranes. These membranes were fabricated by combining the material mentioned above and subsequently tuning their properties for particular applications [19]. In the 1970s two researchers, Paul and Kemp, combined 5A zeolite with polydimethylsiloxane (PDMS) to explore the diffusion lag for gases such as CO_2 and CH_4 [47]. Today, among many polymer types, Udel, Ultem, and Matrimid have been widely used to synthesize mixed matrix membranes by embedding various inorganic fillers to improve the gas separation properties [48–49].

Inorganic fillers are generally classified as porous and nonporous. Their selection depends on the knowledge of its shape, size, and interaction with the polymer and penetrant molecules, as shown in Figure 6.4 [50]. Porous type inorganic fillers

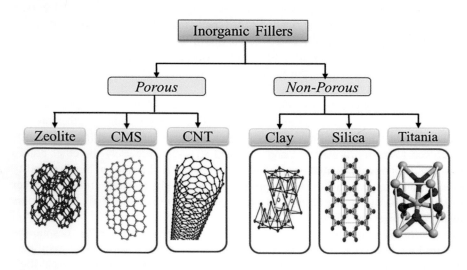

FIGURE 6.4 Classification of inorganic fillers.

include zeolite, carbon molecular sieve, and carbon nanotubes, whereas nano-clay, silica, and metal-organic frameworks are nonporous type inorganic fillers [51–52].

In general, gas molecules are transported in MMM based on their solubility. For CO$_2$/CH$_4$ gas mixture, CO$_2$ is more soluble than CH$_4$ for its lower critical temperature in addition to its lower kinematic diameter (0.33 nm). Solubility is a thermodynamic property and depends on the polarity of incident gas and polymer matrix. Moreover, certain polymers with functional groups like ether, oxygen, nitrile, and acetate have the highest selectivity for CO$_2$ solubility. This feature can be confirmed by comparing the concentration of ether's oxygen in polybutadiene to polytetraethylene and polyethylene oxide. Hence, CO$_2$ solubility also changes in the same order as that of ether's oxygen in the aforementioned polymer types. Similarly, selectivity of CO$_2$/CH$_4$ solubility increases with the concentration of carbonyl and sulfone group [53–54].

Table 6.3 analyzes the performance of MMMs incorporating different fillers found in the literature. Nonporous filler, i.e., Cloisite 15 A incorporated in PEI matrix, showed excellent selectivity of 50 even at high pressure of 15 bars. Zeolite has been extensively included in various polymers to synthesize the MMMs for separating CO$_2$ from CH$_4$ gas [55]. It was found that the addition of Zeolite into PSF/Matrimid hollow fiber membranes resulted in enhanced selectivity up to 50%. Likewise, incorporating the CMS into Matrimid and Ultem resulted in a 45% and 40% improvement in membrane selectivity, respectively, for CO$_2$/CH$_4$ gas mixture [36]. This increase in membranes' performance is due to the porous nature of the incorporated filler, which provides an additional sieving, thus improving the selectivity.

Anjum et al. embedded varying concentrations of carbon silica nanocomposite as a filler in Matrimid® 9725 for gas separation studies [64]. Their study reported a uniform distribution of filler particles due to a better adhesion of the filler particles with polymers. They suggested that the addition of carbon-based filler particles might have provided extra porosity and free volume, which helped to improve the separation efficiency. In their study, the permeability and selectivity achieved for the gas mixture of CO$_2$/N$_2$ at 50:50 feed composition were 27 Barrer and 42.5, respectively.

Limitations of mixed matrix membranes: Literature study depicts that researchers in successful MMM formation face various challenges. This section presents the problems and proposed solutions associated with successful MMM development.

Filler dispersion in the organic phase: One of the key challenges to the MMMs has been the homogeneous and uniform dispersion of inorganic fillers in the organic matrix since these fillers tend to agglomerate in the MMMs [65]. Thus, agglomeration of filler particles is avoided, preventing the formation of stress spots in MMM that might compromise the mechanical stability of succeeding MMM. Typically, this behavior occurs at high filler concentrations.

Polymer–inorganic filler compatibility: The filler compatibility with the polymer matrix is another critical component in the creation of MMM. Compatibilizers are thus included to facilitate chemical interactions between the organic and inorganic filler phases, hence generating enough compatibility between the two phases [66–67].

MMMs surface defects: MMM performance is susceptible to imperfections on the membrane surface; therefore, it is desired to fabricate flawless membranes for improved separation. Asymmetric membranes (i.e., integrally skinned membranes)

TABLE 6.3

CO_2/CH_4 and CO_2/N_2 Separation of Mixed Matrix Membranes

Polymers	Filler	Permeability				Selectivity		Reference
		CO_2	N_2	CH_4	Units	CO_2/N_2	CO_2/CH_4	
PSF	Silica	90.04	2.5	2.75	GPU	36	32.8	[56]
PI	Silica	80	5	2.16	Barrer	16	37	[57]
Matrimid	$Cu_3(BTC)_2$	17	–	0.74	GPU	–	23	[58]
Polyurethane	Alumina	74.67	–	3.18	Barrer	–	23.48	[59]
PEI	Cloisite	0.93	–	0.02	GPU	–	50.0	[60]
CA	Bentonite	6.29	0.64	0.52	Barrer	9.83	13.1	[3]
6FDA-DAM	ZIF-90	720	–	–	Barrer	–	37	[61]
Ultem	Zeolite	6.8	–	0.14	GPU	–	46.9	[62]
Matrimid 9725	CMS	27	–	0.64	Barrer	42.5	–	[62]
PEI	SAPO-34	1.3	–	0.02	Barrer	–	60	[63]

typically contain a top-thin selection layer and a very porous bottom support layer. These two layers, when combined, determine the membrane's overall performance in terms of selectivity and permeability. Notably, the skin layer is responsible for most gas molecule movement over the membrane surface, while the porous support offers mechanical stability. Because the skin layer controls the diffusion and penetration of penetrant gas molecules, a defect-free surface becomes a need for separation efficiency. [68–69]

Plasticization: Due to the polymeric phase included in MMMs, plasticization poses the same danger to the MMM's performance as a polymer membrane. Plasticization causes polymer chains to grow to the point where they become unsuitable for separation due to an abundance of CO_2 molecules around them. Generally, all polymers with polar groups assist this process, as CO_2 is extremely polarizing. The presence of $-OCOCH_3$ and $-COOCH_3$ groups on the surfaces of polymethyl methacrylate and polyethyl methacrylate is a famous example. Thus, the presence of these groups results in the plasticization of these polymers when CO_2 concentrations are increased [70–71].

6.8 ACQUIRED SUSTAINABILITY AND APPLICATIONS OF GAS SEPARATION MEMBRANES

For separation processes, the United States alone consumes 4500 trillion Btu of energy annually, accounting for 22% of total in-plant energy consumption. When it comes to separation procedures, distillation consumes the greatest energy. Nearly 40,000 distillation columns are utilized in the United States for 200 distinct operations, accounting for 49% of total industrial energy required for separation operations [72]. Additional separation techniques include absorption, crystallization, and membrane separation. On the other hand, membranes are more attractive than the

TABLE 6.4

Industrial Applications of Gas Separation Membranes

Gas Pair	Membranes	Applications
CO_2/CH_4	Prism (PSF hollow fiber)	Acid gas treatment
N_2/O_2	Generon (PC hollow fiber)	Nitrogen enrichment
H_2/N_2	Cynara (CA hollow fiber)	Ammonia purge gas recovery
H_2/CH_4	CMS membranes (lab scale)	Refinery gas purification
H_2/CO	Silicon rubber, polyimide	Syngas ratio adjustment
H_2O/air	–	Dehydration
$CO_2/flue$ gas	KEPCO	Coal combustion

other two processes since they do not require sorbents or a thermal driving force to separate gas mixtures.

Membranes are used in the industrial sector to purify refinery gas, remove acid gases, enrich nitrogen, recycle ammonia purge gas, dehydrate, and synthesize oxo-chemicals [73–74]. Table 6.4 illustrates representative gas pairs that must be isolated in these applications.

Merkel et al. demonstrated in their research that a membrane with CO_2 permeance of 4000 GPU and a CO_2/N_2 selectivity of more than 50 could capture CO_2 for less than \$15 per ton. In comparison, the US Department of Energy (DOE) established a CO_2 collection target of \$20/ton [75]. Similarly, Hussain et al. examined the influence of process parameters on energy consumption and cost by utilizing a novel CO_2-selective fixed-site carrier (FSC) membrane and in-house developed membrane software [76]. According to their research, a membrane system comprising high-performance FSC membranes could absorb CO_2 from a flue gas stream with a low CO_2 concentration (10%). Compared to amine absorption, the system could collect more than 90% of CO_2 with a purity of over 95% in the permeate stream.

Peters et al. used HYSYS integrated with an in-house membrane software to perform process design, simulation, and optimization for CO_2 removal from natural gas [77]. They claimed that a two-stage membrane system with a CO_2/CH_4 selectivity of 40 and CO_2 permeance of 0.3 m^3 is comparable to an amine phase. However, the CH_4 purity of 98% for sweet gas is lower than that of the amine-based method (99.5%), which is lower than the gas sales requirement (less than 2% CO_2) in natural gas. Bhide et al. devised a hybrid approach that combines a membrane system for bulk CO_2 removal from the crude natural gas supply with an amine unit for final purification to satisfy the international pipeline requirements [78]. Combining a membrane system with an amine unit might give a low-cost alternative to all-amine or all-membrane plants [79–80].

6.9 NOVEL MATERIALS FOR CO_2 SEPARATION MEMBRANES

Membrane technology has developed new materials and enhanced existing classes of gas separation membranes in past years. Thermally rearranged (TR) polymers, polymers with inherent microporosity (PIMs), polymerized room temperature ionic liquids

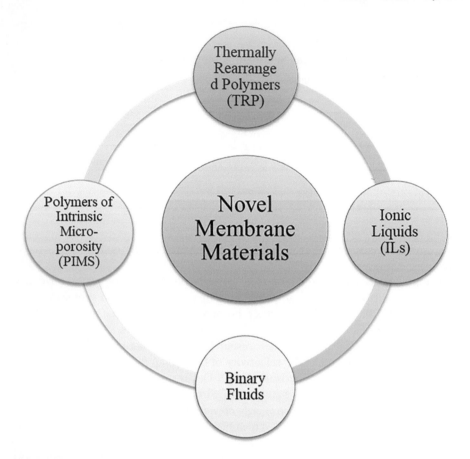

FIGURE 6.5 Novel material trends in CO_2 separation membranes.

(poly (RTIL)s), and perfluoropolymers are among the latest classes of materials that will be explored in this section and depicted in Figure 6.5. In addition, novel fillers for MMMs are also emerging materials for high-performance CO_2 separation membranes.

6.9.1 THERMALLY REARRANGED POLYMERS

A new polymeric membrane called TR polymers was identified for CO_2 capturing [31]. These materials exhibited increased CO_2 permeability and CO_2/CH_4 selectivity, in addition to exceptional resistance to CO_2-induced plasticization. A well-known example of this type of polymer is the TR-1 polymer. It is prepared by reacting a fluorinated diamine with dianhydride. This polymer has a CO_2 permeability of around 2000 Barrer and a CO_2/CH_4 selectivity of around 40, with no evidence of plasticization up to a pressure of 15.2 bars.

TR polymers of polybenzoxazole (PBO) are thought to be formed via molecular thermal rearrangement of aromatic polyimides with ortho hydroxyl groups on the imide ring. One such example is the thermal conversion of aromatic poly(hydroxyimide)s to PBOs when heated to elevated temperatures in an inert environment, which results in the release of CO$_2$. Apart from the TR method's excellent membrane efficiency, another notable benefit is that it overcomes the characteristic insolubility of PBOs by starting with a soluble precursor, enabling industrial processing such as hollow fiber spinning [81]. In Figure 6.6, TR polymers frequently exhibit transport capabilities that exceed the CO$_2$/CH$_4$ upper bound, making them one of the greatest natural gas production products currently accessible.

Following the TR procedure, polyimides with ortho-functional groups other than hydroxyl groups showed much higher permeability than TR polymers made from their poly(hydroxyimide) analogs. Gas permeability rose with the size of the ortho-functional group, although selectivity dropped marginally. This ortho-position group is eliminated as the temperature of the thermal rearrangement increases, resulting in a considerable rise in permeability [82]. As a result, the ortho-position group of the TR polymer precursor may be modified to alter TR polymer permeability and selectivity.

As a result, TR polymers show promise as high-performance membrane materials for gas separation. If the conversion temperature could be decreased to a more

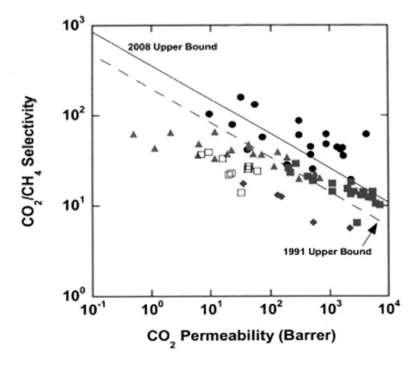

FIGURE 6.6 Separation characteristics of various novel emerging polymer systems for (CO$_2$, CH$_4$) gas pair.

Note: **Separation characteristics of various novel emerging polymer systems for (CO$_2$, CH$_4$) gas pair. TR polymers (•). PIMs (■). Perfluoropolymers (♦). Polyimides (▲). Poly(RTIL)s (□) [72].**

energy-efficient level and the monomers become more readily accessible, their practical applications may be significantly expanded.

6.9.2 Polymers of Intrinsic Microporosity (PIMs)

Numerous attempts have been made to synthesize organic or organic-inorganic composite materials that replicate the structure of zeolites. Following that, Budd *et al.* discovered a new class of nonnetwork polymers with 'intrinsic microporosity' (PIMs). The 'microporous' composition of PIMs is impacted by their very stiff and deformed molecular architectures, which prevent macromolecules from packing efficiently in the solid state [83–84]. PIMs are identified by their kinked backbones and severely limited rotational movements of the backbone. The term 'intrinsic microporosity' refers to the property of PIMs that emerges solely from their molecular characteristics and is indifferent to the polymer's temperature or manufacturing history. The preferred approach for producing a scaffold for PIMs is to combine an aromatic tetra-pyrrol with an active halogen-forming monomer via a double nucleophilic aromatic substitution (SNAr) reaction to generate dibenzodioxane [72]. The reactivity of the aromatic halides greatly influences the reactivity of the double substitution. The aromatic halides' reactivity has a significant impact on the double substitution's reactivity. For CO_2/CH_4 and CO_2/N_2 gas pairings, PIM-1 and PIM-7 both display transport characteristics substantially above the Robeson curves [85]. Aside from that, several PIMs with different backbones and pendant groups have been created, and they have shown increased permeability but lower selectivity.

Tetrazole-containing PIMs (TZPIMs), on the other hand, do not enhance CO_2/CH_4 and CO_2/N_2 mixed gas selectivity in mixed gas settings. TZPIMS demonstrated selectivity of around 17.5 at a CO_2 partial pressure of 10 bars in both pure gas experiments and measurements using a 50:50 CO_2/CH_4 mixed gas stream. Additionally, the CO_2/N_2 mixed gas selectivity was 37.5 at 10 atm CO_2 partial pressure, larger than the CO_2/CH_4 selectivity of 27.5 under the same experimental circumstances for pure gas tests. These selectivity-enhancing characteristics are related to solubility effects [72]. As a result, these PIMs keep their separation efficiency even in a mixed gas environment. However, further study is necessary to understand better the influence of contaminants on mixed gas permeation products. Additionally, physical aging at thicknesses of less than 1 micron, molecular weights, and mechanical activity of PIMs all require more investigation.

Membranes combining PIM-1 and Troger's base polymer were made with a high degree of compatibility [86]. The PIM-1/TB (2/8) has the highest ideal selectivity of 16.2 for the (CO_2, CH_4) gas combination, with a CO_2 flow of 727 Barrer. The compatibility and homogeneity of numerous components are two of the most difficult aspects of the blend membrane production process. PIM-1/Am-PAFEK mix membranes, on the other hand, were made by solution casting. PIM-1/Am-PAFEK membrane systems have optimum selective permeabilities of 26.7 and 24.3 for CO_2/N_2 and CO_2/CH_4. The CO_2 permeability of the membrane was 752 Barrer, which was higher than the upper limits. PIMs have a lot of promise for usage in commercial gas separation membranes as an organic filler [87].

6.9.3 Ionic Liquids as Filler for MMMs

The salts that occur in a liquid state are known as ionic liquids (ILs). Although ILs were discovered a long time ago, their applications were limited until the late 1990s, with only a few examples in electrochemistry and organic chemistry [62]. ILs have several advantages over traditional volatile organic solvents, including the ability to retain liquid form even at lower temperatures (<100 °C). In addition, ILs exhibited low vapor pressure, superior thermal resilience, and nonflammability [88]. High viscosity, possible environmental impact, and high manufacturing costs are the key factors that restrict the applications of ILs. There are two different kinds of ILs. The first is RTILs (room temperature ILs), and the second is TSILs (task-specific ILs) or functionalized ILs [89–90].

Fam *et al.* developed hollow fiber TFC Pebax/(emim)(BF4) gel membranes with an IL loading of 80 wt% on the PTMSP gutter layer and a PVDF support layer using a scalable and easy dip-coating process [91]. From free-standing films to thin films, these membranes have shown remarkable separation characteristics and operation stability. At 35 °C and 3 bars, the CO$_2$ permeance of TFC Pebax/IL 80 gel membranes was 305.8 GPU, with a selectivity of 36.4 and 15.2 for CO$_2$/N$_2$ and CO$_2$/CH$_4$ gas pairs, respectively.

Bara *et al.* synthesized RTIL monomers of various length n-alkyl substituents and processed them into polymer films [92]. The separation of CO$_2$, N$_2$, and CH$_4$ was evaluated using these membranes, and it was observed that introducing n-alkyl substituents increased CO$_2$ permeability in a nonlinear fashion. The efficiency of CO$_2$/N$_2$ separation remained stable as CO$_2$ permeability rose. Their performance was determined using the CO$_2$/N$_2$ Robeson plot, demonstrating that this first-generation poly (RTIL) maintains the chart's upper bond and outperforms a wide variety of other polymeric membranes.

6.9.4 Binary Filler Incorporation in MMMs

As previously mentioned, adding inorganic filler to the base polymer is supposed to increase the system performance and physicochemical properties of polymeric membranes. In most cases, though, the single filler may only improve one of the two properties, permeability or selectivity, but not both [93]. Furthermore, producing MMM with nano-fillers has caused serious agglomeration problems in the host polymers. In general, organo-functionalization of filler particles is used to address this problem, and literature indicates that the newer functionalities have increased filler particle dispersion by reducing intermolecular interaction [4].

However, scaling up presents a problem in controlling the degree of functionalization on the filler surface. Consequently, a new method for combining two different fillers with different properties, morphologies, design, and dimensions has been established. This procedure involves combining two distinct filler particles into a host polymer in different proportions to produce the new MMM. The polymer-filler interaction is thought to be enhanced by this formulation, resulting in less agglomeration and increased membrane efficiency [94].

Zornoza *et al.* developed a binary filler-based MMM for CO_2/CH_4 separation by combining HKUST-1 and ZIF-8 into polysulfone [95]. The combined effect of adding two fillers with different properties was observed to disperse the filler particles more homogeneously, enhancing the polymer matrix's physicochemical properties and gas separation efficiency. At 8/8 wt% of HKUST-1/ZIF-8 filler loadings, the optimum CO_2/CH_4 separation 22.4 was reported at 8.9 Barrer of CO_2 permeability. Sarfraz *et al.* used CNT and ZIF-302 binary filler to create a high-performance MMM. The resulting membrane showed a 177% and 218% increase in CO_2 permeability and CO_2/N_2 separation at a loading of 8/12 wt% CNT/ZIF-302, respectively, crossing the Robeson upper bound boundary [96]. The resulting membrane was thermally and mechanically stable, with uniform filler dispersion content and no polymer-filler interfacial defects.

In contrast to single filler-based MMM, binary filler-based MMM has significantly increased thermal stability and mechanical efficiency [97]. As a result, this technique of combining two fillers in a polymer matrix has a great deal of potential for gas separation membrane progress.

6.10   FUTURE TRENDS AND CONCLUDING REMARKS

Membrane technology is a promising emerging separation technology that can be used for various applications, including CO_2 capture and separation applications. Membrane technology, which uses various types and geometries of membranes, will dominate as a separation technology for improved CO_2 separation performance.

Importantly, the durability of the membrane materials is insufficient to reach industrial standards or implementations. This simple reality has stifled the development of this technology in the separation industry. As a result, one of the essential characteristics of advanced membrane materials is their toughness, which allows them to survive harsh environments such as elevated temperatures and pressures. In addition to the ability to cope with large feed volumes of penetrant gases, durability would be critical in acquiring wider acceptance of these membranes.

It is essential to develop a membrane from porous inorganic materials that can have better overall performance in terms of permeability and selectivity over a long span of membrane life in the presence of undesirable components like water, H_2S, and C_2, C_4 hydrocarbons. Further research in porous inorganic type gas membranes demands higher selectivity at a reduced operating temperature lower than 500°C and permeability better than that of nonporous type membranes. Another hurdle in developing such porous membranes with superior features is the fabrication of defect-free membrane modules.

Similar challenges are being faced by researchers interested in developing mixed matrix membranes that have zero defects, improved separation, higher permeability, and an upgraded interaction between filler and polymer matrix. Improvement in all the aforementioned aspects for mixed matrix membranes could only cement them as a successful alternative to the existing membrane types. For example, MMM containing glassy polymeric membranes face compatibility issues and plasticization. Apart from these, new categories of materials are being proposed to enhance the separation performance of the membranes, such as TR polymers, PIMs, poly(RTIL)s, and binary filler-containing membranes. The new developments have shown tremendous performance so far; however, further insight is required to enhance the membrane performance and thermal stability and to commercialize the potential of the membranes.

REFERENCES

1. J. K. Adewole, A. L. Ahmad, S. Ismail and C. P. Leo, "Current challenges in membrane separation of CO$_2$ from natural gas: A review," *International Journal of Greenhouse Gas Control*, vol. 17, pp. 46–65, 2013.
2. A. Kargari and S. Rezaeinia, "State-of-the-art modification of polymeric membranes by PEO and PEG for carbon dioxide separation: A review of the current status and future perspectives," *Journal of Industrial and Engineering Chemistry*, vol. 84, pp. 1–22, 2020.
3. A. Jamil, O. P. Ching, M. Naqvi, H. A. Aslam Khan and S. R. Naqvi, "Polyetherimide-montmorillonite nano-hybrid composite membranes: CO$_2$ permeance study via theoretical models," *Processes*, vol. 8, 2020.
4. A. Jamil, P. C. Oh and A. M. Shariff, "Polyetherimide-montmorillonite mixed matrix hollow fibre membranes: Effect of inorganic/organic montmorillonite on CO$_2$/CH$_4$ separation," *Separation and Purification Technology*, vol. 206, pp. 256–267, 2018.
5. T. A. Saleh and V. K. Gupta, "Membrane classification and membrane operations," *Nanomaterial and Polymer Membranes*, pp. 55–82, 2016.
6. P. Pandey and R. S. Chauhan, "Membranes for gas separation," *Progress in Polymer Science (Oxford)*, vol. 26, pp. 853–893, 2001.
7. C. W. Jones and W. J. Koros, "Carbon molecular sieve gas separation membranes-I. Preparation and characterization based on polyimide precursors," *Carbon*, vol. 32, pp. 1419–1425, 1994.
8. W. J. Koros and G. K. Fleming, "Membrane-based gas separation," *Membrane Technology*, vol. 2001, p. 16, 2001.
9. M. Knudsen, "The laws of molecular flow and of inner friction flow of gases through tubes," *Journal of Membrane Science*, vol. 100, pp. 23–25, 1995.
10. J. D. Perry, K. Nagai and W. J. Koros, "Polymer membranes for hydrogen separations," *MRS Bulletin*, vol. 31, pp. 745–749, 2006.
11. R. Mahajan and W. J. Koros, "Mixed matrix membrane materials with glassy polymers: Part 2," *Polymer Engineering and Science*, vol. 42, pp. 1432–1441, 2002.
12. E. A. Mason, "From pig bladders and cracked jars to polysulfones: An historical perspective on membrane transport," *Journal of Membrane Science*, vol. 60, pp. 125–145, 1991.
13. D. F. Mohshim, H. B. Mukhtar, Z. Man and R. Nasir, "Latest Development on Membrane Fabrication for Natural Gas Purification: A Review," *Journal of Engineering (United Kingdom)*, vol. 2013, 2013.
14. Z. Y. Yeo, T. L. Chew, P. W. Zhu, A. R. Mohamed and S. P. Chai, "Conventional processes and membrane technology for carbon dioxide removal from natural gas: A review," *Journal of Natural Gas Chemistry*, vol. 21, pp. 282–298, 2012.
15. Z. Dai, L. Ansaloni and L. Deng, "Recent advances in multi-layer composite polymeric membranes for CO$_2$ separation: A review," *Green Energy and Environment*, vol. 1, pp. 102–128, 2016.
16. W. Guo, H. H. Ngo and J. Li, "A mini-review on membrane fouling," *Bioresource Technology*, vol. 122, pp. 27–34, 2012.
17. M. Wessling, S. Schoeman, T. van der Boomgaard and C. A. Smolders, "Plasticization of gas separation membranes," *Gas Separation and Purification*, vol. 5, pp. 222–228, 1991.
18. T. Visser, N. Masetto and M. Wessling, "Materials dependence of mixed gas plasticization behavior in asymmetric membranes," *Journal of Membrane Science*, vol. 306, pp. 16–28, 2007.
19. P. Bernardo, E. Drioli and G. Golemme, "Membrane Gas Separation: A Review/State of the Art," *Ind. Eng. Chem. Res.* 2009, 48, 4638–4663.

20. R. Nasir, H. Mukhtar, Z. Man and D. F. Mohshim, "Material advancements in fabrication of mixed-matrix membranes," *Chemical Engineering and Technology*, vol. 36, pp. 717–727, 2013.

21. M. Mubashir, Y. F. Yeong and K. K. Lau, "Ultrasonic-assisted secondary growth of deca-dodecasil 3 rhombohedral (DD3R) membrane and its process optimization studies in CO_2/CH_4 separation using response surface methodology," *Journal of Natural Gas Science and Engineering*, vol. 30, pp. 50–63, 2016.

22. T. Tomita, K. Nakayama and H. Sakai, "Gas separation characteristics of DDR type zeolite membrane," *Microporous and Mesoporous Materials*, vol. 68, pp. 71–75, 2004.

23. Y. Hasegawa, T. Tanaka, K. Watanabe, B. H. Jeong, K. Kusakabe and S. Morooka, "Separation of CO_2-CH_4 and CO_2-N_2 Systems Using Ion-Exchanged FAU-type Zeolite Membranes with Different Si/Al Ratios," *Korean Journal of Chemical Engineering*, vol. 19, pp. 309–313, 2002.

24. B. S. Li, J. L. Falconer and R. D. Noble, "Improved SAPO-34 Membranes for CO_2/CH_4 Separations," pp. 2601–2603, 2006. [Online]. Available: https://onlinelibrary.wiley.com/doi/abs/10.1002/adma.200601147.

25. M. C. Lovallo, A. Gouzinis and M. Tsapatsis, "Synthesis and characterization of oriented MFI membranes prepared by secondary growth," *Reactors, Kinetics, and Catalysis*, vol. 44, pp. 1903–1913, 1998.

26. M. Ogawa and Y. Nakano, "Separation of CO_2/CH_4 mixture through carbonized membrane prepared by gel modification," *The Journal of Membrane Science*, vol. 173, pp. 123–132, 2000.

27. R. M. D. Vos and H. Verweij, "Improved performance of silica membranes for gas separation," *The Journal of Membrane Science*, vol. 143, 1998.

28. E. Hayakawa and S. Himeno, "Microporous and mesoporous materials synthesis of all-silica ZSM-58 zeolite membranes for separation of CO_2/CH_4 and CO_2/N_2 gas mixtures," *Microporous and Mesoporous Materials*, vol. 291, p. 109695, 2020.

29. J. van den Bergh, W. Zhu and F. Kapteijn, "Separation of CO_2 and CH_4 by a DDR membrane," *Research on Chemical Intermediates*, vol. 34, pp. 467–474, 2008.

30. L. J. P. van den Broeke, W. J. W. Bakker, F. Kapteijn, and J. A. Moulijn, "Transport and separation properties of a silicalite-1 membrane – I: Operating conditions," *Chemical Engineering Science*, vol. 54, 1999.

31. J. Park, N. F. Attia, M. Jung, M. Eun, K. Lee, J. Chung *et al.*, "Sustainable nanoporous carbon for CO_2, CH_4, N_2, H_2 adsorption and CO_2/CH_4 and CO_2/N_2 separation," *Energy*, vol. 158, pp. 9–16, 2018.

32. A. Jamil, O. P. Ching and A. B. M. Shariff, "Current status and future prospect of polymer-layered silicate mixed-matrix membranes for CO_2/CH_4 separation," *Chemical Engineering and Technology*, vol. 39, pp. 1393–1405, 2016.

33. A. Idris, Z. Man, A. S. Maulud and M. S. Khan, "Effects of Phase Separation Behavior on Morphology and Performance of Polycarbonate Membranes," *Membranes*, vol. 7, 2017.

34. A. C. Puleo, D. R. Paul and S. S. Kelley, "The effect of degree of acetylation on gas sorption and transport behavior in cellulose acetate," *Journal of Membrane Science*, vol. 47, pp. 301–332, 1989.

35. C. L. Aitken, W. J. Koros and D. R. Paul, "Effect of structural symmetry on gas transport properties of polysulfones," *Macromolecules*, vol. 25, pp. 3424–3434, 1992.

36. D. Q. Vu, W. J. Koros and S. J. Miller, "Effect of condensable impurity in CO_2/CH_4 gas feeds on performance of mixed matrix membranes using carbon molecular sieves," *Journal of Membrane Science*, vol. 221, pp. 233–239, 2003.

37. W. J. Koros, G. K. Fleming, S. M. Jordan, T. H. Kim, and H. H. Hoehn, "Polymeric membrane materials solution-diffusion based permeation separations membrane science and polymer science have grown synergistically over the past thirty years: Clearly,

today's impressive set of membrane processes and products could not exist," *Progress in Polymer Science*, vol. 13, pp. 339–401, 1988.

38. N. Muruganandam, W. J. Koros and D. R. Paul, "Gas sorption and transport in substituted polycarbonates," *Journal of Polymer Science, Part B: Polymer Physics*, vol. 25, pp. 1999–2026, 1999.

39. S. Sridhar, T. M. Aminabhavi and M. Ramakrishna, "Separation of binary mixtures of carbon dioxide and methane through sulfonated polycarbonate membranes," *Journal of Applied Polymer Science*, vol. 105, 2007.

40. Y. Liu, M. Lin, T. S. Chung and R. Wang, "Effects of amidation on gas permeation properties of polyimide membranes," *Journal of Membrane Science*, vol. 214, pp. 83–92, 2003.

41. N. M. Jose, "Synthesis, characterization, and permeability evaluation of hybrid organic-inorganic films," *Journal of Polymer Science Part B: Polymer Physics*, pp. 4281–4292, 2004.

42. M. Jamshidi, V. Pirouzfar, R. Abedini and M. Z. Pedram, "The influence of nanoparticles on gas transport properties of mixed matrix membranes: An experimental investigation and modeling," *Korean Journal of Chemical Engineering*, vol. 34, pp. 829–843, 2017.

43. A. K. Zulhairun, A. F. Ismail, T. Matsuura, M. S. Abdullah and A. Mustafa, "Asymmetric mixed matrix membrane incorporating organically modified clay particle for gas separation," *Chemical Engineering Journal*, vol. 241, pp. 495–503, 2014.

44. D. T. Clausi and W. J. Koros, "Formation of defect-free polyimide hollow fiber membranes for gas separations," *Journal of Membrane Science*, vol. 167, pp. 79–89, 2000.

45. A. Bos, I. G. M. Pünt, M. Wessling and H. Strathmann, "Plasticization-resistant glassy polyimide membranes for CO$_2$/CO$_4$ separations," *Separation and Purification Technology*, vol. 14, pp. 27–39, 1998.

46. M. Aguilar-Vega and D. R. Paul, "Gas transport properties of polyphenylene ethers," *Journal of Polymer Science Part B: Polymer Physics*, vol. 31, pp. 1577–1589, 1993.

47. D. R. Paul and D. R. Kemp, "Containing Adsorptive Fillers," *Journal of Polymer Science: Polymer Symposia*, vol. 93, pp. 79–93, 1973.

48. C. M. Zimmerman, A. Singh and W. J. Koros, "Tailoring mixed matrix composite membranes for gas separations," *Journal of Membrane Science*, vol. 137, pp. 145–154, 1997.

49. G. Dong, H. Li and V. Chen, "Plasticization mechanisms and effects of thermal annealing of Matrimid hollow fiber membranes for CO$_2$ removal," *Journal of Membrane Science*, vol. 369, pp. 206–220, 2011.

50. P. S. Goh, A. F. Ismail, S. M. Sanip, B. C. Ng and M. Aziz, "Recent advances of inorganic fillers in mixed matrix membrane for gas separation," *Separation and Purification Technology*, vol. 81, pp. 243–264, 2011.

51. A. Jamil, O. P. Ching and A. B. M. Shariff, "Polymer-nanoclay mixed matrix membranes for CO$_2$/CH$_4$ separation: A review," *Applied Mechanics and Materials*, vol. 625, pp. 690–695, 2014.

52. K. K. Youn, B. P. Ho and M. L. Young, "Preparation and characterization of carbon molecular sieve membranes derived from BTDA-ODA polyimide and their gas separation properties," *Journal of Membrane Science*, vol. 255, pp. 265–273, 2005.

53. A. L. Khan, C. Klaysom, A. Gahlaut, A. U. Khan and I. F. J. Vankelecom, "Mixed matrix membranes comprising of Matrimid and -SO$_3$H functionalized mesoporous MCM-41 for gas separation," *Journal of Membrane Science*, vol. 447, pp. 73–79, 2013.

54. D. Bera, P. Bandyopadhyay, S. Ghosh and S. Banerjee, "Gas transport properties of aromatic polyamides containing adamantyl moiety," *Journal of Membrane Science*, vol. 453, pp. 175–191, 2014.

55. Y. Zhang, J. Sunarso, S. Liu and R. Wang, "International Journal of Greenhouse Gas Control Current status and development of membranes for CO$_2$/CH$_4$ separation: A review," *International Journal of Greenhouse Gas Control*, vol. 12, pp. 84–107, 2013.

56. M. F. A. Wahab, A. F. Ismail and S. J. Shilton, "Studies on gas permeation performance of asymmetric polysulfone hollow fiber mixed matrix membranes using nanosized fumed silica as fillers," *Separation and Purification Technology*, vol. 86, pp. 41–48, 2012.

57. C. Hibshman, C. J. Cornelius and E. Marand, "The gas separation effects of annealing polyimide-organosilicate hybrid membranes," *Journal of Membrane Science*, vol. 211, pp. 25–40, 2003.

58. S. Basu, A. Cano-odena and I. F. J. Vankelecom, "MOF-containing mixed-matrix membranes for CO_2/CH_4 and CO_2/N_2 binary gas mixture separations," *Separation and Purification Technology*, vol. 81, pp. 31–40, 2011.

59. E. Ameri, M. Sadeghi, N. Zarei and A. Pournaghshband, "Enhancement of the gas separation properties of polyurethane membranes by alumina nanoparticles," *Journal of Membrane Science*, vol. 479, pp. 11–19, 2015.

60. S. A. Hashemifard, A. F. Ismail and T. Matsuura, "Effects of montmorillonite nano-clay fillers on PEI mixed matrix membrane for CO_2 removal," *Chemical Engineering Journal*, vol. 170, pp. 316–325, 2011.

61. T. H. Bae, J. S. Lee, W. Qiu, W. J. Koros, and C. W. Jones, "MOF membranes a high-performance gas-separation membrane containing submicrometer-sized metal-organic framework crystals," *Angewandte Chemie*, pp. 10059–10062, 2010.

62. M. Chawla, H. Saulat, M. Masood Khan, M. Mahmood Khan, S. Rafiq, L. Cheng *et al.*, "Membranes for CO_2/CH_4 and CO_2/N_2 gas separation," *Chemical Engineering and Technology*, vol. 43, pp. 184–199, 2020.

63. S. Belhaj, A. Takagaki, T. Sugawara, R. Kikuchi and S. T. Oyama, "Mixed matrix membranes using SAPO-34/polyetherimide for carbon dioxide/methane separation," *Separation and Purification Technology*, vol. 148, pp. 38–48, 2015.

64. M. Waqas Anjum, F. de Clippel, J. Didden, A. Laeeq Khan, S. Couck, G. V. Baron *et al.*, "Polyimide mixed matrix membranes for CO_2 separations using carbon-silica nano-composite fillers," *Journal of Membrane Science*, vol. 495, pp. 121–129, 2015.

65. P. C. Oh and N. A. Mansur, "Effects of aluminosilicate mineral nano-clay fillers on polysulfone mixed matrix membrane for carbon dioxide removal," *Jurnal Teknologi (Sciences and Engineering)*, vol. 69, pp. 23–27, 2014.

66. E. Picard, H. Gauthier, J. F. Gérard and E. Espuche, "Influence of the intercalated cations on the surface energy of montmorillonites: Consequences for the morphology and gas barrier properties of polyethylene/montmorillonites nanocomposites," *Journal of Colloid and Interface Science*, vol. 307, pp. 364–376, 2007.

67. V. Goodarzi, S. H. Jafari, H. A. Khonakdar, S. A. Monemian and M. Mortazavi, "An assessment of the role of morphology in thermal/thermo-oxidative degradation mechanism of PP/EVA/clay nanocomposites," *Polymer Degradation and Stability*, vol. 95, pp. 859–869, 2010.

68. A. F. Ismail and P. Y. Lai, "Development of defect-free asymmetric polysulfone membranes for gas separation using response surface methodology," *Separation and Purification Technology*, vol. 40, pp. 191–207, 2004.

69. D. Huang, B. Mu and A. Wang, "Preparation and properties of chitosan/poly (vinyl alcohol) nanocomposite films reinforced with rod-like sepiolite," *Materials Letters*, vol. 86, pp. 69–72, 2012.

70. A. F. Ismail and W. Lorna, "Penetrant-induced plasticization phenomenon in glassy polymers for gas separation membrane," *Separation and Purification Technology*, vol. 27, pp. 173–194, 2002.

71. A. Jamil, O. P. Ching, T. Iqbal, S. Rafiq, M. Zia-ul-Haq, M. Z. Shahid *et al.*, "Development of an extended model for the mitigation of environmentally hazardous CO_2 gas over asymmetric hollow fiber composite membranes," *Journal of Hazardous Materials*, p. 126000, 2021.

72. D. F. Sanders, Z. P. Smith, R. Guo, L. M. Robeson, J. E. McGrath, D. R. Paul et al., "Energy-efficient polymeric gas separation membranes for a sustainable future: A review," Polymer, vol. 54, pp. 4729–4761, 2013.

73. I. Fergus, "Membrane technology," Water, vol. 26, pp. 20–21, 1999.

74. G. Zarca, I. Ortiz and A. Urtiaga, "Gas permeation properties of 1-hexyl-3-methyl-imidazolium chloride supported liquid membranes," Procedia Engineering, vol. 44, pp. 1114–1116, 2012.

75. T. C. Merkel, H. Lin, X. Wei and R. Baker, "Power plant post-combustion carbon dioxide capture: An opportunity for membranes," Journal of Membrane Science, vol. 359, pp. 126–139, 2010.

76. A. Hussain and M. B. Hägg, "A feasibility study of CO$_2$ capture from flue gas by a facilitated transport membrane," Journal of Membrane Science, vol. 359, pp. 140–148, 2010.

77. L. Peters, A. Hussain, M. Follmann, T. Melin and M. B. Hägg, "CO$_2$ removal from natural gas by employing amine absorption and membrane technology – A technical and economical analysis," Chemical Engineering Journal, vol. 172, pp. 952–960, 2011.

78. B. D. Bhide, A. Voskericyan and S. A. Stern, "Hybrid processes for the removal of acid gases from natural gas," Journal of Membrane Science, vol. 140, pp. 27–49, 1998.

79. R. W. Baker and K. Lokhandwala, "Industrial & Engineering Chemistry Research (2008), 47(7), 2109–2121.pdf," Ind. Eng. Chem. Res, vol. 47, pp. 2109–2121, 2008.

80. X. He and M. B. Hägg, "Membranes for environmentally friendly energy processes," Membranes, vol. 2, pp. 706–726, 2012.

81. K. T. Woo, J. Lee, G. Dong, J. S. Kim, Y. S. Do, H. J. Jo et al., "Thermally rearranged poly(benzoxazole-co-imide) hollow fiber membranes for CO$_2$ capture," Journal of Membrane Science, vol. 498, pp. 125–134, 2016.

82. R. Guo, D. F. Sanders, Z. P. Smith, B. D. Freeman, D. R. Paul and J. E. McGrath, "Synthesis and characterization of Thermally Rearranged (TR) polymers: Influence of ortho-positioned functional groups of polyimide precursors on TR process and gas transport properties," Journal of Materials Chemistry A, vol. 1, pp. 262–272, 2013.

83. N. B. Mc Keown and P. M. Budd, "Polymers of intrinsic microporosity (PIMs): Organic materials for membrane separations, heterogeneous catalysis and hydrogen storage," Chemical Society Reviews, vol. 35, pp. 675–683, 2006.

84. P. M. Budd, E. S. Elabas, B. S. Ghanem, S. Makhseed, N. B. McKeown, K. J. Msayib et al., "Solution-processed, organophilic membrane derived from a polymer of intrinsic microporosity," Advanced Materials, vol. 16, pp. 456–459, 2004.

85. P. M. Budd, K. J. Msayib, C. E. Tattershall, B. S. Ghanem, K. J. Reynolds, N. B. McKeown et al., "Gas separation membranes from polymers of intrinsic microporosity," Journal of Membrane Science, vol. 251, pp. 263–269, 2005.

86. S. Zhao, J. Liao, D. Li, X. Wang and N. Li, "Blending of compatible polymer of intrinsic microporosity (PIM-1) with Tröger's base polymer for gas separation membranes," Journal of Membrane Science, vol. 566, pp. 77–86, 2018.

87. L. Hou, Z. Wang, Z. Chen, W. Chen and C. Yang, "PIM-1 as an organic filler to enhance CO$_2$ separation performance of poly (arylene fluorene ether ketone)," Separation and Purification Technology, vol. 242, 2020.

88. D. Mecerreyes, "Polymeric ionic liquids: Broadening the properties and applications of polyelectrolytes," Progress in Polymer Science (Oxford), vol. 36, pp. 1629–1648, 2011.

89. H. Abdul Mannan, T. M. Yih, R. Nasir, H. Muhktar and D. F. Mohshim, "Fabrication and characterization of polyetherimide/polyvinyl acetate polymer blend membranes for CO$_2$/CH$_4$ separation," Polymer Engineering and Science, vol. 59, pp. E293–E301, 2019.

90. M. Zia ul Mustafa, H. bin Mukhtar, N. A. H. Md. Nordin, H. A. Mannan, R. Nasir and N. Fazil, "Recent developments and applications of ionic liquids in gas separation membranes," Chemical Engineering and Technology, vol. 42, pp. 2580–2593, 2019.

91. W. Fam, J. Mansouri, H. Li and V. Chen, "Improving CO_2 separation performance of thin film composite hollow fiber with Pebax®1657/ionic liquid gel membranes," *Journal of Membrane Science*, vol. 537, pp. 54–68, 2017.

92. J. E. Bara, S. Lessmann, C. J. Gabriel, E. S. Hatakeyama, R. D. Noble and D. L. Gin, "Synthesis and performance of polymerizable room-temperature ionic liquids as gas separation membranes," *Journal of Membrane Science*, pp. 5397–5404, 2007.

93. M. Behroozi and M. Pakizeh, "Study the effects of Cloisite15A nanoclay incorporation on the morphology and gas permeation properties of Pebax2533 polymer," *Journal of Applied Polymer Science*, vol. 134, pp. 1–11, 2017.

94. A. Galve, D. Sieffert, C. Staudt, G. Carme and C. Te, "Combination of ordered mesoporous silica MCM-41 and layered titanosilicate JDF-L1 fillers for 6FDA-based copolyimide mixed matrix membranes," *Journal of Membrane Science*, vol. 431, pp. 163–170, 2013.

95. B. Zornoza, B. Seoane, J. M. Zamaro and C. Tøllez, "Combination of MOFs and Zeolites for Mixed-Matrix Membranes," *ChemPhysChem*, vol. 12, no. 15, pp. 2781–2785, 2011.

96. M. Sarfraz, "Combined Effect of CNTs with ZIF-302 into polysulfone to fabricate MMMs for enhanced CO_2 separation from flue gases," *Arabian Journal for Science and Engineering*, vol. 41, no. 7, pp. 2573–2582, 2016.

97. P. S. Murugiah, O. P. Ching and L. K. Keong, "Collegial effect of carbonaceous hybrid fillers in mixed matrix membrane development," *Reactive and Functional Polymers*, vol. 135, pp. 8–15, 2019.

7 Cryogenic CO_2 Capture

Khuram Maqsood, Abulhassan Ali, and Aymn Abdulrahman

CONTENTS

7.1 INTRODUCTION: BACKGROUND AND DRIVING FORCES

Cryogenic separation uses the freezing/thawing properties of components of a fluidic mixture at a certain temperature to separate them in different streams. However, the process becomes tedious if the constituents of the fluidic mixture have physical or chemical interactions. It is common to find that physicochemical properties of one substance do affect the viscosity, melting point, or boiling point of another constituent. In simplest words, for the separation of two gases or liquids with different freezing points, the temperature must be lowered enough to freeze one of the gases/liquids (converting it to near solid) in the mixture. Once separated, the temperature of the frozen constituent is raised again to convert it back to the original state. Cryogenics can help separate a gas like carbon dioxide from a mixture in a highly pure form [1–2].

DOI: 10.1201/9781003162780-7

There are several methods in the literature for capturing carbon dioxide that uses a different combination of physical and chemical processes. Carbon dioxide separation is costly; the storage, processing, and recycling of chemicals leads to many operational and environmental challenges. Several innovative sequestrating technologies are being researched for CO_2 capturing [3]. The economics and process efficiency are the main factors for using technology at the industrial level. Absorption using amines or other suitable chemical solutions is the prevalent process for separating carbon dioxide. However, the economics of current technologies for high carbon dioxide content makes the process less attractive, and new process development becomes crucial.

The cryogenic separation of carbon dioxide involves the liquefaction or solidification of CO_2 with high purity (>90%) at low (sub-zero) temperatures. Cryogenic CO_2 capture provides many benefits compared to other existing processes, as it does not require any solvents (which leads to a major portion in operational costs). Pure carbon dioxide is retrieved from the most cryogenic separation processes, which can be easily transported to the injection site to enhanced oil recovery (EOR) or enhanced coal-bed methane (ECBM) recovery.

Cryogenic based CO_2 separation finds applications in industrial gas treatment, both for pre-combustion (e.g. sweetening of natural gas) and post-combustion (e.g., the capture of carbon dioxide from a flue/exhaust gas). Moreover, cryogenics-based technologies also find application in oxyfuel combustion, where they first provide almost pure oxygen for combustion and then capture the resulting carbon dioxide.

7.2   CRYOGENIC CO_2 TECHNOLOGIES FOR FLUE GAS CLEANING (CO_2 CAPTURE)

Flue gases from fossil fuel-based power generation are the largest volume-based emitter of CO_2 to the atmosphere (their CO_2 content is rather low, at less than 5%). To address this issue, sustainable and renewable fossil fuels with low environmental impact, and cost-effective CO_2 capture technologies to mitigate the effects of conventional power generation, are urgently required [4].

Among the numerous known sources of CO_2 emissions, coal-fired power plants are the major contributor of CO_2 emissions globally and produce nearly 14,766 million tons of CO_2 per year, as shown in Figure 7.1 [5]. In contrast, oil and natural gas are the next on the list. Global CO_2 emissions have risen rapidly in recent decades, especially in developing countries. The largest source of CO_2 emissions comes from the electricity and power sectors, followed by the transport and the manufacturing sectors, as depicted in Figure 7.2 [5]. Carbon dioxide capture, utilization, and storage (CCUS) is a viable method for reducing CO_2 emissions by capturing it from significant point sources and then either utilizing or storing it underground in specific porous rocks [6]. Scientists have thoroughly studied various CO_2 capture and separation technologies; the application of the technologies depends on the process's economics and efficiency.Fossil fuels are burned to produce energy and

CO$_2$ (million tonnes) emmision by fuel (2018)

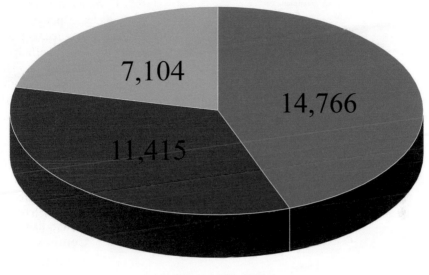

FIGURE 7.1   CO$_2$ emission sources.

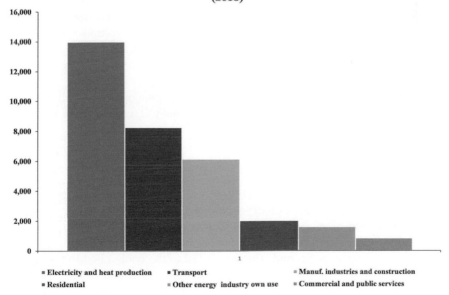

FIGURE 7.2   Significant sources of CO$_2$ emission.

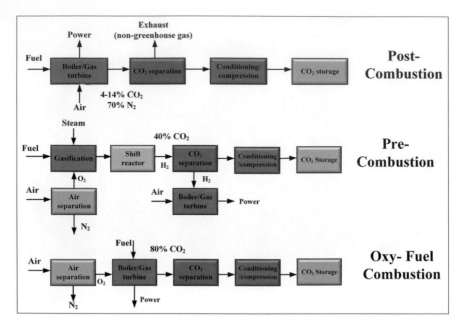

FIGURE 7.3 CO_2 capture technologies from exhaust gases.

carbon dioxide. The capturing of carbon dioxide from the process has been done in three different ways, namely, pre-combustion capture, post-combustion capture, and oxyfuel combustion capture. The selection of a CO_2 capture technology is primarily based on the CO_2 emission conditions (e.g., the concentration of CO_2 in the flue gas, the pressure of the gas stream). The process description of these technologies is illustrated in Figure 7.3.

Several techniques and methods based on condensation or desublimation properties of CO_2 have already been reported for cryogenic CO_2 capture from flue gases. The description of these processes is given below.

7.2.1 CRYOGENIC PACKED BED

A process based on cryogenic packed beds has been developed for carbon capture by Tuinier *et al.* [7–8]. The cryogenic packed bed process is based on the periodic operation of multiple beds. The process consists of cooling, capture, and recovery steps. First, the bed is cooled to below the freezing temperature of CO_2 in the cooling step. Second, flue gas is fed to the cooled bed in the capture step. The lower temperature facilitates the condensation and desublimation of water and carbon dioxide on the packed material. In the third and final step, water and carbon dioxide is recovered from the bed by introducing air for water and carbon dioxide for desublimated CO$_2$. To ensure that the process runs continuously, all three beds must be turned on simultaneously. The process is shown in Figure 7.4 (a).

7.2.2 Anti-Sublimation (AnSU)

An efficient process of cryogenic CO$_2$ capture is developed by Clodic *et al.* [9–10], where flue gases are first cleaned before the CO$_2$ desublimation onto surfaces of heat exchangers. It is preceded by a regeneration step at high pressures to obtain liquid CO$_2$ (Figure 7.4 [b]). However, to avoid clogging, the water from the flue gases should be removed. Moreover, the moisture in the flue gas should be removed from the flue gases to prevent clogging. The frosted layer of carbon dioxide on the heat exchanger walls has a considerable effect on the efficiency of the process.

7.2.3 Stirling Cooler System

Cryogenic CO$_2$ separation by the Stirling cooler is given by Song *et al.* [11–12]. Stirling coolers are highly efficient, dependable, and compact. A pre-freezing tower, the main freezing tower, and a storage tower are all part of the process Figure 7.4 (c). The Stirling cooler-1 separates water in a pre-freezing tower. The Stirling cooler-2 eliminates extra heat from the main freeze tower until the temperature is low enough to capture CO$_2$ by anti-sublimation. The de-sublimated CO$_2$ is collected from the storage tower, cooled by the Stirling cooler-3, whereas the other gases are released to the atmosphere. The process is still tested at a laboratory scale with a binary mixture of nitrogen-carbon dioxide.

7.2.4 External Cooling Loop Cryogenic Carbon Capture (CCCECL)

A hybrid cryogenic carbon capture system via an external cooling loop (CCCECL) by utilizing waste cooling is proposed by Baxter *et al.* [13]. Following are the steps for the CCCECL process: (1) the flue gas is dried and cooled, (2) gas is further compressed and cooled slightly above the CO$_2$ solidification temperature, (3) the gas is expanded to reduce the temperature, (4) CO$_2$ is further solidified, (5) CO$_2$ is pressurized, and (6) CO$_2$ is heated with the help of flue gas feed. The process is described in Figure 7.4 (d). The final form of CO$_2$ is liquid, whereas the rest of the mixture (mainly N$_2$) is discharged.

7.3 CRYOGENIC CO$_2$ TECHNOLOGIES FOR SWEETENING OF NATURAL GAS (CO$_2$ REMOVAL)

The presence of carbon dioxide (CO$_2$) and other sour gases in different concentrations in various natural gas reserves has encouraged many researchers and organizations to invent various innovative technologies [3]. It is crucial to have a thermodynamic

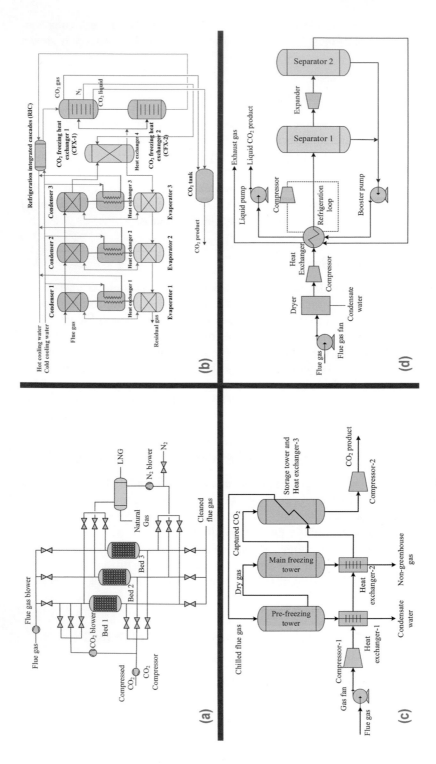

FIGURE 7.4 Cryogenic CO_2 capture technologies from flue gases: (a) cryogenic packed bed; (b) anti-sublimation CO_2 capture process (AnSU); (c) Stirling cooler system; and (d) external cooling loop cryogenic carbon capture (CCCECL).

analysis of carbon dioxide and natural gas mixture to develop the process design properly. A comprehensive review of the thermodynamic data for the methane-carbon dioxide system is presented by Babar *et al.* [14].

A modern approach to capturing, processing, and transporting carbon dioxide simultaneously carries immense potential for industrial applications. Cryogenic CO$_2$ capture does not need any chemicals and solvents, leading to a reduction in consumable costs. Also, no process heating and water are required, and carbon dioxide can be available at high pressure. Furthermore, when process integration is used and the required cooling duty is obtained at a low cost from a liquefied natural gas (LNG) terminal, cryogenic CO$_2$ capture becomes very attractive economically. High-content CO$_2$ natural gas reserves are present in different countries in the world. Indonesia has the highest amount of sour natural gas reserves, followed by Thailand and Malaysia. The distribution of high-content CO$_2$ reserves is presented in Figure 7.5.

7.3.1 CONVENTIONAL SIMPLE CRYOGENIC DISTILLATION

The operation of cryogenic distillation at low temperatures and high pressures can separate CO$_2$ and other components according to their varying boiling temperatures. The output of the cryogenic distillation column is pure and liquid CO$_2$, which can easily be transported. The process is very suitable when the concentration of CO$_2$ in the feed is very high. The process is not economically viable for low carbon dioxide

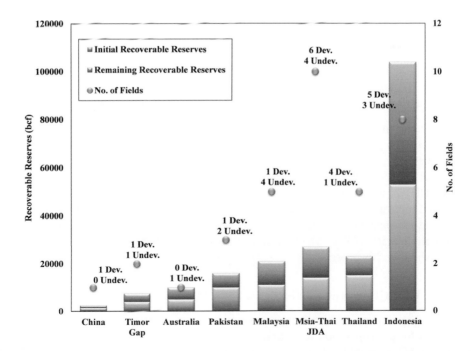

FIGURE 7.5　Sour CO$_2$ reserves in the world [15].

content. There is a high possibility of carbon dioxide solidification at lower pressure in the top part of the distillation column. This solid formation can be avoided by using extractive distillation.

7.3.2 EXTRACTIVE CRYOGENIC DISTILLATION

Extractive distillation is the process of adding chemicals to facilitate the separation in the distillation column. An extractive distillation process was developed by Holmes *et al.* [16–17] by adding heavier hydrocarbon (typically C_3–C_5) to avoid carbon dioxide solidification distillation column, as shown in Figure 7.6 (a). Helium, an additive, is used by Valencia *et al.* [18] to avoid CO_2 solidification. Multiple distillation columns with variable pressure were introduced by Atkinson *et al.* [19] to capture carbon dioxide from natural gas. Berstad *et al.* [2] presented a three-column distillation column with pentane as an additive. An extractive distillation network using butane as an additive is presented by Maqsood *et al.* [20].

7.3.3 MULTIPLE CRYOGENIC PACKED BEDS

A multiple cryogenic packed bed concept was given by Abulhassan and Nor Syahera [21]. A high-content carbon dioxide feed (70% CO_2) was used for the experimentation. The countercurrent packed bed concept was used for carbon dioxide capture. The process comprises three steps. In the first cooling step, the temperature of the packed bed is reduced to below the freezing point of CO_2 (around $-85°C$ to $-100°C$). The second step is the capture step: a mixture of CH_4–CO_2 is put into the packed bed and because of the low temperature, CO_2 is desublimated on the packing. This process continuous until the bed is saturated with desublimated CO_2. In the third step, recovery, the sublimated CO_2 is recovered with the help of some inert gas.

7.3.4 CO_2 SEPARATION TECHNOLOGIES BY USING COMPRESSION-REFRIGERATION

A cryogenic multi-stage process was designed to separate carbon dioxide from a high CO_2 feed [22]. A high CO_2 content (71%) from Natuna gas field in Indonesia was selected. A series of separators was used to separate carbon dioxide from methane. CANMET Energy Technology Centre proposed a cryogenic process (CO_2CCU) based on several compression units for CO_2. The key process steps included moisture removal, compression stages with separators and coolers, dryers, and heat exchangers. A final purity of 95% carbon dioxide is reported, while the feed contains 50 vol% or higher carbon dioxide feed. A pilot-scale unit was also established for this process.

7.3.5 CO_2 CAPTURE BY GAS HYDRATES

Carbon dioxide capture using hydrate formation is a promising technology. Clathrate hydrates for gas separation were first explored in 1930 by Nikitin to separate rare gases by SO_2 hydrate formation [23–24]. Happel *et al.* [25] utilized the same techniques to separate specific gas from a multi-component gas mixture by hydrate

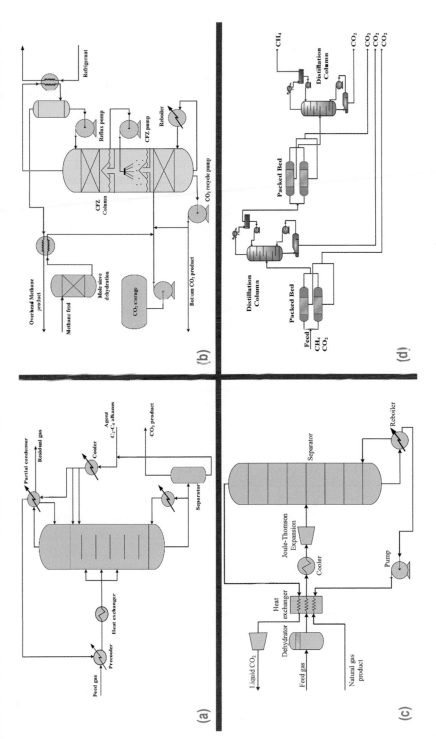

FIGURE 7.6 CO₂ capture technologies from natural gas: (a) extractive distillation; (b) Ryan Holmes CFZ; (c) cryocell technology; and (d) hybrid cryogenic network.

formation. Clathrate hydrates crystalline solid compounds formed by hydrogen-bonded water molecules that use van der Waals forces to trap small guest molecules (10 Å) in the cavities of the hydrates. The most common guest molecules that form a stable hydrate are CH_4 (methane), C_2H_6 (ethane), C_3H_8 (propane), i-C_4H10, n-C_4H_{10} (iso and normal butane), CO_2, H_2S, and N_2. A suitable pressure and temperature combination depending on gas composition is a key control parameter for hydrate formation. Hydrate formation is usually facilitated by high pressure and low temperature. Lee and Kang [26] explored the hydrate formation process to capture CO_2. An extensive research effort was carried out to capture CO_2 using hydrate formation [27–30]. Hydrate formation is done at high pressure that leads to high operating cost [31]. Some researchers add promoter molecules tetra-n-butylammonium bromide (TBAB) [32–34] and tetrahydrofuran (THF) [35–37] to reduce the hydrate formation pressure. Other promoters include cyclopentane [38] and propane [39]. A detailed discussion on the use of gas hydrates for CO_2 capture can be found in Chapter 4 of this book.

7.3.6 CONTROLLED FREEZE ZONE TECHNOLOGY

A cryogenic process of CO_2 separation was developed by ExxonMobil [40]. The research, named "controlled freeze zone" (CFZ) technology for cryogenic carbon dioxide separation, was performed on a pilot plant. CFZ technology was a new design for the conventional distillation column. The liquid stream in the upper rectification section enters the CFZ section operating at a lower temperature (– °80 to –90 °C) to solidify the CO_2. In contrast, methane is separated in the gas phase. Once the CO_2 is solidified, it drops on the melting section, and later sent to the stripping section. The buildup of solid CO_2 on the heat exchanging surfaces can cause a decrease in the heat transfer rate. The process is presented in Figure 7.6(b). ExxonMobil has set up an ExxonMobil's Shute Creek Treatment Facility close to its LaBarge commercial demonstration plant (CDP). The process was tested for different feed concentrations of carbon dioxide and hydrogen sulfide in the natural gas. A wide range of CO_2 concentration feed (8%–71%) was tested, and the results showed that the product streams contained less than 2% CO_2. The plant was also tested for H_2S removal, with promising results [41].

7.3.7 CRYOCELL-BASED SEPARATION

Cryocell technology is tested for natural gas contain a high concentration of CO_2 [1]. In the cryocell process, the feed is first cooled to a temperature just above the CO_2 freezing point. The mixture is then converted into a liquid by applying pressure. A constant enthalpy flash is carried out in the final stage using a Joule-Thomson valve, and the mixture splits into vapor, fluid, and solid phases. CO_2 is converted into a solid phase with high purity. While the gas stream is compressed to market specifications, solid CO_2 is melted and moved to the bottom of the separator. The process is shown in Figure 7.6(c). Cryocell demonstration plants were established, and feed gases with compositions of 3.5 to 60 mol% CO_2 content were tested. Unfortunately, the available literature on this method does not provide modeling or fundamental data or other essential information because of commercial reasons.

7.3.8 CONDENSED CONTAMINANT CENTRIFUGAL SEPARATION

Some cryogenic CO$_2$ separation technologies have been reported involving mechanical methods to purify natural gas with high content CO$_2$. A process of C3 separation (condensed contaminant centrifugal separation) is presented in which CO$_2$ is condensed into droplets and then separated from the natural gas using rotational separators [42]. Condensed contaminant centrifugal separation combines the benefit of low temperature with mechanical separation to capture carbon dioxide. The principle of separation is based on the difference in the boiling points. The feed is cooled in specially designed turbo-expanders, and different components of a gas mixture are converted into small droplets in the range of 1–10 μm. A cyclone separation mechanism was suggested to separate the small droplets from the vapor phase.

7.3.9 HYBRID CRYOGENIC NETWORK

A hybrid cryogenic network is presented by Maqsood *et al.* [43–45]. The hybrid cryogenic network helps to combine the benefits of V-L (distillation) and V-S (cryogenic packed bed) CO$_2$ separation in a single unit. One type of configuration is illustrated in Figure 7.6(d). Initially, process simulation using Aspen Plus was conducted [44]. The process was compared with simple extractive distillation concerning cost, methane losses, and methane purity for a high content CO$_2$ natural gas feed. The studies were further enhanced by adding the cryogenic packed bed at different locations [45]. In a recent study, the process concept and simulation studies are compared with the experimental data to show the effectiveness of the process [43]. The experimental results and the simulation results, and the experimental data show that the methane purity of 90%–97% can be achieved with a different arrangement of separators.

7.4 ADVANTAGES OF CRYOGENIC TECHNOLOGIES

One of the main advantages of cryogenic CO$_2$ capture is the availability of highly pure and high-pressure CO$_2$ due to desublimation or liquefaction. This benefit aids in the transportation of CO$_2$ for long-distance storage and makes it very convenient to use in processes such as steam reforming or chemical synthesis. Another benefit of cryogenic capture is the potential reuse of low-temperature CO$_2$ as an industrial energy source of cold energy, especially LNG plants.

7.5 KEY CHALLENGES

Cryogenic technologies have great potential and advantages, as discussed earlier. However, specific challenges are associated with these technologies that should be overwhelmed to use these technologies for large-scale CO$_2$ capture. A comparison of these technologies is presented in Table 7.1. It is evident from the table that there are

TABLE 7.1

Comparison and Key Challenges of the Different Technologies

Technology name	Feed	CO_2 capture efficiency	Required energy	Key challenges
Conventional cryogenic distillation [46]	Natural gas with high CO_2 content	>98%	High	High energy cost and possible plugging of solid CO_2
Extractive distillation [47]	CO_2–ethane	>96%	High, 1.394×10^6 MJ/hr	The operation and design is the difficulty for the distillation columns
Cryogenic packed bed concept [7–8]	Exhaust gases	99%	1.8 MJ/kg of CO_2 0.66–0.8 MJ/kg CO_2	The process was used for low-pressure gases
Stirling coolers [11–12]	Exhaust gases	96%	1.5 MJ/kg of CO_2	Separation needs a long time Heat losses due to frost CO_2 layer on heat exchanger surface costly
Gas hydrate formation [48]	Exhaust gas 17% and 65% CO_2			A high-pressure process that requires temperature control because of the exothermic reaction
CFZ technology [40–41]	Natural gas 15%–71% CO_2	>90%		The process is applicable only for high CO_2, concentrations, and the process is very complicated.
Cryocell process [1]	Natural gas 13%–60%	81%	0.864 MJ/kg of CO_2	Higher compression power requirements and separation percentage was low
Condensed contaminated centrifugal separation [49] [42]	Exhaust gases and natural gas 21%–70% CO_2	>70%–90%	0.842 MJ/kg of CO_2	High operating and maintenance cost due to the rotational equipment
Hybrid cryogenic distillation network [20, 43]	Natural gas	>97%	1527 kJ/kg of CO_2	High maintenance cost and complex arrangement of equipment

three key challenges for using cryogenic technologies as a replacement for conventional technologies. These challenges are:

1. Capital and operational cost
2. Efficiency
3. Operation complexities.

Due to the low-temperature operation, the capital cost of cryogenic equipment is high due to the expensive construction material. Second, the operational cost is high because of the cooling energy requirement. The process efficiency improves as the temperature drops but, on the contrary, the energy efficiency improves as the temperature rises. The problem can be addressed by using multi-objective optimization. The arrangement and operation of the equipment are complicated and should be addressed in future research.

REFERENCES

1. A. Hart and N. Gnanendran, "Cryogenic CO$_2$ capture in natural gas," *Energy Procedia*, vol. 1, pp. 697–706, 2009.
2. D. Berstad, P. Nekså and R. Anantharaman, "Low-temperature CO$_2$ removal from natural gas," *Energy Procedia*, vol. 26, pp. 41–48, 2012.
3. IPCC, *Policymaker's Summary of the Scientific Assessment of Climate Change; Report to IPCC from Working Group*, Meteorological Office, 1990.
4. C. Song, Q. Liu, S. Deng, H. Li, and Y. Kitamura, "Cryogenic-based CO$_2$ capture technologies: State-of-the-art developments and current challenges," *Renewable and Sustainable Energy Reviews*, vol. 101, pp. 265–278, 2019.
5. I. S. D. Service, "CO$_2$ Emissions from Fuel Combustion 2020: Highlights," 2020. [Online]. Available: http://wds.iea.org/wds/pdf/Worldco2_Documentation.pdf
6. T. Ajayi, J. S. Gomes and A. Bera, "A review of CO$_2$ storage in geological formations emphasizing modeling, monitoring and capacity estimation approaches," *Petroleum Science*, vol. 16, pp. 1028–1063, 2019.
7. M. J. Tuinier, M. van Sint Annaland, G. J. Kramer and J. A. M. Kuipers, "Cryogenic CO$_2$ capture using dynamically operated packed beds," *Chemical Engineering Science*, vol. 65, pp. 114–119, 2010.
8. M. J. Tuinier, H. P. Hamers and M. van Sint Annaland, "Techno-economic evaluation of cryogenic CO$_2$ capture – A comparison with absorption and membrane technology," *International Journal of Greenhouse Gas Control*, vol. 5, pp. 1559–1565, 2011.
9. D. Clodic and M. Younes, "A new method for CO$_2$ Capture: Frosting CO$_2$ at atmospheric pressure," in *Greenhouse Gas Control Technologies – 6th International Conference*, J. Gale and Y. Kaya, Eds. Pergamon, 2003, pp. 155–160.
10. D. Clodic, R. El Hitti, M. Younes, A. Bill and F. Casier, "CO$_2$ capture by anti-sublimation Thermo-economic process evaluation," Fourth Annual Conference on Carbon Capture & Sequestration, Alexandria, 2005, pp. 2–5.
11. C. F. Song, Y. Kitamura and S. H. Li, "Evaluation of Stirling cooler system for cryogenic CO$_2$ capture," *Applied Energy*, vol. 98, pp. 491–501, 2012.
12. C. F. Song, Y. Kitamura, S. H. Li and K. Ogasawara, "Design of a cryogenic CO$_2$ capture system based on Stirling coolers," *International Journal of Greenhouse Gas Control*, vol. 7, pp. 107–114, 2012.
13. M. Jensen, "Energy process enabled by cryogenic carbon capture," PhD, Brigham Young University Provo, 2015.
14. M. Babar, M. A. Bustam, A. Ali, A. Shah Maulud, U. Shafiq, A. Mukhtar *et al.*, "Thermodynamic data for cryogenic carbon dioxide capture from natural gas: A review," *Cryogenics*, vol. 102, pp. 85–104, 2019.
15. B. Cahill, "Regional Economic and Energy Outlook: Perspectives on Malaysia," presented at the 3rd Energy Forum, Kuala Lumpur, Malaysia, 2011.
16. A. S. Holmes and J. M. Ryan, "Cryogenic distillative separation of acid gases from methane," 4,318,723, 1982.

17. A. S. Holmes, B. C. Price, J. M. Ryan and R. E. Styring, "Pilot tests prove out cryogenic acid-gas/hydrocarbon separation processes," *Oil and Gas Journal*, vol. 81, pp. 85–86, 1983.

18. J. A. Valencia and R. D. Denton, "Method of separating acid gases, particularly carbon dioxide from Methane by the addition of a light gas such as Helium," 1983. [Online]. Available: https://www.osti.gov/biblio/5247359-method-separating-acid-gases-particularly-carbon-dioxide-from-methane-addition-light-gas-helium.

19. T. D. Atkinson, J. T. Lavin and D. T. Linnett, "Separation of gaseous mixtures," 4,759,786, 1988.

20. K. Maqsood, J. Pal, D. Turunawarasu, A. J. Pal and S. Ganguly, "Performance enhancement and energy reduction using hybrid cryogenic distillation networks for purification of natural gas with high CO_2 Content," *Korean Journal of Chemical Engineering*, 2014.

21. A. Abulhassan, K. Maqsood, N. Syahera, Azmi B. M. Shariff and S. Ganguly, "Energy minimization in cryogenic packed beds during purification of natural gas with high CO_2 content," *Chemical Engineering and Technology*, vol. 37, no. 10, pp. 1675–1685, 2014. https://doi.org/10.1002/ceat.201400215.

22. I. P. Suarsana, "Producing high CO_2 gas content reservoirs in pertamina Indonesia using multi stage cryogenic process," SPE Asia Pacific Oil and Gas Conference and Exhibition, 2010.

23. T. E. Rufford, S. Smart, G. C. Y. Watson, B. F. Graham, J. Boxall, J. C. Diniz da Costa *et al.*, "The removal of CO_2 and N_2 from natural gas: A review of conventional and emerging process technologies," *Journal of Petroleum Science and Engineering*, vol. 94–95, pp. 123–154, 2012.

24. E. Sloan, K. Dendy and A. Carolyn, *Clathrate hydrates of natural gases*. Boca Raton, Fla.: CRC Press, 2008.

25. J. Happel, M. A. Hnatow and H. Meyer, "The study of separation of nitrogen from methane by hydrate formation using a novel apparatus," *Annals of the New York Academy of Sciences*, vol. 715, pp. 412–424, 1994.

26. H. Lee and S. P. Kang, "Method for separation of gas constituents employing hydrate promoter," ed: Google Patents, 2003.

27. P. Linga, R. Kumar and P. Englezos, "Gas hydrate formation from hydrogen/carbon dioxide and nitrogen/carbon dioxide gas mixtures," *Chemical Engineering Science*, vol. 62, pp. 4268–4276, 2007.

28. N. H. Duc, F. Chauvy and J. M. Herri, "CO_2 capture by hydrate crystallization – a potential solution for gas emission of steelmaking industry," *Energy Conversion and Management*, vol. 48, pp. 1313–1322, 2007.

29. J. Zhang, P. Yedlapalli and J. W. Lee, "Thermodynamic analysis of hydrate-based precombustion capture of," *Chemical Engineering Science*, vol. 64, pp. 4732–4736, 2009.

30. R. Kumar, P. Linga and P. Englezos, "Pre and post combustion capture of carbon dioxide via hydrate formation," *EIC Climate Change Technology*, 2006, pp. 1–7.

31. H. Tajima, A. Yamasaki and F. Kiyono, "Energy consumption estimation for greenhouse gas separation processes by clathrate hydrate formation," *Energy*, vol. 29, pp. 1713–1729, 2004.

32. H. Y. Acosta, P. R. Bishnoi and M. A. Clarke, "Experimental Measurements of the Thermodynamic Equilibrium Conditions of Tetra-n-butylammonium Bromide Semiclathrates Formed from Synthetic Landfill Gases," *Journal of Chemical and Engineering Data*, vol. 56, pp. 69–73, 2010.

33. Q. Sun, X. Guo, A. Liu, B. Liu, Y. Huo and G. Chen, "Experimental study on the separation of CH_4 and N_2 via hydrate formation in TBAB solution," *Industrial and Engineering Chemistry Research*, vol. 50, pp. 2284–2288, 2010.

34. X. S. Li, C. G. Xu, Z. Y. Chen and H. J. Wu, "Tetra-n-butyl ammonium bromide semiclathrate hydrate process for post-combustion capture of carbon dioxide in the presence of dodecyl trimethyl ammonium chloride," *Energy*, vol. 35, pp. 3902–3908, 2010.

35. S. P. Kang and H. Lee, "Recovery of CO$_2$ from flue gas using gas hydrate: Thermodynamic verification through phase equilibrium measurements," *Environmental Science and Technology*, vol. 34, pp. 4397–4400, 2000.
36. Y. Seo, S.-P. Kang, S. Lee and H. Lee, "Experimental measurements of hydrate phase equilibria for carbon dioxide in the presence of THF, propylene oxide, and 1,4-dioxane," *Journal of Chemical and Engineering Data*, vol. 53, pp. 2833–2837, 2008.
37. C. Y. Sun, G. J. Chen and L. W. Zhang, "Hydrate phase equilibrium and structure for (methane plus ethane plus tetrahydrofuran plus water) system," *The Journal of Chemical Thermodynamics*, vol. 42, pp. 1173–1179, 2010.
38. S. Li, S. Fan, J. Wang, X. Lang and Y. Wang, "Clathrate Hydrate Capture of CO$_2$ from Simulated Flue Gas with Cyclopentane/Water Emulsion," *Chinese Journal of Chemical Engineering*, vol. 18, pp. 202–206, 2010.
39. R. Kumar, H. J. Wu and P. Englezos, "Incipient hydrate phase equilibrium for gas mixtures containing hydrogen, carbon dioxide and propane," *Fluid Phase Equilibria*, vol. 244, pp. 167–171, 2006.
40. P. S. Northrop and J. A. Valencia, "The CFZ™ process: A cryogenic method for handling high-CO$_2$ and H$_2$S gas reserves and facilitating geosequestration of CO$_2$ and acid gases," *Energy Procedia*, vol. 1, pp. 171–177, 2009.
41. S. Kelman, D. Maher, A. Nagavarapu and J. Valencia, "The controlled freeze zone technology for the commercialization of sour gas resources," presented at the International Petroleum Technology Conference, Doha, Qatar, 2014.
42. G. P. Willems, M. Golombok, G. Tesselaar and J. J. H. Brouwers, "Condensed rotational separation of CO$_2$ from natural gas," *AIChE Journal*, vol. 56, pp. 150–159, 2010.
43. K. Maqsood, A. Ali, R. Nasir, A. Abdulrahman, A. B. Mahfouz, A. Ahmed *et al.*, "Experimental and simulation study on high-pressure V-L-S cryogenic hybrid network for CO$_2$ capture from highly sour natural gas," *Process Safety and Environmental Protection*, vol. 150, pp. 36–50, 2021.
44. K. Maqsood, J. Pal, D. Turunawarasu, A. Pal and S. Ganguly, "Performance enhancement and energy reduction using hybrid cryogenic distillation networks for purification of natural gas with high CO$_2$ content," *Korean Journal of Chemical Engineering*, vol. 31, pp. 1120–1135, 2014.
45. K. Maqsood, A. Ali, A. B. M. Shariff and S. Ganguly, "Process intensification using mixed sequential and integrated hybrid cryogenic distillation network for purification of high CO$_2$ natural gas," *Chemical Engineering Research and Design*, vol. 117, pp. 414–438, 2017.
46. M. Garner, "Chevron Buckeye CO$_2$ plant treating of natural gas using the Ryan/Holmes separation process," University of Texas, 2008.
47. F. Lastari, V. Pareek, M. Trebble, M. O. Tade, D. Chinn, N. C. Tsai *et al.*, "Extractive distillation for CO$_2$-ethane azeotrope separation," *Chemical Engineering and Processing*, vol. 52, pp. 155–161, 2012.
48. C. G. Xu, Z. Y. Chen, J. Cai and X. S. Li, "Study on pilot-scale CO$_2$ separation from flue gas by the hydrate method," *Energy & Fuels*, vol. 28, pp. 1242–1248, 2013.
49. R. J. Van Benthum, H. P. Van Kemenade, J. J. H. Brouwers and M. Golombok, "Condensed rotational separation of CO$_2$," *Applied Energy*, vol. 93, pp. 457–465, 2012.

8 Advance Reforming Technologies for Biofuel Production from Carbon Capture

Ain Syuhada, Muhammad Izham Shahbudin, and Mohammad Tazli Azizan

CONTENTS

8.1 INTRODUCTION: BACKGROUND

In recent years, the increase in carbon dioxide (CO_2) emissions to the environment has become a serious environmental problem, as it leads to global warming and climate change. The increase in CO_2 emissions is mainly caused by the combustion of fossil fuels, coal, and natural gas as the main energy sources for various applications, such as transportation and electricity generation in industry and residential areas. The International Energy Agency (IEA) reported that approximately 80% of the energy generation comes from the of combustion oil, coal, and natural gas. The high dependency on these nonrenewable energy resources as the main energy source should be reduced by substituting these resources with other renewable resources (e.g., biomass, wind, thermal) to avoid more serious environmental problems.

DOI: 10.1201/9781003162780-8

Figure 8.1 shows the US energy consumption by sources and energy-related CO_2 emission by sources in 2019 reported by the Energy Information Administration (EIA). The data shows that petroleum is the largest energy source, which contributes approximately 37% of the total energy consumption. The second highest energy consumption is from natural gas sources (32%), followed by nonfossil sources (20%). The least energy consumption is from coal sources, contributing only 11% of the total energy consumption. Meanwhile, for the energy-related CO_2 emission, burning petroleum fuels contributed the highest percentage of 46% from the total CO_2 emission. The second-highest CO_2 emission is from burning natural gas (33%) and the least CO_2 emission from burning coal (21%).Moreover, in 2018, the EIA reported that the combustion of fossil fuels to generate energy contributed approximately 75% of the total US anthropogenic greenhouse gas emission. This high emission of CO_2 was contributed by various end-use sectors such as transportation, industrial, residential, and commercial sectors. In 2019, EIA reported that the transportation sector emits the highest amount of CO_2 (1902 million metric tons) followed by the industrial sector (1453 million metric tons), the residential sector (964 million metric tons), and from commercial sectors (839 million metric tons; Figure 8.2).

In December 2015, the UN Climate Change Conference held in Paris acknowledged the long-term goal of managing average global warming below 2 °C. This can be achieved by considering two long-term emission goals: peaking emissions as soon as possible and net greenhouse gas neutrality (expressed as a balance between anthropogenic emissions by sources and removals by sinks) in the second half of this century. Therefore, it is expected that these emission goals will help to reduce the CO_2 emission caused by human activities.

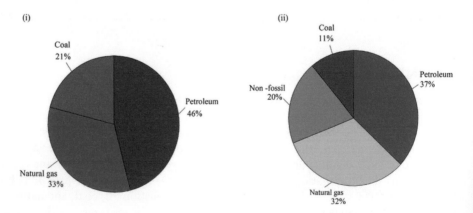

FIGURE 8.1   US energy consumption by (i) sources and (ii) energy-related CO_2 emission by sources in year 2019.

Source: **Data from EIA [1].**

On the other hand, over the next decade, it is expected that the increase in the world population will increase the demand for fossil fuels as energy sources and, thus, will increase CO_2 emissions. Therefore, one possible way to decrease the CO_2 emission receiving much interest is through carbon capture and utilization (CCU) and carbon capture and sequestration (CCS) technologies.

The concepts of CCU and CCS are different, as they are targeted for different applications, as summarized in Figure 8.3; thus they should be evaluated separately. The CCU technology involves the capture of CO_2, and it can be recycled in a reforming reaction as a renewable feed to be converted into biofuel and various chemical products via physical, chemical, or biological processes or can be directly used for the enhanced oil recovery process. Meanwhile, the CCS focuses on removing CO_2 from gas streams, which are then compressed into a supercritical condition and transported to the storage site. Between these two technologies, CCU technology, which involved the conversion of CO_2, is preferred for CO_2 recovery as it promotes economic growth and reduces the utilization of other renewable resources.

The CO_2 capture technology had been demonstrated using various routes, such as absorption [3–6], membrane [7–10], adsorption [11–13], and chemical looping

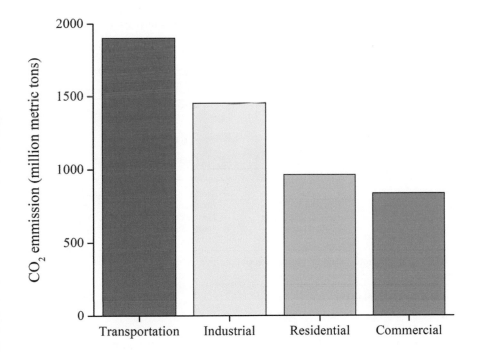

FIGURE 8.2 CO_2 emission by end-use sectors in 2019.

Source: **Data from EIA [1].**

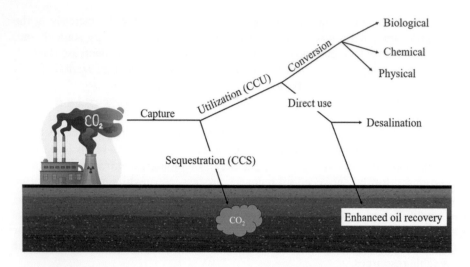

FIGURE 8.3 The concepts of carbon capture and utilization (CCU) and carbon capture and sequestration (CCS).

Source: **Adapted from [2].**

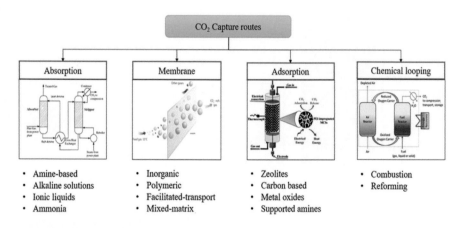

FIGURE 8.4 Various carbon capture technologies used for industries and lab-scale application.

Source: **Adapted from [18].**

[14–17] as summarized in Figure 8.4. For each industry, the selected CO_2 capture route varies depending on the type of industry producing the CO_2. For example, the chemical industry producing ethanol that produced high-concentration CO_2 would require more energy for the CO_2 captured than thermal power plants, which produced less CO_2 and required less energy for the CO_2 capture process.

The conversion of captured CO_2 to syngas, consisting of hydrogen (H_2) and carbon monoxide (CO) via reforming technology, was identified as the most feasible utilization of the captured CO_2. In the future, syngas will become the most promising renewable energy for electricity generation in industry and a residential area. Besides that, syngas is also used as a feed in Fischer-Tropsch process to synthesize petrochemical products and fuel for car engines.

A few studies have proved that CO_2 reforming was identified as the most suitable technology for syngas production. Zhan et al. [19] investigated the performance of Ni-Mg-Al catalysts for syngas production via biogas (CH_4 and CO_2) reforming of CO_2 at 750°C with GHSV of 28,000 mL g^{-1} h^{-1}. It was found that the Mg compositions significantly affect the CH_4 and CO_2 conversion to syngas. Another study [20] also demonstrated syngas production via biogas reforming over Ni/γ-Al_2O_3 catalysts conducted at a temperature range of 300°C to 1000°C with atmospheric pressure. They reported that at a low level of CO_2, the highest yield of H_2 (49%) and CO (45%) was obtained.

Catalyst based on noble metals such as Pt, Ir, Rh, and Ru has been extensively reported for syngas production via reforming technologies due to their high catalytic activity and high resistance towards carbon deposition during the reaction. However, high cost and low availability limit their use in large-scale applications. Therefore, the development of catalysts based on non-noble metals has attracted much attention. Ni-based catalysts have come to interest due to high availability, lower cost, and high activity for reforming reactions [21]. For the support selection, alkaline earth metal oxides were identified as the potential support for CO_2 reforming to increase the basic sites of the catalyst, which improve the CO_2 adsorption ability and consequently inhibit the catalyst deactivation. Son et al. [22] and Ryi et al. [23] investigated the performance of MgO-promoted Co-Ni/ Al_2O_3 catalyst for CO_2 reforming. It was found that the conversion of CO_2 over MgO-promoted Co-Ni/Al_2O_3 (96.7%) catalysts is higher compared to Co-Ni/Al_2O_3 (92.1%) catalysts. Therefore, this shows that metal oxide–promoted catalysts demonstrate excellent performance compared to catalysts without metal oxide promoters during the reaction.

Many challenges have not been fully addressed in catalysis and reforming of CO_2, such as reactor design and its performance and determining the reaction mechanism and kinetics for syngas production via CO_2 reforming. In terms of technical aspects, the challenge that needs to be tackled during the catalytic reaction includes catalyst deactivation due to coke formation and metal sintering at high temperatures.

This chapter will critically discuss previous work carried out by other researchers focusing on syngas production utilizing CO_2 as renewable feedstock via respective reforming technologies and catalysts development.

8.2   REFORMING TECHNOLOGIES

Reforming is a well-developed technology commonly used to synthesize syngas from renewable feedstocks such as hydrocarbons and natural gas. This reforming technology in an endothermic reaction which carried out at relatively high temperature between 400°C to 800°C. However, reactant and suitable metal-based catalysts are very important for this reforming technology as they may lead to coke deposition, which may cause catalyst deactivation. To date, several developments in selecting metal-based catalysts and reactants have been carried out to improve the yield of the desired products. A few studies on syngas production via CO_2 reforming have been reported previously. These studies have proved that this technology can convert CO_2 and hydrocarbon, mainly methane, into syngas.

8.2.1 SYNGAS PRODUCTION VIA CO_2 REFORMING

In the petrochemical industry, steam-reforming technology is the most common technology used to convert natural gas to hydrogen or syngas. This reaction is an endothermic reaction usually carried out at a high temperature in the range of 400°C to 800°C and ambient pressure. However, the selection of the operating temperature is highly dependent on the type of reactant used. For example, hydrocarbons with longer carbon chains would require a higher reaction temperature than a hydrocarbon with a shorter carbon chain for better conversion. Besides that, this technology also requires high energy input to sustain the reaction, makes this reaction costly. In addition, this technology also possesses a great challenge, such as the catalyst may experience sintering effects and coking at high temperature, leading to catalyst deactivation and thus reducing the activity of the catalyst during the reaction. Therefore, the researchers are currently working on catalyst development to improve catalyst stability by doping with other noble metals, using basic supports or using multiple or mixed oxides as support.

The highest application of the captured CO_2 is used to produce syngas via the combination of steam and CO_2 reforming of methane, as reported by [24]. Authors reported that these processes were the most promising method for syngas production as they can control the desired H_2/CO by manipulating the feedstock ratio of CO_2 and H_2O (Table 8.1). Equations 1 to 3 show the reactions involved during the CO_2 reforming of methane reaction, which includes steam reforming of methane, CO_2 reforming of methane and water gas shift (WGS) or reverse water gas shift (RWGS) reaction, respectively, as proposed by [24].

$$CH_4 + H_2O \rightarrow 3H_2 + CO \ (\Delta H°_{298K} = +206 \ kJ \ mol^{-1}) \tag{1}$$
$$CH_4 + CO_2 \rightarrow 2H_2 + 2CO \ (\Delta H°_{298K} = +247 \ kJ \ mol^{-1}) \tag{2}$$
$$CO + H_2O \rightarrow H_2 + CO_2 \ (\Delta H°_{298K} = -41.2 \ kJ \ mol^{-1}) \tag{3}$$

Another study [25] also conducted a study on syngas production conducted in a quartz fixed-bed reactor via CO_2 reforming and CO_2-steam reforming of methane utilizing Ni/Ce-SBA-15 catalysts. Their research mainly focused on the effect of the steam feed ratio and CO_2 feed ratio on CH_4 and CO_2 conversion and H_2 and CO yield. It was reported that the increase in the steam ratio from 10 to 20 kPa increased the CH_4 conversion, H_2 yield, and CO yield from 60% to 70%, 35% to 50%, and 36% to 47%, respectively. Meanwhile, the CO_2 conversion slightly decreased from 85 to 65%. The presence of excess H_2O mainly causes this enhanced CH_4 reforming reaction, resulting in a high H_2/CO ratio. For the effect of CO_2 feed ratios, it was reported that the increased in the CO_2 feed ratio from 10 to 20 kPa decreased the CH_4 conversion, CO_2 conversion, and H_2 yield from 76% to 60%, 60% to 40%, and 62% to 45%, respectively. Meanwhile, a slight increase in CO yield from 45% to 55% was observed with an increased CO_2 feed ratio from 10 to 20 kPa. This might be caused by the reduction in the CO_2 steam reforming of methane reaction due to the adsorption and decomposition of CH_4 on the catalyst surface.

Mirzaei et al. [26] also reported on syngas production via carbon dioxide reforming of methane over a series of Co-MgO mixed oxide catalysts. The reaction was performed at a temperature of 550 to 700°C in steps of 50°C for 30 minutes at each temperature, with gas hourly space velocity (GHSV) of 12,000 ml/(h $g_{catalyst}$) in a fixed-bed continuous flow reactor loaded with 200 mg catalyst. It was reported that 10 CoMgO catalyst showed the highest catalytic activity with H_2/CO of 0.85, H_2 yield of 60.70%, and CO yield of 71.38%. Meanwhile, for the effect of temperature on CO_2 conversion, the increased reaction temperature from 550°C to 700°C increased the CO_2 conversion from 30% to 78%. The higher CO_2 conversion is mainly contributed by the RWGS reaction ($CO_2 + H_2O \rightarrow H_2 + CO$), which simultaneously occurs during the reforming reaction.

Thermodynamic equilibrium study of syngas production via dry reforming of methane (CH_4) with CO_2 and water (H_2O) was reported by [27]. The study aimed to observe the effect of temperature on the CH_4 and CO_2 conversion and molar ratio of H_2/CO. For dry reforming of methane (DRM) with CO_2, it was reported that syngas production was more favored at a high temperature above 850 °C. This was proved by the increase in CH_4 and CO_2 conversion from 96% to 99% and 93% to 99%, respectively, with increased temperatures from 850°C to 1000°C. Meanwhile, the H_2/CO decreased from 1.1 to 1.0 with an increase in temperature from 850°C to 1000°C, which showed that H_2 and CO production were also favored at high temperatures. This shows that dry reforming of methane was favored at temperatures above 850°C. For dual reforming of CH_4 with CO_2 and H_2O, it was reported that at temperatures above 850°C, H_2 and CO were produced as the main product during the reaction. with a maximum H_2/CO molar ratio of 1.7.

Meanwhile, the maximum CH_4 and CO_2 conversions of 100% and 96% were obtained at the maximum reaction temperature of 1000°C. Comparing the H_2/CO molar ratio of DRM (1.0) and dual reforming of CH_4 with CO_2 and H_2O (1.7), it was reported that the syngas produced via dual reforming of CH_4 with CO_2 and H_2O is more favorable to be used as feedstock for the synthesis of valuable chemical via the Fisher-Tropsch reaction. Therefore, it can be concluded that the syngas production via DRM and dual reforming of methane with CO_2 and H_2O was more favored at temperatures above 850°C.

Dan et al. [28] also reported on the combined steam and dry reforming of methane process, which utilized CO_2 as the main feedstock for syngas production using Ni/Al_2O_3 catalyst. The reaction was conducted in a micro-activity reference unit equipped with a stainless-steel reactor at atmospheric pressure and temperature varying from 600°C to 700°C. It was reported that the increase in the steam to carbon ratio (S/C) by increasing water content in the feed leads to an increase in the CH_4 conversion and decreased the CO_2 conversion due to the formation of larger quantity

TABLE 8.1

Summary on the Utilization of CO_2 as Feedstock for Syngas Production via Various Reforming Technologies

Catalyst	Process	Operating conditions				Results		Ref
		T (°C)	P (kPa)	GHSV (mL g^{-1} h^{-1})	CH$_4$/CO$_2$ ratio	CO$_2$ conv. (%)	H$_2$/ CO ratio	
NiM$_0$A	BRCD	750	101	28,000	1	70	0.98	[19]
NiM$_5$A	BRCD	750	101	28,000	1	86	0.98	[19]
NiM$_{10}$A	BRCD	750	101	28,000	1	90	0.96	[19]
NiM$_{20}$A	BRCD	750	101	28,000	1	84	0.96	[19]
NiM$_{30}$A	BRCD	750	101	28,000	1	80	0.92	[19]
Ni/Al$_2$O$_3$	DRM	550	101	1450	1.5	40	0.82	[20]
Ni/Al$_2$O$_3$	DRM	600	101	1,450	1.5	30	0.86	[20]
Ni/Al$_2$O$_3$	DRM	650	101	1450	1.5	20	0.92	[20]
Ni/Al$_2$O$_3$	DRM	700	101	1,450	1.5	15	0.96	[20]
Ni/Al$_2$O$_3$	DRM	550	101	1150	1	30	0.62	[20]
Ni/Al$_2$O$_3$	DRM	600	101	1,150	1	30	0.73	[20]
Ni/Al$_2$O$_3$	DRM	650	101	1150	1	15	0.84	[20]
Ni/Al$_2$O$_3$	DRM	700	101	1,150	1	10	0.90	[20]
CoNi/Al$_2$O$_3$	CDRM	700	101	40,000	1	65	0.40	[23]
CoNi/Al$_2$O$_3$	CDRM	750	101	40,000	1	74	0.70	[23]
CoNi/Al$_2$O$_3$	CDRM	800	101	40,000	1	84	0.82	[23]
CoNi/Al$_2$O$_3$	CDRM	850	101	40,000	1	90	0.90	[23]
MgCoNi/Al$_2$O$_3$	CDRM	700	101	40,000	1	75	0.84	[23]
MgCoNi/Al$_2$O$_3$	CDRM	750	101	40,000	1	84	0.88	[23]
MgCoNi/Al$_2$O$_3$	CDRM	800	101	40,000	1	92	0.92	[23]
MgCoNi/Al$_2$O$_3$	CDRM	850	101	40,000	1	95	1.00	[23]
Nickel membrane	CSDRM	650	101	30,000	CO$_2$/H$_2$O = 1.0	10	2.20	[24]
Nickel membrane	CSDRM	700	101	30,000	CO$_2$/H$_2$O = 1.0	20	2.10	[24]
Nickel membrane	CSDRM	750	101	30,000	CO$_2$/H$_2$O = 1.0	22	2.00	[24]
Hongce lignite	CRM	1000	101	216	1.0	98	0.90	[30]
Shenmu bituminous coal	CRM	1000	101	216	1.0	94	0.80	[30]
Ni/MgAl$_2$O$_4$	CSCRM	650	101	530,000	2.5	32	2.30	[31]
Ni-2.5Ce/ MgAl$_2$O$_4$	CSCRM	650	101	530,000	2.5	36	2.35	[31]
Ni-5 Ce/ MgAl$_2$O$_4$	CSCRM	650	101	530,000	2.5	42	2.20	[31]
Ni-10Ce/ MgAl$_2$O$_4$	CSCRM	650	101	530,000	2.5	34	2.25	[31]
Ni/MgAl$_2$O$_4$	CSCRM	600	101	530,000	2.5	8	3.00	
Ni-2.5Ce/MgAl$_2$O$_4$	CSCRM	600	101	530,000	2.5	12	2.80	[31]
Ni-5 Ce/ MgAl$_2$O$_4$	CSCRM	600	101	530,000	2.5	16	2.70	[31]
Ni-10Ce/ MgAl$_2$O$_4$	CSCRM	600	101	530,000	2.5	15	2.85	[31]
Ni/Al$_2$O$_3$	CSDRM	600	101	12,000	2.1	25	2.70	[28]
Ni/Al$_2$O$_3$(M)	CSDRM	600	101	12,000	2.1	30	3.00	[28]

TABLE 8.1 (Continued)

Ni/Al_2O_3	CSDRM	650	101	12,000	2.1	43	2.50	[28]
Ni/Al_2O_3(M)	CSDRM	650	101	12,000	2.1	48	2.80	[28]
Ni/Al_2O_3	CSDRM	700	101	12,000	2.1	50	2.30	[28]
Ni/Al_2O_3(M)	CSDRM	700	101	12,000	2.1	55	2.40	[28]
$Ni_{0.5}Mg_{2.5}AlO_{4.5}$	CDRM	800	101	7,200	1.0	80	1.10	[32]
$Ni_{0.5}Mg_{2.5}Al_{0.96}$ $La_{0.04}O_{4.5}$	CDRM	800	101	7,200	1.0	82	1.15	[32]
$Ni_{0.5}Mg_{2.5}Al_{0.9}$ $La_{0.1}O_{4.5}$	CDRM	800	101	7,200	1.0	86	1.16	[32]
$Ni_{0.5}Mg_{2.5}Al_{0.85}$ $La_{0.15}O_{4.5}$	CDRM	800	101	7,200	1.0	84	1.18	[32]
$Sr_{0.8}La_{0.2}Ni_{0.3}$ $Al_{0.7}O_{2.6}$	CDRM	800	101	3,000	1.0	88.6	1.10	[33]
$Ir/Ce_{0.75}Zr_{0.25}O_2$ (550)	CDRM	700	101	10,000	$C_2H_6O:CO_2 = 1:1$	54	1.20	[34]
$Ir/Ce_{0.75}Zr_{0.25}O_2$ (700)	CDRM	700	101	10,000	$C_2H_6O:CO_2 = 1:1$	48	1.21	[34]
$Ir/Ce_{0.75}Zr_{0.25}O_2$ (850)	CDRM	700	101	10,000	$C_2H_6O:CO_2 = 1:1$	37	1.25	[34]
Ni(12)-Mo(1)	CSCRM	900	101	40,000	0.67	99	1.7	[35]
Ni(12)-Mo(1)-Sb(1)	CSCRM	900	101	40,000	0.67	91	1.7	[35]

Abbreviations: BRCD = Biogas reforming of carbon dioxide; CRM = carbon dioxide reforming of methane; CSCRM = combined steam and carbon dioxide reforming of methane; CSDRM = combined steam and carbon dioxide reforming of methane; DRM = dry reforming of methane.

of CO_2. The high quantity of CO_2 produced was contributed by the steam reforming of methane ($CH_4 + 2H_2O \rightarrow 4H_2 + CO_2$) and WGS reactions ($CO + H_2O \rightarrow H_2 + CO_2$), both enhanced by the presence of high-water content in the feed. Therefore, to increase the CO_2 conversion, the H_2O content in feed should be kept lower to enhance the dry reforming of methane (Eq. 4) at both studied temperatures. Akbari *et al.* [29] investigated the effect of CeO_2-promoted Ni-MgO-Al_2O_3 catalysts for the CO_2 reforming of methane. The CO_2 reforming of methane reaction was carried out in a quartz reactor at atmospheric pressure, and the temperature varied from 550°C to 700°C. It was reported that the CO_2 conversion was higher than the CH_4 conversion due to the simultaneous RWGS reaction ($H_2 + CO_2 \rightarrow CO + H_2O$) that occurred during the CO_2 reforming of methane. However, this RWGS reaction significantly affects the H_2/CO molar, with the maximum value obtained less than 0.9. Besides that, it was also reported that the reactant conversion increased with increased in temperature from 550°C to 700°C due to the endothermic properties of the CO_2 reforming of methane reaction. It was also proved that the CeO_2 promoted Ni-MgO-Al_2O_3 catalysts showed higher stability and high resistance against coke deposition due to the improvement of Ni dispersion and modification of the metal-support interaction.

8.3   CATALYST DEVELOPMENT

The biggest challenge in any process that utilizes or produces CO_2 is to develop the suitable catalysts in optimum operating conditions to avoid carbon formation. It is known that most transition metals can be applied as catalysts in the CO_2 reforming [36–44]. Moreover, noble metals are also considered catalysts for the CO_2 reforming process because of their high activity and stability during the reforming of CO_2 [43, 45–49]. One of the most important criteria in choosing the catalysts for CO_2 utilization processes is to choose a catalyst that can prevent coking, such as transition metals like Co, Ce, Ni, and Mo [50–53], and noble metals.

8.3.1   SELECTION OF METAL

The selection of suitable metal catalysts is vital to prevent coke formation on the surface of the catalysts and sintering effects at the high reaction temperature, leading to catalyst deactivation.

8.3.1.1   Noble Metals

Noble metals such as Rh, Ru, Pd, and Pt are greatly active when used in carbon reforming and highly resistant to carbon formation compared to other metals [45, 54–57]. The main disadvantage of using noble metals is that they are costly and hard to find [57], but their performance in terms of high activity and coke resistance cannot be compared with other conventional transition metals used. Li *et al.* [58], in their research, used Ru (2%) on various supports such as Al_2O_3, $MgAl_2O_4$, $Mg_3(Al)$ O, and MgO and got excellent results in terms of CO_2 conversion (90%) and also coke resistance. The catalysts were prepared via incipient wetness impregnation, which gave high dispersion of Ru on the supports, especially $Mg_3(Al)O$ where there is no sintering of Ru metal on the support. High dispersion of Ru gave better stability and activity of the catalysts, as the authors concluded. No coke was seen after 30 hours of reaction, but 8% of the catalysts had been deactivated, which is possibly due to other reasons instead of coke formation. While Chen and colleagues [59] synthesized, Ru (0.5, 1.0, 1.5, and 3.0 wt%) supported on $Ce_{0.75}Zr_{0.25}O_2$ by using the co-precipitation method gave a better result in terms of CO_2 conversion (98%) and also coke formation, where only 0.06 wt% was observed after 100 hours of run. They concluded that because of the Ru, which is highly dispersed on the support, it improved the interaction between the metal and support, leading to high stability, activity, and great anti-carbon deposition.

Reddy *et al.* [49] used the deposition-precipitation method to prepare their 1% Pt supported on ZrO_2/SiO_2. Their results were remarkable, where the volume of coke deposited on the catalyst was below the TGA detection limit. Moreover, the conversion of CO_2 was high, which is 96% at the temperature of 800 °C. Some research had been done on bimetallic noble metals to improve their stability and performance. Du *et al.* [46] used incipient wetness impregnation, synthesizing $Pt-Ru/Al_2O_3$ at the temperature of 800 °C and converted CO_2 at 96%. They determined that the bimetallic noble metals suppress the formation of carbon on the catalyst for more than 500 hours.

8.3.1.2 Transition Metals Catalysts

Nickel is one of the most popular metal to be utilized as a catalyst for reforming technology. This is because nickel is cheaper, widely available, and comparable in terms of performance to noble metals such as Pt, Pd, and Ru [38, 40, 43, 60–68]. The drawback of using nickel compared to noble metals is that nickel disintegrates easy and is less stable. To overcome this drawback, much research was done to support nickel with other compounds, such as zeolites and oxides. In terms of CO_2 reforming, basicity of the support plays a major role in speeding up the reaction rate by boosting the CO_2 activation. Vafaeian et al. [69] used Ni with ZSM-5 in CO_2 reforming of methane to improve nickel performance. They reported that zeolite has potential as a support in CO_2 reforming, as zeolite has excellent properties in terms of high surface area, high affinity for CO_2 adsorbent, good micropore structure, and high thermal stability, hence increasing the activity and stability of catalyst in the CO_2 reforming [69–70]. They utilize the sonochemical method to synthesize the catalyst. Ultrasound irradiation was reported to improve the dispersion of metals and prevent agglomeration during the synthesis process. Their study shows that the conversion of CH_4 and CO_2 is more than 70% at the temperature of 800 °C, comparable to the study of [71] by using Pt as the catalyst.

Cobalt is also one of the most promising transition metals used for reforming CO_2 and showed quite comparable results with nickel catalysts. Tang et al. [72] used sol-gel 10% Co-Al at 700 °C and got a result of the CO_2 conversion rate of 93% with the coking rate of 0.0025 g (carbon)/g (cat) h, which is less than 0.01 after 6 hours of reaction. Another notable result using Co catalyst was achieved by Ruckenstein and Wang [73]; they calculated a conversion of CO_2 at 93% by using 9%Co/Al_2O_3. The catalyst only deactivates for 2.1% after 50 hours of reaction. Furthermore, it is also reported that adding another metal to the catalyst would enhance the stability, improve reducibility, and decrease the coke forming on the catalysts [74–77]. Bimetallic catalysts are frequently better in performance because both metals functioned and were active during the reaction. Turap et al. [74] stated that adding Co to Ni improved the activity and stability. The method that had been applied for their research is incipient wet impregnation synthesis. The stability of the catalyst improved mainly because of the carbon removal efficiency resulting from the promotion of surface oxygen adsorption on the Co. Besides, they indicated that the Co-Ni alloy had weak chemisorption towards hydrogen, hence preventing the RWGS reaction, decreasing the carbon formation on the catalyst even after 10 hours of reaction. From their study, the conversion rate of CH_4 and CO_2 is more than 80% at 800 °C, and nearly 100% at 850 °C by using the Co-Ni/CeO_2, which is better than the monometallic metal that they compare with, Ni/CeO_2. Gonzalez et al. [78] and Feng et al. [79] also researched bimetallic catalysts, where they utilized Ni-Co/ZrO_2 and Ni-Co/Si_3N_4, respectively. They also reported that there was less carbon formation by using bimetallics, and the catalysts gave better activity and performance. Another study that agreed with the statement was performed by Zhang and teammates [80], where they observed no carbon deposited on the catalyst even after 250 hours of run. The catalyst that they used was 0.02 Ni-0.03 Co synthesized via the co-precipitation method.

Other than that, some research had been done on trimetallic catalysts of nickel to improve the capabilities of the catalysts further. Yong *et al.* [81] synthesized $LaNi_{0.34}Co_{0.33}Mn_{0.33}O_3$ and ran it with dry reforming of methane. From their finding, the reaction rate was improved by the addition of Co, and the addition of Mn promotes the stability of the catalysts because of the synergistic effect of the tri-metals. As for Bhavani [82], they did the study on tri-metals of NiCoMn supported on ZrO_2. Their study found out that the doped $NiCoMn/ZrO_2$ with other alkaline metals (Ce, La, Ca, K) gave a high conversion of CO_2 (95.5%) and high selectivity towards H_2. The doped catalyst gave a high performance in terms of activity and stability where there is close to negligible carbon formation. The tri-metals affect the catalysts in many ways, such as reducing the crystal size, greatly improving dispersion of the metals, and increasing the availability of active sites.

8.3.1.3 Combination of Transition and Noble Metals

By combining noble and transition metals, some of the disadvantages of each part can be overcome. For example, the lack of activity and low stability of transition metals can be overcome by adding a few noble metals. On the other hand, the pricey and rare noble metals can be covered by adding more percentage of transition metals to the catalysts. In addition, the bimetallic catalyst had shown a promising catalyst activity and high resistance over coke formation during the CO_2 reforming reaction. Figure 8.5 shows the simplified mechanism of bimetallic catalysts to improve the catalyst activity and high resistance over coke formation during reforming the CO_2 response.

Singha *et al.* [84] combined Pd with CeO_2 by using the surfactant-induced method and observed that the catalyst exhibited activity at low temperatures (300–500 °C) due to the good dispersion of metals. The catalyst also improved coke and sintering resistance during the reaction and gave a high conversion of CO_2 (95%). Ni-Au-Pt catalyst supported on alumina was prepared by Wu [85] by using the sol-gel + impregnation method. The result was that the composite proved to be better than the monometallic of each metal. The addition of Au and Pt to the Ni demonstrated synergistic interaction, which enhanced the catalyst in terms of coke resistance, stability, and performance (conversion of CO_2 = 93%) by facilitating NiO reduction. Aramouni

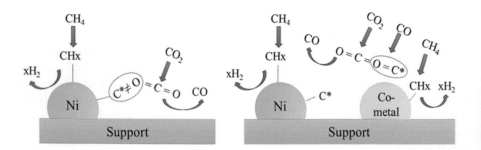

FIGURE 8.5 Simplified mechanism of bimetallic catalyst for catalytic activity improvement and high resistance over coke formation.

Source: **Adapted from [83].**

TABLE 8.2
CO_2 Reforming Using Various Catalysts at Their Optimized Reaction Conditions

Catalyst	Reaction condition		Preparation method	Carbon formation/ stability	CO_2 conversion (%)	Ref
	T (°C)	GHSV (ml g^{-1} h^{-1})				
CoAl	700	40,000	Sol-gel	Rate = 0.0025 g (carbon)/g (cat) h	93	[72]
Co (12 wt%)/MgO	900	60,000	conventional impregnation	N/A	94	[87]
Co(9 wt%)/Al$_2$O$_3$	900	60,000	conventional impregnation	2.1% deactivate after 50 hours	93	[73]
Ni(2.8–58.6 wt.%)/ MgO Al$_2$O$_3$	750	20,000	co-precipitation	Rate = 0.0002 g (carbon)/g (cat) h	92	[88]
Ni/Al$_2$O$_3$	800	6,545,000	atomic layer deposition	steady activity after 50 hours	97	[89]
Ru/Mg$_3$(Al)O	750	60,000	incipient wetness impregnation	8 wt % deactivated after 30 hours	90	[58]
1.5 wt% Ru/CeO	900	12,000	co-precipitation method	0.06 wt% after 100 hours	98	[59]
(1 wt%) Pt/ZrO$_2$/ SiO$_2$	800	400,000	Deposition precipitation method	Below detection range of TGA	96	[73]
(0.5 wt%) Pt/ZrO$_2$	852	25,584	wet impregnation	0.0002 g (carbon)/g (cat) h	83	[90]
Co$_2$Ni$_2$Mg$_2$Al$_2$	800	32,000	co-precipitation	N/A	95	[91]
4.0 Ni-3.6Co/ Si$_3$ N$_4$	800	12,000	reactions between silicon nitride and metal halides	No carbon formation after 100 hours	90	[79]
La/Ni	750	30,000	situ one-pot hydrothermal method	Resist coke formation for 22 hours	96	[92]
Ni-3Ce$_2$Ca	800	36,000	co-impregnation	High stable activity for 14 hours	95	[93]
B/Ni	750	36,000	sequential incipient wetness impregnation	1.3 wt% loss after 10 hours	92	[94]
Gd/Ni	600	39,000	co-impregnation	no weight loss	84	[95]
NiCoMn/ZrO2 doped with Ce	800	243,000	co-precipitation	0.05 wt%	95.5	[82]
Ni@HSS	800	144,000	facile one-pot micelle method	no carbon formation	95	[96]
40 LaNi$_{0.75}$Fe$_{0.25}$O$_3$/ SiO^2	500	600,000	Sol-gel	Most stable and effective for 15 hours DRM reaction	96	[97]
Ni-ZrO$_2$@SiO$_2$	800	18,000	one-pot synthesis	no coke formation	93	[98]

et al. [86] developed the material with zero carbon formation and zero activity loss, even after 47 hours of reaction using their catalyst, Ni-Co-Ru/MgO-Al$_2$O$_3$ synthesized via the neutral sol-gel method. They concluded that the main reason for the good performance is because of the small pore size of their catalyst. Al-Doghachi *et al.* [47] calculated a 99% conversion of CO$_2$ by using their trimetallic catalyst, which is Pt, Pd, Ni/MgO, a trimetallic promoted by Ce (III). The catalyst was very active at 900 °C and had low carbon formation and high thermal stability for 200 hours.

By selecting the best catalyst with the suitable synthesis method, CO$_2$ reforming can be developed to its highest potential to yield the best results from the reaction. Furthermore, it can be suggested that by introducing a minor quantity of noble metals into the transition metal catalysts, the performance and stability of the catalyst can be enhanced and perfected for the utilization of CO$_2$. Some of the research of excellent catalysts done on CO$_2$ reforming are shown in Table 8.2.

8.4 CONCLUSIONS

Several conclusions can be drawn from the critical analysis and our understanding of the literature. The CO$_2$ reforming process was identified as the most feasible technology for capturing CO$_2$ for syngas production. The syngas production utilizing CO$_2$ as feedstock had been demonstrated via various reforming processes, such as biogas reforming of carbon dioxide, dry reforming of methane, combined steam and dry reforming of methane, carbon dioxide reforming of methane, and combined steam and carbon dioxide reforming of methane. However, during the reforming process, other side reactions such as WGS or RWGS reactions simultaneously occurred. These WGS or RWGS reaction significantly affect syngas production and the conversion of the reactants. Besides that, the other challenges during the CO$_2$ reforming reaction are the catalysts' sintering effects at high reaction temperature and coke formation, which led to catalyst deactivation. Therefore, selecting suitable metal-based catalysts is necessary to improve the stability of the catalyst at high temperatures and high resistance over coke formation. Nickel-based catalysts is the commonly used catalyst for CO$_2$ reforming; however, some modifications can be carried out to improve its performance by adding other metals as promoters. It was determined that bimetallic catalysts showed higher catalytic activity and high resistance over coke formation during the CO$_2$ reforming reaction. It is believed that a more extensive pilot-scale study is needed for the utilization of the captured CO$_2$ for biofuel production via reforming processes.

REFERENCES

1. U. E. I. Administration, *International Energy Outlook, 2010*. Government Printing Office, 2010.
2. S. Nanda, S. N. Reddy, S. K. Mitra and J. A. Kozinski, "The progressive routes for carbon capture and sequestration," *Energy Science & Engineering*, vol. 4, no. 2, pp. 99–122, 2016.
3. F. Isa *et al.*, "An overview on CO$_2$ removal via absorption: Effect of elevated pressures in counter-current packed column," *Journal of Natural Gas Science and Engineering*, vol. 33, pp. 666–677, 2016.

4. L. C. Barbosa, O. d. Q. F. Araújo and J. L. de Medeiros, "Carbon capture and adjustment of water and hydrocarbon dew-points via absorption with ionic liquid [Bmim] [NTf2] in offshore processing of CO_2-rich natural gas," *Journal of Natural Gas Science and Engineering*, vol. 66, pp. 26–41, 2019.
5. L. C. Barbosa, M. V. d. C. Nascimento, F. A. Ofélia de Queiroz and J. L. de Medeiros, "A cleaner and more sustainable decarbonation process via ionic-liquid absorption for natural gas with high carbon dioxide content," *Journal of Cleaner Production*, vol. 242, p. 118421, 2020.
6. L. Han *et al.*, "Enhanced hydrogen production via catalytic toluene reforming with in situ carbon dioxide capture: Effects of a hybrid iron-calcium composite prepared by impregnation," *Energy Conversion and Management*, vol. 214, p. 112834, 2020/06/15/ 2020, doi:10.1016/j.enconman.2020.112834.
7. M. A. Abdulhameed *et al.*, "Carbon dioxide capture using a superhydrophobic ceramic hollow fibre membrane for gas-liquid contacting process," *Journal of Cleaner Production*, vol. 140, pp. 1731–1738, 2017.
8. A. Huang, L.-H. Chen, C.-H. Chen, H.-Y. Tsai and K.-L. Tung, "Carbon dioxide capture using an omniphobic membrane for a gas-liquid contacting process," *Journal of Membrane Science*, vol. 556, pp. 227–237, 2018.
9. L. Liu, W. Qiu, E. S. Sanders, C. Ma and W. J. Koros, "Post-combustion carbon dioxide capture via 6FDA/BPDA-DAM hollow fiber membranes at sub-ambient temperatures," *Journal of Membrane Science*, vol. 510, pp. 447–454, 2016.
10. H.-C. Wu, Z. Rui and J. Y. Lin, "Hydrogen production with carbon dioxide capture by dual-phase ceramic-carbonate membrane reactor via steam reforming of methane," *Journal of Membrane Science*, vol. 598, p. 117780, 2020.
11. X. Zhu, Y. Shi and N. Cai, "Integrated gasification combined cycle with carbon dioxide capture by elevated temperature pressure swing adsorption," *Applied Energy*, vol. 176, pp. 196–208, 2016.
12. L. A. Darunte, K. S. Walton, D. S. Sholl and C. W. Jones, "CO_2 capture via adsorption in amine-functionalized sorbents," *Current Opinion in Chemical Engineering*, vol. 12, pp. 82–90, 2016.
13. R. Zhao, L. Liu, L. Zhao, S. Deng, S. Li and Y. Zhang, "A comprehensive performance evaluation of temperature swing adsorption for post-combustion carbon dioxide capture," *Renewable and Sustainable Energy Reviews*, vol. 114, p. 109285, 2019.
14. T. Xu *et al.*, "Syngas production via chemical looping reforming biomass pyrolysis oil using NiO/dolomite as oxygen carrier, catalyst or sorbent," *Energy Conversion and Management*, vol. 198, p. 111835, 2019.
15. Z. Wang, L. Li and G. Zhang, "Life cycle greenhouse gas assessment of hydrogen production via chemical looping combustion thermally coupled steam reforming," *Journal of Cleaner Production*, vol. 179, pp. 335–346, 2018.
16. R. Pérez-Vega *et al.*, "Coal combustion via chemical looping assisted by oxygen uncoupling with a manganese-iron mixed oxide doped with titanium," *Fuel Processing Technology*, vol. 197, p. 106184, 2020.
17. H. Bahzad *et al.*, "Development and techno-economic analyses of a novel hydrogen production process via chemical looping," *International Journal of Hydrogen Energy*, vol. 44, no. 39, pp. 21251–21263, 2019.
18. A. Al-Mamoori, A. Krishnamurthy, A. A. Rownaghi and F. Rezaei, "Carbon capture and utilization update," *Energy Technology*, vol. 5, no. 6, pp. 834–849, 2017.
19. Y. Zhan *et al.*, "Biogas reforming of carbon dioxide to syngas production over Ni-Mg-Al catalysts," *Molecular Catalysis*, vol. 436, pp. 248–258, 2017, doi:10.1016/j.mcat.2017.04.032.
20. V. Rathod and P. V. Bhale, "Experimental investigation on biogas reforming for syngas production over an alumina based nickel catalyst," *Energy Procedia*, vol. 54, pp. 236–245, 2014.

21. G. Garbarino *et al.*, "A study of Ni/Al2O3 and Ni-La/Al$_2$O$_3$ catalysts for the steam reforming of ethanol and phenol," *Applied Catalysis B: Environmental*, vol. 174–175, pp. 21–34, 2015, doi:10.1016/j.apcatb.2015.02.024.

22. K. Jabbour, P. Massiani, A. Davidson, S. Casale, and N. El Hassan, "Ordered mesoporous 'one-pot' synthesized Ni-Mg (Ca)-Al$_2$O$_3$ as effective and remarkably stable catalysts for combined steam and dry reforming of methane (CSDRM)," *Applied Catalysis B: Environmental*, vol. 201, pp. 527–542, 2017.

23. I. H. Son, S. J. Lee and H.-S. Roh, "Hydrogen production from carbon dioxide reforming of methane over highly active and stable MgO promoted Co-Ni/γ-Al$_2$O$_3$ catalyst," *International Journal of Hydrogen Energy*, vol. 39, no. 8, pp. 3762–3770, 2014.

24. S.-K. Ryi, S.-W. Lee, J.-W. Park, D.-K. Oh, J.-S. Park and S. S. Kim, "Combined steam and CO$_2$ reforming of methane using catalytic nickel membrane for gas to liquid (GTL) process," *Catalysis Today*, vol. 236, pp. 49–56, doi:10.1016/j.cattod.2013.11.001.

25. J. S. Tan, H. T. Danh, S. Singh, Q. D. Truong, H. D. Setiabudi and D. V. N. Vo, "Syngas Production from CO$_2$ Reforming and CO$_2$-steam Reforming of Methane over Ni/Ce-SBA-15 Catalyst," *IOP Conference Series: Materials Science and Engineering*, vol. 206, p. 012017, 2017, doi:10.1088/1757-899x/206/1/012017.

26. F. Mirzaei, M. Rezaei, F. Meshkani and Z. Fattah, "Carbon dioxide reforming of methane for syngas production over Co-MgO mixed oxide nanocatalysts," *Journal of Industrial and Engineering Chemistry*, vol. 21, pp. 662–667, 2015, doi:10.1016/j.jiec.2014.03.034.

27. D. P. Minh, T. S. Pham, D. Grouset and A. Nzihou, "Thermodynamic equilibrium study of methane reforming with carbon dioxide, water and oxygen," *Journal of Clean Engineering Technologies*, vol. 6, no. 4, pp. 309–313, 2018.

28. M. Dan, M. Mihet and M. D. Lazar, "Hydrogen and/or syngas production by combined steam and dry reforming of methane on nickel catalysts," *International Journal of Hydrogen Energy*, vol. 45, no. 49, pp. 26254–26264, 2020, doi:10.1016/j.ijhydene.2019.12.158.

29. E. Akbari, S. M. Alavi and M. Rezaei, "CeO$_2$ Promoted Ni-MgO-Al$_2$O$_3$ nanocatalysts for carbon dioxide reforming of methane," *Journal of CO$_2$ Utilization*, vol. 24, pp. 128–138, 2018, doi:10.1016/j.jcou.2017.12.015.

30. F. Guo, Y. Zhang, G. Zhang and H. Zhao, "Syngas production by carbon dioxide reforming of methane over different semi-cokes," *Journal of Power Sources*, vol. 231, pp. 82–90, 2013, doi:10.1016/j.jpowsour.2013.01.003.

31. K. Y. Koo, S.-h. Lee, U. H. Jung, H.-S. Roh and W. L. Yoon, "Syngas production via combined steam and carbon dioxide reforming of methane over Ni-Ce/MgAl$_2$O$_4$ catalysts with enhanced coke resistance," *Fuel Processing Technology*, vol. 119, pp. 151–157, 2014, doi:10.1016/j.fuproc.2013.11.005.

32. X. Yu, N. Wang, W. Chu and M. Liu, "Carbon dioxide reforming of methane for syngas production over La-promoted NiMgAl catalysts derived from hydrotalcites," *Chemical Engineering Journal*, vol. 209, pp. 623–632, 2012, doi:10.1016/j.cej.2012.08.037.

33. M. K. Nikoo and N. A. S. Amin, "Thermodynamic analysis of carbon dioxide reforming of methane in view of solid carbon formation," *Fuel Processing Technology*, vol. 92, no. 3, pp. 678–691, 2011, doi:10.1016/j.fuproc.2010.11.027.

34. F. Qu *et al.*, "Syngas Production from Carbon Dioxide Reforming of Ethanol over Ir/Ce$_{0.75}$Zr$_{0.25}$O$_2$ Catalyst: Effect of Calcination Temperatures," *Energy & Fuels*, vol. 32, no. 2, pp. 2104–2116, 2018, doi:10.1021/acs.energyfuels.7b03945.

35. H. Ryoo, B. C. Ma and Y. C. Kim, "Syngas Production via Combined Steam and Carbon Dioxide Reforming of Methane over Ni-Mo-Sb/Al$_2$O$_3$ Catalysts," *Journal of Nanoscience and Nanotechnology*, vol. 19, no. 2, pp. 988–990, 2019.

36. C. Crisafulli, S. Scirè, S. Minicò and L. J. Solarino, "Ni-Ru bimetallic catalysts for the CO$_2$ reforming of methane," *Applied Catalysis A: General*, vol. 225, no. 1–2, pp. 1–9, 2002.

37. A. Birot, F. Epron, C. Descorme and D. J. Duprez, "Ethanol steam reforming over Rh/ CexZr1-xO₂ catalysts: Impact of the CO-CO₂-CH₄ interconversion reactions on the H₂ production," *Applied Catalysis B: Environmental*, vol. 79, no. 1, pp. 17–25, 2008.

38. P. Frontera *et al.*, "Bimetallic zeolite catalyst for CO₂ reforming of methane," *Topics in Catalysis*, vol. 53, no. 3–4, pp. 265–272, 2010.

39. S. Gheno, S. Damyanova, B. Riguetto and C. M.P. Marques, "CO₂ reforming of CH₄ over Ru/zeolite catalysts modified with Ti," *Molecular Catalysis*, vol. 198, no. 1–2, pp. 263–275, 2003.

40. N. Gokon, Y. Yamawaki, D. Nakazawa and T. Kodama, "Ni/MgO-Al₂O₃ and Ni-Mg-O catalyzed SiC foam absorbers for high temperature solar reforming of methane," *International Journal of Hydrogen Energy*, vol. 35, no. 14, pp. 7441–7453, 2010.

41. M. Yang, H. Guo, Y. Li and Q. Dang, "CH₄-CO₂ reforming to syngas over Pt-CeO₂- ZrO₂/MgO catalysts: Modification of support using ion exchange resin method," *Journal of Natural Gas Chemistry*, vol. 21, no. 1, pp. 76–82, 2012.

42. S. S.-Y. Lin, H. Daimon and S. Y. Ha, "Co/CeO₂-ZrO₂ catalysts prepared by impregnation and coprecipitation for ethanol steam reforming," *Applied Catalysis A: General*, vol. 366, no. 2, pp. 252–261, 2009.

43. Ş. Özkara-Aydınoğlu and A. E. Aksoylu, "CO₂ reforming of methane over Pt-Ni/Al₂O₃ catalysts: Effects of catalyst composition, and water and oxygen addition to the feed," *International Journal of Hydrogen Energy*, vol. 36, no. 4, pp. 2950–2959, 2011.

44. S. Tanaka *et al.*, "Preparation of highly dispersed silica-supported palladium catalysts by a complexing agent-assisted sol-gel method and their characteristics," *Applied Catalysis A: General*, vol. 229, no. 1–2, pp. 165–174, 2002.

45. D. Pakhare and J. Spivey, "A review of dry (CO₂) reforming of methane over noble metal catalysts," *Chemical Society Reviews*, vol. 43, no. 22, pp. 7813–7837, 2014.

46. J. Du *et al.*, "Carbon dioxide reforming of methane over bimetallic catalysts of Pt-Ru/γ- Al₂O₃ for thermochemical energy storage," *Journal of Central South University*, vol. 20, no. 5, pp. 1307–1313, 2013.

47. F. A. Al-Doghachi, U. Rashid and Y. H. Taufiq-Yap, "Investigation of Ce(iii) promoter effects on the tri-metallic Pt, Pd, Ni/MgO catalyst in dry-reforming of methane," *RSC Advances*, vol. 6, no. 13, pp. 10372–10384, 2016.

48. M. C. Bradford and M. A. Vannice, "Metal-support interactions during the CO₂ reforming of CH₄ over model TiOx/Pt catalysts," *Catalysis Letters*, vol. 48, no. 1–2, pp. 31–38, 1997.

49. G. K. Reddy, S. Loridant, A. Takahashi, P. Delichère and B. M. Reddy, "Reforming of methane with carbon dioxide over Pt/ZrO₂/SiO₂ catalysts – Effect of zirconia to silica ratio," *Applied Catalysis A: General*, vol. 389, no. 1–2, pp. 92–100, 2010.

50. M. D. Porosoff, S. Kattel, W. Li, P. Liu and J. G. Chen, "Identifying trends and descriptors for selective CO₂ conversion to CO over transition metal carbides," *Chemical Communications*, vol. 51, no. 32, pp. 6988–6991, 2015.

51. J. H. Oh *et al.*, "Importance of exsolution in transition-metal (Co, Rh, and Ir)-doped LaCrO₃ perovskite catalysts for boosting dry reforming of CH₄ using CO₂ for hydrogen production," *Ind. Eng. Chem. Res.*, vol. 58, no. 16, pp. 6385–6393, 2019.

52. P. Djinović and A. Pintar, "Stable and selective syngas production from dry CH₄-CO₂ streams over supported bimetallic transition metal catalysts," *Applied Catalysis B: Environmental*, vol. 206, pp. 675–682, 2017.

53. C. Kunkel, F. Vines and F. J. E. Illas, "Transition metal carbides as novel materials for CO₂ capture, storage, and activation," *Energy & Environmental Science*, vol. 9, no. 1, pp. 141–144, 2016.

54. J. Kehres *et al.*, "Dynamical properties of a Ru/MgAl₂O₄ catalyst during reduction and dry methane reforming," *J. Phys. Chem. C*, vol. 116, no. 40, pp. 21407–21415, 2012.

55. K. Sutthiumporn and S. Kawi, "Promotional effect of alkaline earth over Ni-La₂O₃ catalyst for CO₂ reforming of CH₄: Role of surface oxygen species on H₂ production

and carbon suppression," *International Journal of Hydrogen Energy*, vol. 36, no. 22, pp. 14435–14446, 2011.

56. M. Usman, W. W. Daud, H. F. J. R. Abbas and S. E. Reviews, "Dry reforming of methane: Influence of process parameters – a review," *Renewable and Sustainable Energy Reviews*, vol. 45, pp. 710–744, 2015.

57. R. Singh, A. Dhir, S. K. Mohapatra, S. K. and Mahla, S. K., "Dry reforming of methane using various catalysts in the process," vol. 10, no. 2, pp. 567–587, 2020.

58. D. Li, R. Li, M. Lu, X. Lin, Y. Zhan and L. Jiang, "Carbon dioxide reforming of methane over Ru catalysts supported on Mg-Al oxides: A highly dispersed and stable Ru/Mg (Al) O catalyst," *Applied Catalysis B: Environmental*, vol. 200, pp. 566–577, 2017.

59. J. Chen, C. Yao, Y. Zhao and P. J. Jia, "Synthesis gas production from dry reforming of methane over $Ce_{0.75}Zr_{0.25}O_2$-supported Ru catalysts," *International Journal of Hydrogen Energy*, vol. 35, no. 4, pp. 1630–1642, 2010.

60. B. Fidalgo, A. Arenillas and J. J. Menéndez, "Mixtures of carbon and Ni/Al_2O_3 as catalysts for the microwave-assisted CO_2 reforming of CH_4," *Fuel Processing Technology*, vol. 92, no. 8, pp. 1531–1536, 2011.

61. M.-S. Fan, A. Z. Abdullah and S. J. Bhatia, "Hydrogen production from carbon dioxide reforming of methane over $Ni-Co/MgO-ZrO_2$ catalyst: Process optimization," *International Journal of Hydrogen Energy*, vol. 36, no. 8, pp. 4875–4886, 2011.

62. L. Huang, F. Zhang, N. Wang, R. Chen and A. T. J. Hsu, "Nickel-based perovskite catalysts with iron-doping via self-combustion for hydrogen production in auto-thermal reforming of ethanol," *International Journal of Hydrogen Energy*, vol. 37, no. 2, pp. 1272–1279, 2012.

63. D. Liu, R. Lau, A. Borgna and Y. J. Yang, "Carbon dioxide reforming of methane to synthesis gas over Ni-MCM-41 catalysts," *Applied Catalysis A: General*, vol. 358, no. 2, pp. 110–118, 2009.

64. D. San-José-Alonso, J. Juan-Juan, M. Illán-Gómez and M. J. Román-Martínez, "Ni, Co and bimetallic Ni-Co catalysts for the dry reforming of methane," *Applied Catalysis A: General*, vol. 371, no. 1–2, pp. 54–59, 2009.

65. L. Chen, Q. Zhu and R. J. Wu, "Effect of Co-Ni ratio on the activity and stability of Co-Ni bimetallic aerogel catalyst for methane Oxy-CO_2 reforming," *International Journal of Hydrogen Energy*, vol. 36, no. 3, pp. 2128–2136, 2011.

66. M. Crişan et al., "Sol-gel based alumina powders with catalytic applications," *Applied Surface Science*, vol. 258, no. 1, pp. 448–455, 2011.

67. A. Al-Fatesh and A. J. Fakeeha, "Effects of calcination and activation temperature on dry reforming catalysts," *Journal of Saudi Chemical Society*, vol. 16, no. 1, pp. 55–61, 2012.

68. S. Damyanova, B. Pawelec, K. Arishtirova and J. J. Fierro, "Ni-based catalysts for reforming of methane with CO_2," *International Journal of Hydrogen Energy*, vol. 37, no. 21, pp. 15966–15975, 2012.

69. Y. Vafaeian, M. Haghighi, S. Aghamohammadi, "Ultrasound assisted dispersion of different amount of Ni over ZSM-5 used as nanostructured catalyst for hydrogen production via CO_2 reforming of methane," *Energy Conversion and Management*, vol. 76, pp. 1093–1103, 2013.

70. A. N. Pinheiro, A. Valentini, J. M. Sasaki and A. C. Oliveira, "Highly stable dealuminated zeolite support for the production of hydrogen by dry reforming of methane," *Applied Catalysis A: General*, vol. 355, no. 1–2, pp. 156–168, 2009.

71. B. Sarkar et al., "Pt nanoparticles supported on mesoporous ZSM-5: A potential catalyst for reforming of methane with carbon dioxide," *Indian Journal of Chemistry*, vol. 51, 2012.

72. S. Tang, L. Ji, J. Lin, H. Zeng, K. Tan and K. J. Li, "CO_2 reforming of methane to synthesis gas over sol-gel-made $Ni/\gamma-Al_2O_3$ catalysts from organometallic precursors," *Journal of Catalysis*, vol. 194, no. 2, pp. 424–430, 2000.

73. E. Ruckenstein and H. J. Wang, "Carbon deposition and catalytic deactivation during CO_2 reforming of CH_4 over Co/γ-Al_2O_3 catalysts," *Journal of Catalysis*, vol. 205, no. 2, pp. 289–293, 2002.

74. Y. Turap, I. Wang, T. Fu, Y. Wu, Y. Wang and W. J. Wang, "Co-Ni alloy supported on CeO_2 as a bimetallic catalyst for dry reforming of methane," *International Journal of Hydrogen Energy*, vol. 45, no. 11, pp. 6538–6548, 2020.

75. N. A. Abd Ghani, A. Azapour, S. A. F. a. S. Muhammad, N. M. Ramli, D.-V. N. Vo and B. J. Abdullah, "Dry reforming of methane for syngas production over Ni-Co-supported Al_2O_3-MgO catalysts," *Applied Petrochemical Research*, vol. 8, no. 4, pp. 263–270, 2018.

76. Y.-M. Dai, C.-Y. Lu and C.-J. J. Chang, "Catalytic activity of mesoporous Ni/CNT, Ni/SBA-15 and (Cu, Ca, Mg, Mn, Co) – Ni/SBA-15 catalysts for CO_2 reforming of CH_4," *RSC Advances*, vol. 6, no. 77, pp. 73887–73896, 2016.

77. K. Y. Koo, S. H. Lee, U. H. Jung, H. S. Roh and W. L. Yoon, "Syngas production via combined steam and carbon dioxide reforming of methane over Ni – Ce/$MgAl_2O_4$ catalysts with enhanced coke resistance," *Journal of Fuel Processing Technology*, vol. 119, pp. 151–157, 2014.

78. V. M. Gonzalez-delaCruz, R. Pereniguez, F. Ternero, J. P. Holgado and A. J. T. Caballero, "In situ XAS study of synergic effects on Ni-Co/ZrO_2 methane reforming catalysts," *J. Phys. Chem. C*, vol. 116, no. 4, pp. 2919–2926, 2012.

79. T. C. Feng, W. T. Zheng, K. Q. Sun and B. Q. J. C. C. Xu, "CO_2 reforming of methane over coke-resistant Ni – Co/Si_3N_4 catalyst prepared via reactions between silicon nitride and metal halides," *Catalyst Communications*, vol. 73, pp. 54–57, 2016.

80. J. Zhang, H. Wang and A. K. J. Dalai, "Effects of metal content on activity and stability of Ni-Co bimetallic catalysts for CO_2 reforming of CH_4," *Applied Catalysis A: General*, vol. 339, no. 2, pp. 121–129, 2008.

81. W. Y. Kim, J. S. Jang, E. C. Ra, K. Y. Kim, E. H. Kim and J. S. Lee, "Reduced perovskite $LaNiO_3$ catalysts modified with Co and Mn for low coke formation in dry reforming of methane," *Applied Catalysis A: General*, vol. 575, pp. 198–203, 2019. [Online]. Available: https://ur.booksc.eu/book/82057791/8d50e1.

82. A. G. Bhavani, W. Y. Kim, J. Y. Kim and J. S. Lee, "Improved activity and coke resistance by promoters of nanosized trimetallic catalysts for autothermal carbon dioxide reforming of methane," *Applied Catalysis A: General*, vol. 450, pp. 63–72, 2013.

83. M. A. A. Aziz, H. D. Setiabudi, L. P. Teh, M. Asmadi, J. Matmin and S. Wongsakulphasatch, "High-performance bimetallic catalysts for low-temperature carbon dioxide reforming of methane," *Chemical Engineering & Technology*, vol. 43, no. 4, pp. 661–671, 2020.

84. R. K. Singha, A. Yadav, A. Shukla, M. Kumar and R. Bal, "Low temperature dry reforming of methane over Pd-CeO_2 nanocatalyst," *Catalysis Communications*, vol. 92, pp. 19–22, 2017. [Online]. Available: https://www.infona.pl/resource/bwmeta1.element.elsevier-6fc67136-5790-3724-b64c-b1ac129bf7e9.

85. H. Wu et al., "Bi-and trimetallic Ni catalysts over Al_2O_3 and Al_2O_3-MO_x (M = Ce or Mg) oxides for methane dry reforming: Au and Pt additive effects," *Applied Catalysis B: Environmental*, vol. 156, pp. 350–361, 2014. [Online]. Available: https://researchportal.unamur.be/en/publications/bi-and-trimetallic-ni-catalysts-over-alsub2subosub3sub-and-alsub2

86. N. A. K. Aramouni, J. Zeaiter, W. Kwapinski, J. J. Leahy and M. N. J. Ahmad, "Eclectic trimetallic Ni-Co-Ru catalyst for the dry reforming of methane," *International Journal of Hydrogen Energy*, vol. 45, 2020.

87. J. Li, J. Li and Q. J. Zhu, "Carbon deposition and catalytic deactivation during CO_2 reforming of CH_4 over Co/MgO catalyst," *Chinese Journal of Chemical Engineering*, vol. 26, no. 11, pp. 2344–2350, 2018.

88. E. Akbari, S. M. Alavi and M. Rezaei, "Synthesis gas production over highly active and stable nanostructured NiMgOAl$_2$O$_3$ catalysts in dry reforming of methane: Effects of Ni contents," *Fuel*, vol. 194, pp. 171–179, 2017.

89. E. Baktash, P. Littlewood, R. Schomäcker, A. Thomas and P. C. Stair, "Alumina coated nickel nanoparticles as a highly active catalyst for dry reforming of methane," *Applied Catalysis B: Environmental*, vol. 179, pp. 122–127, 2015.

90. K. Nagaoka, K. Seshan, K.-i. Aika and J. A. Lercher, "Carbon deposition during carbon dioxide reforming of methane – comparison between Pt/Al$_2$O$_3$ and Pt/ZrO$_2$," *Journal of Catalysis*, vol. 197, no. 1, pp. 34–42, 2001.

91. C. Tanios *et al.*, "Syngas production by the CO$_2$ reforming of CH$_4$ over Ni-Co-Mg-Al catalysts obtained from hydrotalcite precursors," *International Journal of Hydrogen Energy*, vol. 42, no. 17, pp. 12818–12828, 2017.

92. A. Abdulrasheed, A. Jalil, M. Hamid, T. Siang and T. Abdullah, "Dry reforming of CH$_4$ over stabilized Ni-La@ KCC-1 catalyst: Effects of La promoter and optimization studies using RSM," *Journal of CO$_2$ Utilization*, vol. 37, pp. 230–239, 2020.

93. Y. Sun, G. Zhang, J. Liu, Y. Xu and Y. J. Lv, "Production of syngas via CO$_2$ methane reforming process: Effect of cerium and calcium promoters on the performance of Ni-MSC catalysts," *International Journal of Hydrogen Energy*, vol. 45, no. 1, pp. 640–649, 2020.

94. S. Singh *et al.*, "Boron-doped Ni/SBA-15 catalysts with enhanced coke resistance and catalytic performance for dry reforming of methane," *Journal – Energy Institute*, vol. 93, no. 1, pp. 31–42, 2020.

95. A. S. Al-Fatesh *et al.*, "CO$_2$ reforming of CH$_4$: Effect of Gd as promoter for Ni supported over MCM-41 as catalyst," *Renewable Energy*, vol. 140, pp. 658–667, 2019.

96. Y. Lu, D. Guo, Y. Ruan, Y. Zhao, S. Wang and X. Ma, "Facile one-pot synthesis of Ni@ HSS as a novel yolk-shell structure catalyst for dry reforming of methane," *Journal of CO$_2$ Utilization*, vol. 24, pp. 190–199, 2018.

97. P. K. Yadav and T. Das, "Production of syngas from carbon dioxide reforming of methane by using LaNixFe1–xO$_3$ perovskite type catalysts," *International Journal of Hydrogen Energy*, vol. 44, no. 3, pp. 1659–1670, 2019, doi:10.1016/j.ijhydene.2018.11.108.

98. W. Liu, L. Li, X. Zhang, Z. Wang, X. Wang and H. Peng, "Design of Ni-ZrO$_2$@ SiO$_2$ catalyst with ultra-high sintering and coking resistance for dry reforming of methane to prepare syngas," *Journal of CO$_2$ Utilization*, vol. 27, pp. 297–307, 2018.

9 Blue/Bio-Hydrogen and Carbon Capture

Md. Abdus Salam, Md. Tauhidul Islam, and Nasrin Papri

CONTENTS

9.1 INTRODUCTION TO HYDROGEN ECONOMY

The presence of hydrogen in nature is scarce and too small to be of any economic significance. Hence, all hydrogen has to be synthesized and is thus termed a synthetic energy. It requires energy conversion from other processes. Therefore, the much-awaited transition to a hydrogen economy will affect energy generation and distributions patterns worldwide. The success of the hydrogen economy lies

DOI: 10.1201/9781003162780-9

within the mass scale production and cost-effectiveness of hydrogen as an alternative fuel. This will require concerted research and development (R&D) efforts and a key change in our socio-economic fabric regarding the use of fuel. Long-term commitment and an increased pace of implementation are required, which demands a global investment in R&D for the hydrogen economy. With the possibility that fossil fuel production could not meet the increased demand for energy or the efforts to mitigate climate change are implemented, a move to an environmentally sustainable fuel must be carried out, which must also be affordable and available in large quantities.

It is expected that fossil fuels will remain in the competition for energy supply (and may have the leading role) due to the well-established supply chain and low cost. But their contribution to air pollution and global warming is a bottleneck for their prolonged use [1]. A global consensus exists to replace fossil fuels with hydrogen by renewable energy, dependent on hydric production such as solar photochemical and biological photo water decomposition, water electrolysis coupled with photovoltaic cells, or wind turbines.

Although hydrogen is the most abundant element in the universe, pure hydrogen is not available on the Earth in its molecular form [2]. Most of the hydrogen available on Earth is in the form of water (when hydrogen bonds with oxygen). Moreover, it exists in reaction with carbon, forming countless hydrocarbons. The use of hydrogen as a fuel is not a novel concept. Henry Cavendish first mentioned its use as fuel after successfully isolating it in 1766, seconded by W.R. Grove in 1839.

Various countries across the globe have opted for the development and implementation of the hydrogen economy. Japan was the forerunner in 2017 to implement a hydrogen policy, and plans to become a 'hydrogen society' while aiming to achieve a cost parity against conventional (fossil fuels) like natural gas for electricity production. In January 2019, a hydrogen economy roadmap was revealed by the South Korean government. The country targets to manufacture 6.2 million fuel cell electric vehicles by the year 2040, supported by 1200 fueling stations. Moreover, a 15 GW installed capacity of hydrogen fuel cells will be completed for power generation.

Also in 2019, Australia presented its national hydrogen strategy under the name of 'H$_2$ under 2' with funding of AU\$370 million as a part of Australia's Technology Investment Roadmap. The program envisages producing hydrogen gas under AU\$2 per kilogram of hydrogen. In June 2020, Germany planned a 5 GW production of hydrogen-based energy by 2030 using a 200-fold increase in their electrolyzer capacity under their national hydrogen plan.

The majority of upcoming development of hydrogen production is dependent on the method of steam methane reforming (SMR). The process produces a significant amount of CO_2 due to the reaction. Such a process can release 13.7 kg CO_2 (equivalent) per kg of total hydrogen, accounting up to 77.6% of the greenhouse gases released by the process [3]. A medium-sized SMR plant can generate up to 1 million cubic meters of carbon dioxide in the atmosphere per day. This amount can be doubled when using coal gasification for hydrogen production. Such huge quantities of carbon dioxide are a constraint towards the adoption of SMR-based hydrogen as a hydrogen production method. Hence, researchers have suggested three major pathways to reduce/mitigate the carbon dioxide emissions from SMR-based hydrogen production:

1. Adopt water-splitting thermochemical cycles.
2. Disassociate thermal hydrocarbon through hydrogen and carbon.
3. Sequester CO_2 in an SMR hydrogen plant.

The last option, the capture and storage/utilization of carbon dioxide, is a promising option as the technology is relatively mature compared to other two options and eliminates the risk of any process complexities and unknowns. Successful implementation can prevent the process-based CO_2 emissions from reaching the atmosphere.

9.2   IMPORTANCE OF BLUE AND BIO-HYDROGEN IN THE HYDROGEN ECONOMY

9.2.1   IMPORTANCE OF BLUE HYDROGEN

The future implementation of blue hydrogen looks far promising than other hydrogen contenders. Although the research and development in green hydrogen should be a long-term gain in terms of sustainability, blue hydrogen will play its role as a starter (or somewhat enabler) in the hydrogen economy. This will certainly generate a large cloud of CO_2, which if sequestered can provide a large-scale and affordable energy option. The technology for blue hydrogen (that for hydrogen generation and carbon dioxide sequestration) is already quite mature; the only obstacle is the cost of capturing carbon dioxide, which is quite high. However, with the introduction of the hydrogen economy and increased consumption of hydrogen, it is expected that new hydrogen production and carbon dioxide sequestration facilities will develop. Since the process chain for blue hydrogen is more mature than other hydrogen options, most problems are found downstream, mostly related to transport and storage. However, there is potential for improvement in the upstream, such as choice of a superior catalyst and reducing the reaction/processing temperatures. This is much smaller than other colored hydrogen concepts. For instance, the upstream of green hydrogen is marred by technological issues and weak cost-benefit and scale of implementation. Hence, by selecting immediate development of blue hydrogen, ample time will be available for other hydrogen production methods to gain ground and develop to higher TRLs, alongside developing supply chain and infrastructure for the hydrogen economy. This transition might be slow and gradual but will enable a smoother integration of green hydrogen in the future.

9.2.2 Importance of Bio-Hydrogen

Biologically sourced hydrogen is another renewable concept for producing large, clean, and sustainable quantities of hydrogen, employing basic and already developed technologies. In theory, they have a higher potential than the chemical hydrogen production of today. Hence, bio-hydrogen can contribute significantly to the clean energy demand of the future. The technology is suitable for converting a wide range of biological resources that include wastes and by-products from agriculture and industry, natural biomass as a feedstock, and other bio-wastes. Such sources are mostly free and available in abundance [4]. However, the viability of this concept is still nascent and is connected directly to social, economic, and environmental impacts that are still not well understood. Figure 9.1 depicts how blue and bio-hydrogen will satisfy the growing need for an alternative to fossil fuels.

9.2.3 The Hydrogen Paradigm Shift

The 'hydrogen economy' forms the cornerstone of the paradigm societal shift towards decarbonization [5]. Society now seriously considers the method of energy generation and its effect on the environment, especially climate change. Scientific data shows that the excessive use of fossil fuels has contributed adversely to global warming. Although the effect of electricity generation on the environment was not significant during the first and second industrial revolutions, the impact is now much visible. The scenario has changed significantly as the concept of climate change mitigation gains ground. Today, a search is underway for a fuel that is inexpensive and carbon-neutral, affordable, and capable of mass-scale production. Blue and bio-hydrogen, coupled with carbon capture, utilization, and storage (CCUS)

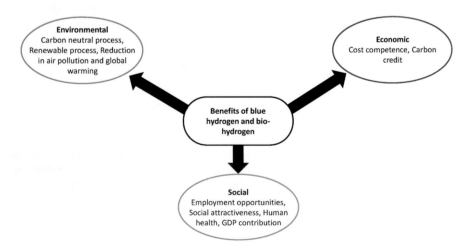

FIGURE 9.1 Benefits of blue and bio-hydrogen as a sustainable fuel to replace fossil fuels.

systems, is among many potential candidates for this transformation. A technologically sound and cost-effective CCUS technology is essential to capture CO_2 from the blue and bio-hydrogen sector to achieve carbon-neutral and carbon-negative emissions scenarios.

9.3 BLUE HYDROGEN

9.3.1 BLUE HYDROGEN

Carbon capture systems have several technological applications in the sectors of pre-combustion capture, post-combustion capture, oxy-combustion, chemical looping, and others [6,7]. The restraint remains in the utilization and storage technologies that are technologically less developed (at low technology readiness levels) [7].

Steam methane reforming produces the major share of hydrogen worldwide. The process also produces carbon dioxide as a by-product. The process is termed 'gray hydrogen' if CO_2 is not sequestered. Contrarily, if the CO_2 is captured, the process is called 'blue hydrogen.' The latter concept is an economical and suitable pathway for industries relying on hydrogen as a fuel, since the CO_2 emissions are nearly zero. The concept is seen as a solution to the short-term downstream hydrogen challenges and making a pathway for the hydrolytic green hydrogen. At the moment, blue hydrogen enjoys the benefits of low energy consumption and the availability of mature technology over the green hydrogen concept.

9.3.2 SOURCES OF BLUE HYDROGEN

Natural gas, oil, coke, coal, and biomass are some of the fuel resources that can be used to produce blue hydrogen via a SMR process coupled with a CCUS system, as presented in Table 9.1. Methane and ethane are the major components of natural gas.

TABLE 9.1
Feed Sources of Blue Hydrogen

Source	Method	Remarks	Ref.
Natural gas or methane	Steam methane reforming (SMR)	Most applied and economic process	[8]
	Partial oxidation	The fuel-air or fuel-oxygen mixture is partially combusted	[9]
	Methane pyrolysis	The process was conducted at a higher temperature	[10]
Liquid hydrocarbon (oil)	Plasma reforming	48% H_2, 40% activated carbon and 10% steam; no CO_2 is produced	[11]
Coal Petroleum coke	Coal gasification	Conversion of solid coal/coke into a gaseous phase	[12]
Biomass	Biomass gasification	Renewable organic source	[13]

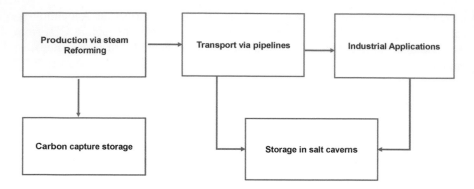

FIGURE 9.2   Process diagram of blue hydrogen chain.

Other light paraffins like propane, butane, pentane, and hexane are found in sparingly small quantities. Heavier feeds like oil, coke, coal, and biomass are also among feed options, but they have to be broken to simpler/smaller molecules for conversion to hydrogen. This brings in its own plethora of problems, which are beyond the scope of this chapter.

9.3.3   Blue Hydrogen Chain

The blue hydrogen chain consists of five basic steps: production, carbon sequestration, transport, daily/seasonal storage (and its fluctuations), and industrial applications, as presented in Figure 9.2.

9.3.4   Hydrogen Production

Natural gas or another fossil fuel (e.g., oil, coal, coke) is converted to syngas (a mixture of CO and H_2) in a conventional SMR or auto thermal reformer (ATR). The reaction is carried out in a temperature range of 500–950 °C at 20–35 atm pressure and steam-to-carbon (S/C) ratios of 2.5–3.0 over a nickel catalyst [14]. A higher temperature and steam-to-carbon ratio in conjunction with low pressure favor the formation of syngas (despite the fact a low S/C ratio is more economical). Due to the use of a nickel catalyst, pre-treatment and removal of sulfur compounds (especially hydrogen sulfide) is mandatory. For feed containing heavier hydrocarbons, a pre-reformer (sometimes termed as a cracker) is required to increase the carbon to hydrogen ratio. Such pre-reformers operate at 400–500 °C [15] and may produce additional CO_2 emissions.

$$CH_4 + H_2O \leftrightarrow CO + 3H_2 \ \Delta H_{rx} = 206 \text{ kJ/mol}$$

In the ATR process, noncatalytic partial oxidation and steam methane reforming is carried out. The methane reacts simultaneously with oxygen and steam in the same reactor.

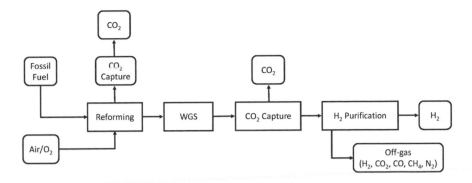

FIGURE 9.3 Flow diagram of steam methane reforming with CO_2 capture.

$$CH_4 + \tfrac{1}{2} H_2O + \tfrac{1}{4} O_2 \leftrightarrow CO + \tfrac{5}{2} H_2 \ \Delta H_{rx} = 84 \text{ kJ/mol}$$

As the reaction equation shows, part of the heat of reaction required for SMR reaction is provided by the partial oxidation reaction. The optimal steam to oxygen ratio is maintained to maximize the conversion to hydrogen and carbon dioxide, with temperature varying between 900–1150 °C and pressure within a wide range of 1–80 bars (depending on the purity of feed gas and required conversion) [16].

For the pure SMR process, a water gas shift (WGS) reaction is necessary. The reaction is carried out in another section of the plant, where steam reacts with syngas to produce H_2 and CO_2. The reaction can occur at a high temperature (320–360 °C, 10–60 bars), called a high-temperature WGS reaction, or at a low temperature (190–250 °C, 10–40 bars), called a low-temperature WGS reaction. The choice between high- and low-temperature WGS depends on the rate of reaction (which is higher in high-temperature WGS) and conversion needed (which is higher in low-temperature WGS) [17].

$$CO + H_2O \leftrightarrow CO_2 + H_2 \ \Delta H_{rx} = -41 \text{ kJ/mol}$$

The product stream may also contain inert, partially converted, or unreacted compounds like CH_4, CO, H_2S, N_2, and Ar. Hence, the by-product CO_2 is captured by a CO_2 separation unit (that might be an extension or a retrofit) in the plant, creating three process streams: purified H_2, purified CO_2, and off-gas containing all other compounds, as shown in Figure 9.3. Notably, the reaction is endothermic (as explained later), which requires the external combustion of a part of the fuel feed to provide energy for the reaction. Nearly 40% of the CO_2 emissions in the SMR reactor comes from the heating source and requires a specialized unit for carbon dioxide separation as in post-combustion carbon capture, which will cause a considerable adverse effect on the capital and operational costs.

Pressure swing adsorption, membranes, and their hybrids form the choice of commercially available technologies for CO_2 capture in the SMR- or ATR-based hydrogen production processes. Carbon capture technique is selected based on hydrogen purity and product stream pressure. Traditionally, amine absorption or K_2CO_3 absorption systems have been used, but these are being rapidly phased out in favor of the aforementioned technologies. Nevertheless, new technology options are always challenging the commercial-scale applications.

9.3.5 CHALLENGES FOR WIDE APPLICATION

Although the technology for blue hydrogen is mature and readily available, hydrogen transition is still a daunting task. Major limitations are the large CO_2 emissions from hydrogen production and the relative immaturity of downstream carbon storage and utilization technologies, which still require research and time for development. Many SMR-WGS plants for hydrogen production will require a large investment for retro-fitting two carbon capture systems: First, removing carbon dioxide from the product gas (H_2 and CO_2); and second, removing carbon dioxide from the flue gas emanating from the combustion to provide process heat/energy. Switching to auto-thermal reformers can reduce the need for a second (flue gas–cleaning) carbon capture unit but still require a larger retrofit price tag for replacing SMRs. Second, the auto-ther-mal reformers are more susceptible to low hydrogen yields and system failures than SMRs. Some theorists point out the use of captured CO_2 for enhanced oil recovery, but this would not be a plausible option once the fossil fuel consumption reduces, and it is not certainly a sustainable solution to decarbonization practices.

9.4 BIO-HYDROGEN

9.4.1 BIO-HYDROGEN

Hydrogen produced from biological resources is termed bio-hydrogen. It can be either produced on a micro scale, probably by a micro-biological organism (e.g., bacte-ria, archaea, algae) or on a macro scale by conversion of bio-wastes and resources. Hence, any supply chain consisting of bio-hydrogen will enjoy the benefits of sus-tainability, waste mitigation, and carbon neutrality and still be counted as a biofuel [18]. Moreover, the process is seemingly attractive due to its low production cost and renewable nature. Microorganisms offers another area of development, while the process remains renewable and carbon-neutral as CO_2 produced during the process is reutilized by the next generation of biomass feedstock [19]. Although the global hydrogen output has surpassed 1 billion m^3 per day, none of the hydrogen comes from the bio-resources but is sourced from the electrolyzer-based processes (4%) and fossil fuels (96%) [20]. Nevertheless, hydrogen production via a biological route (from sev-eral kinds of biomass) remains a potential option for development in the near future.

9.4.2 SOURCES OF BIO-HYDROGEN

Contrary to the feed options for conventional fossil fuel–based hydrogen production, the feed sources for bio-hydrogen are diversified. They can be segregated based on phase, molecular complexity, pretreatment/feed preparation, the energy needed for hydrogen manufacture, or the cost of production itself. Traditionally, combustible biomass resources are preferred for use due to their resemblance to fossil fuels; hence, fossil fuel–based processes can be modified for bio-hydrogen production. Some unconventional sources include very simple inorganic mixtures that contain one to three hydrogen atoms per molecule (e.g., hydrochloric acid, ammonia, hydrogen sulfide). However, development in this area is still at a low level of technology advancement. It is more likely that conventional fossil fuels (oil and coal) may only be exchanged by conventional biomass resources; however, the confusion remains that, if any, substitution will compete with the current food chains, as the biological H_2 production may use waste material and domestic wastes as a feed source that may contain, proteins, carbohydrates (like starch enriched crops, sugar-containing crops, hemicelluloses, glucose, cellulose, sucrose), fats, dairy waste, manure, and wastewater from food/beverage industries [21].

9.4.3 METHODS FOR BIO-HYDROGEN PRODUCTION

The pure biological and thermochemical routes for bio-hydrogen production are the two diverse yet overlapping fields of development that may use many possible pathways of utilizing biomass for the purpose, as illustrated in Figure 9.4 [22].

Bio photolysis uses sunlight to biologically split water into H_2 and O_2 molecules. Usually, bacterial strains like cyanobacteria and green microalgae are used. These strains contain photosynthetic biocatalysts that produce hydrogen and oxygen via the photo-lytic reaction [23]. When seen on the sustainability scale, sunlight and water are abundantly available, which points to the theoretical superiority of the process. Based on how the hydrogen is produced, photolysis is further segregated into two further sub-categories, namely direct and indirect bio photolysis. The details of both processes are given in the following subsections.

Direct bio photolysis: Direct bio photolysis is a system similar to what is found in commonly known algal, plant, and/or tree photosynthesis. The sun provides the energy required for the process, which is then directly used to convert water to hydrogen and oxygen molecules via a chain of photosynthetic reactions. The main reaction is given below:

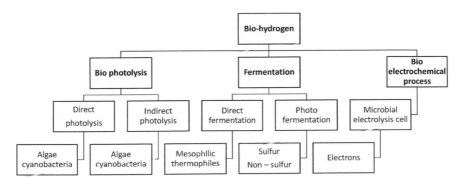

FIGURE 9.4 Bio-hydrogen production process chart.

$$2H_2O \xrightarrow{\text{light energy}} 2H_2 + O_2$$

The process is independent of the intermediate carbon fixation process [21]. The additional benefit is that process can occur at approximately 1 atm partial pressure of oxygen, removing stringent process requirements, opposite to the conventional fossil fuel–based processes. However, the oxygen sensitivity shifts the forward equilibrium, reducing productivity [22].

Indirect bio photolysis: Indirect bio photolysis is a two-phase, two-step process. Organic compounds like glycogen in cyanobacteria and starch in algae act as the electron source for the process. The first step of the process involves a conventional photosynthesis reaction, involving a carbon-fixation step, and a product is released in the form of a complex carbohydrate. In the second step, a fermentation reaction produces hydrogen by breaking down the complex carbohydrate [23] but also releases carbon dioxide, which can be reabsorbed by the process for further processing. Hence, the overall process remains carbon neutral. The reaction steps are shown below.

Step 1: $12 \; H_2O + 6 \; CO_2 + $ 'light energy'$\rightarrow C_6H_{12}O_6 + 6 \; O_2$
Step 2: $C_6H_{12}O_6 + 12 \; H_2O + $ 'light energy'$\rightarrow 12 \; H_2 + 6 \; CO_2$

Fermentation process: The fermentation process involves a fermentation step, where an organic molecule (technically termed as a substrate) is broken down to hydrogen and other by-products (essentially containing carbon dioxide). Many forms of bacteria, protozoa, and yeast-like cells provide enzymes (biocatalysts) for this process. The process can either occur in the dark (subsequently called dark fermentation) or promoted by sunlight (called photo fermentation). Refined sugar, corn stover, wastewater, and raw biomass can be used as a feed for bio-hydrogen production. The subcategories are explained below.

Dark fermentation: As the name suggests, dark fermentation occurs in the absence of sunlight or solar activation. Hydrogen is produced from carbon-rich compounds using anaerobic microorganisms under anaerobic conditions [23]. The process enjoys better reaction rates as compared to photo fermentation and bio photolysis. However, the process suffers from large production of volatile fatty acids as an intermediate, which not only lowers the production of hydrogen but also poses waste-handling issues.

Photo fermentation: The photo fermentation system is deemed as an exceptionally promising option by many researchers in the field. The process enjoys the benefits of operation at mild reaction conditions using sunlight (solar energy) as an energy source. The process also offers waste mitigation as it can transform any organic waste into hydrogen. The equilibrium of the process is always forward (thanks to photo activation), but the rates of conversion are slower than dark fermentation. The following equation shows the conversion of acetic acid to hydrogen and carbon dioxide, which can be absorbed by the next stage of plants, rendering the process carbon neutral.

$$CH_3COOH + 2H_2O \rightarrow 4H_2 + 2CO_2$$

Another restriction for the process is the pre-treatment of the feed, as any of the process variables (pressure, temperature, nutrients, pH) can detrimentally affect production. Moreover, any biomass also has to be pretreated with a proper concentration of organic acid and sugars and diluted accordingly for optimal production [24–25].

9.4.4 BIO-ELECTROCHEMICAL PROCESS

Bio-electrochemical processes are an innovation in bio-hydrogen production. The process combines different enzymes or microorganisms with an electrochemical process to produce electricity and hydrogen. The microbial electrolysis cell (MEC) is a technological innovation for hydrogen production that uses a variety of organic substrates [26]. MECs enhance the pre-established microbial fuel cells [27] and are sometimes referred to as electro-fermentation technology. The technology concept activates the microorganisms electrochemically to produce carbon dioxide (CO_2), electrons (e^-), and proton/hydrogen ions (H^+) at the anode by oxidizing the organic substrate. The concept is represented in Figure 9.5.

Most MECs enjoy very high conversion efficiencies, sometimes above 90%, which is a significant improvement over the mere 33% efficiency seen in dark fermentation processes. Mostly, the organic substrate/feed for the MEC process is the wastewater from the domestic use or food/beverage industry, hence providing a potential pathway for the hydrogen economy based on non–fossil fuels [28].

9.4.5 COMPARISON BETWEEN THE BIO-HYDROGEN PRODUCTION PROCESSES

A hydrogen economy based on bio-hydrogen is now advocated as an alternative to fossil fuel–based hydrogen production. Moreover, bio-hydrogen systems are exquisitely appropriate for small-scale and regional systems. They can be easily multiplied as per requirements and can sustain modular development. Combined with agronomic and benefits like waste mitigation [29], the technology concept will have the least socio-economic challenges as it is being sourced from biological feedstocks and have minimal waste handling issues. The advantages and technical challenges of various technologies/pathways for bio-hydrogen are presented in Table 9.2.

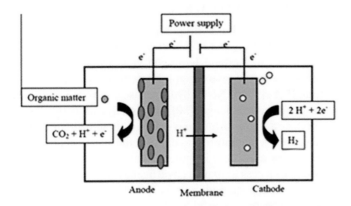

FIGURE 9.5 Schematic of typical two-chamber MEC construction and operation.

TABLE 9.2

Comparison of Bio-Hydrogen Production Process [23,29]

Sr. no.	Methods	Advantages	Challenges
01.	Direct bio photolysis	Direct use of solar irradiation to produce H_2 without any by-products.	Very small-scale hydrogen output, as conversion proficiency is quite low. Custom-made bioreactors are mandatory. Too much sensitive to oxygen that propagates a backward reaction.
02.	Indirect bio photolysis	Producing hydrogen from water can fix N_2 from the air.	Limited hydrogen production due to slow hydrogenase conversion. Light conversion proficiency is extremely low.
03.	Dark fermentation	Independent of a light source. Simultaneous hydrogen production. No oxygen limitation problem. Butyric acid and acetic acid can be valuable products.	Low-scale hydrogen production rate. Small amount organic substrate conversion rates. Not feasible in thermodynamic terms. Difficult to separate hydrogen from other by-products. Production of volatile fatty acids (VAcs) as intermediate, which lowers production.
04.	Photo fermentation	Visible light energy is the main source of energy. Different waste materials can be used as feedstock.	Lower production rate of hydrogen. Light conversion ability is very poor.
05.	Microbial electrolysis cells	Advanced and innovative technology offering the use of wastewater as a feedstock. 90% hydrogen production rate. Microorganisms act as biocatalysts.	Produces carbon dioxide. Low power density. Dependence on Pt.

9.4.6 CHALLENGES FOR WIDE APPLICATION OF BIO-HYDROGEN

Improving the efficiency of the bio-hydrogen process is the greatest challenge faced by the bio-hydrogen production sector. The field requires major technical advancements and improvements to increase hydrogen yield through photo- and dark-fermentation processes. For example, oxygen inhibition reduces the conversion rate of main enzymes (like [FeFe]$^-$ and [NiFe]-hydrogenases) in photo-biological reactions, which remains a daunting challenge [30]. Thermodynamically, the hydrogenase enzyme cannot produce a stable oxygen molecule. Although attempts are being made to resolve these enzymes' vulnerability to oxygen, the success is limited so far.

Similarly, the oxygen absorbers in green algae (namely glucose-oxidase) have been used with limited success to thawart hydrogenase inactivation caused by simultaneous oxygen production during photolysis [31]. Moreover, other limiting factors for the sector are the low hydrogen yield (less than 10%) and the requirement of complex photo-bioreactors [30, 32]. Areas of improvement include developing strategies and systems that can increase the light absorption and conversion efficiency, refining the bioreactor design, reducing structural/capital costs, and decreasing the requirements for agitation, cooling, and so forth, which will cut down the operational expenses.

9.5 OPPORTUNITIES FOR CO_2 CAPTURE IN BLUE AND BIO-HYDROGEN SECTOR

9.5.1 AREAS OF CARBON CAPTURE

Carbon dioxide emissions from fossil fuels are one of the major contributors to climate change. The use of carbon capture, utilization, and storage (CCUS) systems can significantly lower carbon dioxide emissions from conventional electricity and hydrogen generation units. This will allow these systems to have a longer life in a clean energy future. Eventually, fossil fuels will be replaced by alternative fuels in the coming decades [33–36].

CCUS systems target a reduction in the new anthropogenic CO_2 emissions to the atmosphere and trap/preserve them in carbon sinks for an indefinite period. In case fossil fuels have to be used for powering the world, CCUS systems are inevitable for mitigating carbon dioxide emissions. According to the Global CCS Institute, large-scale CCUS systems are becoming a norm. However, the scale is not yet global (as per 2012, only eight CCUS were operational, and only the other nine are under construction). However, the numbers are far better in 2020, pointing out to growing confidence in the technology. Notably, none of the CCUS plants has been installed for the hydrogen production units, owing to energy penalty and high-end operating costs. Nevertheless, the following sections present a quick technical roundup of applying CCUS to the SMR processes. It must be acknowledged that the numbers are prone to change due to demographic locations, climate, and the levels of economic activity and cannot be copied without incorporating changes.

Steam methane reforming with CCUS: Estimates by Scholz show that CO_2 emissions from the SMR operation are 0.44 Nm^3 CO_2/Nm^3 H_2 (or 9.7 kg CO_2/kg H_2) [37]. As mentioned earlier, with a global warming potential (GWP) [38] of 13.7 kg CO_2 (equivalent) per kg of net hydrogen, a modern SMR hydrogen plant can produce nearly 0.4 million m^3 of CO_2, which needs active carbon sequestration. Research efforts do point to a possibility of using CCUS effectively for carbon dioxide mitigation [39]. Notably, 60% of the emissions come from the product stream of a SMR plant, while the rest is from the fuel combusted to provide heat of reaction [40]. Our analysis suggests that using a pressure swing adsorption (PSA) unit is suitable with a theoretical potential of removing all CO_2 and other off-gases, producing 99.99% pure hydrogen. Another (sequential) CO_2 purifying PSA step is required to separate the off-gases from the CO_2, which are eventually combusted to cater the energy needs for this process.

Hydrogen production from coal with CCUS: Hydrogen production from coal gasification is one of the advanced yet rapidly depleting technologies. Finding applications over a century ago, the process has now left its share to natural gas-based hydrogen production. Limited works from Kreutz *et al.* (2005) [41], Gray and Tomlinson (2003) [42], DOE/NETL (2010) [43], and Garcia Cortes *et al.* (2009) [44] show the analysis of using CCUS for capturing carbon dioxide emissions from coal-based hydrogen production. Kreutz *et al.* studied the production of hydrogen from high-sulfur bituminous coal. They managed to produce 1,070 MWth H_2 (with a sulfur content of 3.4%) with an entrained coal gasifier operating at a pressure of 7 MPa. The H_2S and CO_2 were separated from the syngas using the Selexol process (a physical process that separates acid gases). The CO_2 was dried and compressed to 150 atm for pipeline and underground storage. A high-purity H_2 (99.999%) was achieved, with 1.4 kilograms of carbon dioxide released to the atmosphere for each kilogram of hydrogen produced, which is nearly 92% lower than if the carbon capture would not have been employed.

The researchers from NETL have also studied a coal-to-H_2 plant with a capacity to capture and store of 617 kg CO_2/day. With a coal feed rate of 5.3 tons/day, the GE energy radiant gasifier coupled with a PSA reduced the CO_2 emissions by nearly 90%.

9.5.2 TECHNO-ECONOMIC ANALYSIS

A techno-economic evaluation of hydrogen production from a carbon-dependent source combined with CCUS is a topic of interest in recent history. The work of Audus *et al.* presents an idea of combining a SMR-based hydrogen plant with a CCUS unit, using a water reservoir as a sink. The study shows that 25%–30% of the CO_2 produced from the hydrogen process can be disposed off in such a way [39].

Main outputs from various models used for different CO_2 capture technology scenarios are presented in Table 9.3. Conventionally, a CCUS scheme is deemed techno-economically viable as it reaches a net present value (NPV) greater than zero, while the internal rate of return (IRR) falls below the reduced rate. Moreover, the payback period (PBP) should be lower than the actual lifetime of the gas field [46].

A physical absorption solvent is used in the Selexol process. The process has a positive NPV (US$367.37 million), an IRR of 15%, and the lowest PBP among other

TABLE 9.3

Economic Results of the CO_2 Capture Technologies [45]

parameter	Polymeric membrane	Chemical absorption	Physical absorption	Cryogenic distillation
Capital cost in USD million	639.90	838.69	590.51	817.19
Abandonment in USD million	289.96	364.51	271.44	365.49
Net revenue in USD million	1923.06	1753.59	1965.17	1771.83
Payback period	8.55	10.68	7.94	10.45
Net present value	292.94	−6.67	367.39	25.58
Internal rate of return	13%	8%	15%	8%

commercial carbon capture technologies [47]. A polymeric membrane process shows slightly inferior results with a NPV of US$292.94 million, an IRR of 13% (higher than the specified discount rate), and a PBP of 8.55 years. Cryogenic distillation takes the third position with a NPV of USD 25.58 million, an IRR of 8% (which just matches the specified discount rate), and a PBP of 10.45 years [48]. The worst option was chemical absorption-based carbon capture. This process showed a negative NPV of US$6.67 million, an IRR of 8%, and the highest PBP at 10.68 years among the contenders. Although physical absorption is economically attractive, its physicochemical limitations make it technologically unattractive [49].

9.6 CURRENT APPLICATIONS OF CARBON CAPTURE IN HYDROGEN PROJECTS

The current technology focus for producing hydrogen is to use fossil fuels, preferably natural gas, with or without a CCUS system [50]. Although the CO_2 from coal (gasification) based hydrogen production is nearly twice that of natural gas as a feed, the current CCUS systems are flexible enough to capture up to 85% of the CO_2 emitted [51]. This aligns with the hydrogen economy's goal to lower/cancel CO_2 emissions and maintain an affordable supply of energy [52].

However, the burden of carbon dioxide emission reduction is massive. For example, a detailed estimate shows that 100 million metric tons (Mt) of hydrogen will be required each year if hydrogen-based fuel cell vehicles (FCVs) dominate the market in 2050 [53]. Suppose all of this hydrogen comes from the blue hydrogen (which is possible in case green hydrogen path is delayed). In that case, the amount of carbon dioxide to be sequestered will be around 255 million tons of carbon if sourced from natural gas and 518 million tons of carbon if sourced from coal [54]. At the current level of development, CCUS can remove a mere 16% of this carbon dioxide load (no competing method can remove 100% of CO_2) [54], hence leaving a large burden of managing CO_2 emissions emanating from such a technology change [55]. As US vehicle-based emissions [56] amount to nearly 20% of the national CO_2 emissions, a blue hydrogen option can reduce a significant chunk of carbon dioxide contents while contributing to other larger point-source CO_2 contributors. This means any CO_2 storage effort should be a thousand times larger than the first storage facility developed in Sleipner, Norway [57]. The problem will be compounded for the European natural gas-based hydrogen production units, where the inherent contents of carbon dioxide are four times higher than their counterparts in the United States. This will present a further burden for CCUS efforts at mitigation [58].

9.7 CHALLENGES FOR CO₂ CAPTURE FOR APPLICATION IN THE HYDROGEN SECTOR

Of the total hydrogen being produced globally, 96% comes from fossil fuels. Nearly half of this hydrogen (48%) is produced from natural gas, followed by oil/naphtha reforming (30%). A decreasing percentage of hydrogen is produced by coal gasification (18%). All of these hydrogen production methods are producing large quantities of carbon dioxide in the atmosphere. The much cleaner (or theoretically, CO_2-free) but meager portion of 4% of the total global hydrogen is sourced from the electrolysis of water (green hydrogen) [59]. However, it is not clear how much of this energy for electrolysis is sourced from renewable sources. This highly fossil fuel–centered production is channeled by economic factors. In case (and being optimistic), the cost of renewable power sources (wind, PV solar, or thermochemical solar) will drop down one day to make them financially competitive.

At the moment, fossil fuels are expected to control hydrogen production for the foreseeable future [60]. European Commission estimates that by the year 2050, nearly a quarter or even higher (30%) of the Europe's hydrogen production may come from coal, which will replace natural gas steam reforming, attributed to a 13-plus favorable commodity price concept [61]. Another 15% of the hydrogen would be sourced from nuclear energy, while a larger portion of hydrogen will be generated by carbon-neutral energy resources, probably renewables [62]. In the longer run, it is expected that renewables will control the market's share of hydrogen generation.

Today, the renewable-based hydrogen production schemes are developing areas of science jumping from medium technology readiness levels, yet their economic and social impact is not well understood. Scientists believe a tough transition period might have to be faced where the hydrogen production may dwindle as the technology supply chain is developed [63].

Another technology concept called methane deburation (methane cracking) is receiving active research importance. The concept divides methane into its nuclear components (carbon and hydrogen), resulting in the production of almost pure hydrogen without any CO_2 emissions. However, this area is still at a low technology readiness and may take another 15 to 20 years before reaching commercial maturity. Although methane cracking does not generate CO_2 emissions, the by-product carbon aggregates would cause further waste handling issues and/or finding use as a resource [64]. This area, combined with other decarbonization and mitigation strategies, must be developed to avert any climatic catastrophe [65].

Lord Turner's prediction shows that the use of hydrogen will see a rapid rise, prompted by its ability to decarbonize areas of the society, industry, and energy, where the renewables are not applicable owing to the hydrogen's combustion product being water. Opposite to what is believed, implementing a hydrogen-based power supply will require minor changes, as the supply chain is already well established for natural gas. This is not the case with electricity-based domestic heating in many European economies [66]. Several pilot studies show that blends of natural gas and hydrogen (the latter up to 20% concentration) can be run smoothly without any infrastructure change.

Hydrogen can play a major role in industrial applications. Steel production seems to be the biggest beneficiary where iron reduction can be carried out directly by using hydrogen rather than coke or natural gas. For instance, Primetals Technologies (a part of MHI Group) is investigating a technology concept that uses hydrogen for providing heat and reducing gas for steel production. However, the extent of this technology change will rely on the carbon-neutral nature of the sourced hydrogen [67]. If the hydrogen is produced via a blue hydrogen concept, CCUS will play a defining role. Similarly, the Shell's Lynch process is another process area that shows the CCUS is necessary irrespective of how hydrogen is sourced.

Although carbon capture methods are technologically mature, the end-use of captured carbon dioxide is not much clear. As mentioned in Sections 9.1 and 9.2, the commercial use of any CCUS strategy will only be as good as the end-use of carbon dioxide or its storage longevity. This will involve major hurdles that must be compared to other technology options. Some notable issues include the effect of CO_2 purity on the end-use/storage or scaling up for a larger industrial unit, perhaps managing 10–20 million tons of carbon dioxide per year [68]. For such large-scale applications, the design of capture equipment will need significant design changes, and the knowledge in that area is still under development.

Additionally, issues like a retrofit, alteration, and integration are other technology constraints. Many of the issues connected with scaling up and integrating technology might be solved by novel instrumentation. Apart from that, equipment size (that directly controls the capital costs) is another area of improvement. Reducing the column diameter and height of an absorption column by using a better solvent, increasing the purity of oxygen for oxy-combustion process, a control design strategy of PSA/fluidized beds, or pellet size of an adsorbent are some areas that can be optimized to yield innovative and groundbreaking new designs [69]. Such areas though not connected directly to an operational CCUS system can still result in substantial improvements over the current technology level.

Another area of possible improvement is the pre-treatment section. The presence of undesired contaminants can adversely reduce the system's efficiency in separating carbon dioxide and/or increase operational costs. Research has shown that the removal of NO_x, SO_x, and any oxygen can improve the operational efficiency and product purity of a CO_2 separation unit [70].

For the end-user development of carbon dioxide, known carbon sinks (e.g., underground storage or industrial use of carbon dioxide in oil recovery, beverage sector) are too small to cater to a large amount of sequestered carbon dioxide. Circular economies involving construction materials, alkaline carbonate products, reverse conversion to methane, and mineralization are some of the promising pathways that must be developed for large scale CO_2 use.

Post-combustion carbon capture can be actively retrofitted to existing fossil fuel power plants with the least economic effort compared to other available options [71]. Chemical absorption using amines is the choice of the process in this case. However, such systems still find minimal application due to problems in large-scale deployment, resulting in a large energy penalty and an overall high operational expense. Moreover, the safety risks of holding such significant amounts of liquid and gaseous carbon dioxide are another technology factor. A nominally designed amine capture system can handle the capture of 800 tons of CO_2 per day, while a 500 MW coal-fired power station can release up to ten times that volume of carbon dioxide [72]. It is quite understandable that the application of carbon capture systems to blue hydrogen production will face the same issues, where magnitudes of CO_2 production are much larger.

Currently, the pre-combustion technologies have found a limited application in the IGCC facilities (apart from their natural gas sweetening processes). The application is marred by large capital and operating expenses, which further limits its application for carbon dioxide emission reduction. Hence, these technologies are not expected to find any quick application in the blue-hydrogen sector; the same stands for the oxy-combustion carbon capture systems. Current setups are not intended to manage high temperatures in oxygen-rich atmospheres with combustion [73]. With the current level of development in materials of construction for oxy-combustion equipment, it is an expensive pathway for collecting carbon dioxide. Since nitrogen in the product gas stream does not pose any problems whether the hydrogen process is being used for fertilizer production or something else, the oxy-combustion pathway is a dead end for the blue hydrogen production at the present level of research. At the current state of the knowledge and discussed earlier, pressure swing adsorption seems the most promising way of removing carbon dioxide from the hydrogen streams. Until any process breakthrough is achieved in the areas of membranes or solvent capture, the pressure swing adsorption will continue to be the choice of carbon capture method for hydrogen production systems, only being sparingly challenged by the Selexol process.

REFERENCES

1. T. N. Veziroglu, "Hydrogen technology for energy needs of human settlements," *International Journal of Hydrogen Energy*, vol. 12, no. 99, 1987.
2. Debeni G, Marchetti C. Hydrogen, key to the energy market. *Eurospectra* 1970;9:46–50.
3. Spath P, Mann M. Life cycle assessment of hydrogen production via natural gas steam reforming. Technical report NREL/TP-570–27637, NREL, 2000.
4. S. V. Mohan, "Waste to renewable energy: A sustainable and green approach towards production of bio-hydrogen by acidogenic fermentation," in *Sustainable Biotechnology*, O. V. Singh and S. P. Harvey, Eds. Springer, 2010, pp. 129–164.
5. T. Kuhn, *The Structure of Scientific Revolution*. University of Chicago, 1962.
6. Figueroa JD, Fout F, Plasynski S, McIlvired H, Srivastava R. Advances in CO_2 capture technology—The U.S. Department of Energy's Carbon Sequestration Program. Int J Greenhouse Gas Control 2008;2:9–20.
7. Intergovernmental Panel on Climate Change (IPCC), "Special report: Carbon dioxide capture and storage," 2010. [Online]. Available: www.ipcc.ch.
8. N. Muradov, "Low-carbon production of hydrogen from fossil fuels," *Compendium of Hydrogen Energy*, pp. 489–522, 2015.
9. Z. Salameh, "Energy storage," *Renewable Energy System Design*, pp. 201–298, 2014.
10. D. C. Upham, "Catalytic molten metals for the direct conversion of methane to hydrogen and separable carbon in a single reaction step commercial process (at potentially low-cost). This would provide no-pollution hydrogen from natural gas, essentially forever," *ScienceMag.org, American Association for Advancement of Science* (accessed Oct. 31, 2020).
11. Huang, D.-Y., Jang, J.-H., Tsai, W.-R., & Wu, W.-Y. (2016). Improvement in Hydrogen Production with Plasma Reformer System. *Energy Procedia*, 88, 505–509.
12. Gnanapragasam, N. V., Reddy, B. V., & Rosen, M. A. (2010). Hydrogen production from coal gasification for effective downstream CO_2 capture. *International Journal of Hydrogen Energy*, 35(10), 4933–4943
13. Shayan, E., Zare, V., & Mirzaee, I. (2018). Hydrogen production from biomass gasification; a theoretical comparison of using different gasification agents. *Energy Conversion and Management*, 159, 30–41.
14. V. Subramani, P. Sharma, L. Zhang and K. Liu, "Catalytic steam reforming technology for the production of hydrogen and syngas," in *Hydrogen and Syngas Production and purification Technologies*, K. Liu, C. Song and V. Subramani, Eds. Wiley, 2010, pp. 14–126.

15. Ritter JA, Ebner AD. State-of-the-Art adsorption and membrane separation processes for hydrogen production in the chemical and petrochemical industries. *Sep Sci Technol* 2007;42(6):1123–93.

16. K. Liu, G. D. Deluga, A. Bitsch-Larsen, L. D. Schmidt and L. Zhang, "Catalytic partial oxidation and autothermal reforming," in *Hydrogen and Syngas Production and Purification Technologies*, K. Liu, C. Song and V. Subramani, Eds. John Wiley & Sons, 2010, pp. 127–155.

17. A. Platon and Y. Wang, "Water-gas shift technologies," in *Hydrogen and Syngas Production and Purification Technologies*, K. Liu, C. Song and V. Subramani, Eds. Wiley, 2010, pp. 311–328.

18. P. Bakonyi, J. Peter and N. Nemestóthy, "Feasibility study of polyetherimide membrane for enrichment of carbon dioxide from synthetic bio-hydrogen mixture and subsequent utilization scenario using microalgae," *International Journal of Energy Research*, pp. 1–8, 2020.

19. J. Wang and Y. Yin, "Fermentative hydrogen production using pretreated microalgal biomass as feedstock." *Microbial Cell Factories*, vol. 17, p. 22, 2018.

20. M. S. Venkata and A. Pandey, *Bio-Hydrogen Production: An Introduction*, Elsevier, 2013, pp. 1–24.

21. Sinharoy, A. and Pakshirajan, K. 2020. A novel application of biologically synthesized nanoparticles for enhanced bio-hydrogen production and carbon monoxide bioconversion. *Renewable Energy* 147:864–873.

22. Manish, S. and Banerjee, R. 2008. Comparison of bio-hydrogen production processes. *International Journal of Hydrogen Energy* 33.1: 279–286.

23. Kuppam, C., Pandit, S., Kadier, A., Dasagrandhi, C., & Velpuri, J. (2017). *Bio-hydrogen Production: Integrated Approaches to Improve the Process Efficiency. Microbial Applications* Vol.1, 189–210.

24. Revah, S. and Morales, M. 2015. Hydrogen production by an enriched photoheterotrophic culture using dark fermentation effluent as substrate: Effect of flushing method, bicarbonate addition, and outdoor-indoor condition photoheterotrophic culture using dark flushing meth. *International Journal of Hydrogen Energy* 40: 9096–9105.

25. Kayahan, E., Eroğlu. I. and Koku, H. 2016. Design of an outdoor stacked–tubular reactor for biological hydrogen production. *International Journal of Hydrogen Energy* 41: 19357–19366.

26. Moghadamtousi, S., Fadaeinasab, M., Nikzad, S., Mohan, G., Ali, H., & Kadir, H. (2015). *Annona muricata* (Annonaceae): A Review of Its Traditional Uses, Isolated Acetogenins and Biological Activities. *International Journal of Molecular Sciences*, 16(7), 15625–15658.

27. K. Chandrasekhar, G. Kumar, S. V. Mohan, A. Pandey, B. H. Jeon, M. Jang and S. H. Kim, "Microbial electro-remediation (MER) of hazardous waste in aid of sustainable energy generation and resource recovery." *Environmental Technology & Innovation*, p. 100997, 2020.

28. Chandrasekhar, K., Yong-Jik, L. and Dong-Woo, L. 2015. Bio-hydrogen production: Strategies to improve process efficiency through microbial routes. *International Journal of Molecular Sciences* 16.4: 8266–8293.

29. Kotay, Meher, S. and Das, D. 2008. Bio-hydrogen as a renewable energy resource – prospects and potentials. *International Journal of Hydrogen Energy* 33.1: 258–263.

30. Levin DB, Pitt L, Love M. Bio-hydrogen production: Prospects and limitations to practical application. *International Journal of Hydrogen Energy* 2004;29:173–185.

31. Pandu K, Joseph S. Comparisons and limitations of bio-hydrogen production processes: A review. *Int J Adv Eng Technol.* 2012;2:342–356.

32. Das D, Veziroglu TN. Hydrogen production by biological processes: A survey of literature. *International Journal of Hydrogen Energy* 2001;26:13–28.

33. H. Herzog, E. Drake and E. Adams, *CO₂ Capture, Reuse, and Storage Technologies for Mitigating Global Climate Change: A White Paper.* Massachusetts Institute of Technology Energy Laboratory, 1997.

34. IPCC. *Climate change 2001: Mitigation.* Cambridge University Press, 2001.

35. Pacala S, Socolow R. Stabilization wedges: Solving the climate problem for the next 50 years with current technologies. *Science* 2004;305:968–72.

36. UNDP, *World Energy Assessment: Energy and the Challenge of Sustainability.* United Nations Development Programme, 2004.

37. Scholz, W. (1993). Processes for industrial production of hydrogen and associated environmental effect. *Gas Separation & Purification*, 7, 131–139.

38. P. Spath and M. Mann, "Life cycle assessment of hydrogen production via natural gas steam reforming," Technical Report. NREL/TP-570-27637, Golden, CO, 2000.

39. H. Audus, O. Kaarstad and M. Kowal, "Decarbonization of fossil fuels: Hydrogen as an energy vector," Proceedings of 11th World Hydrogen Energy Conference, Stuttgart, Germany, 1996.

40. G. Collodi, "Hydrogen production via steam reforming with CO₂ capture," *Chemical Engineering Transactions*, vol. 19, 2010 (accessed Feb. 10, 2013).

41. Kreutz, T., Williams, P., Chiesa, P., & Consonni, S. (2005). Co-production of hydrogen, electricity and CO₂ from coal with commercially ready technology. *International Journal of Hydrogen Energy*, 30, 769–784.

42. D. Gray and G. Tomlinson, "Hydrogen from coal," Mitretek Technical Paper MTR-2003-13, 2003 (prepared for US DOE NETL).

43. DOE/NETL, "Assessment of hydrogen production with CO₂ capture," In *Baseline state-of-the-art plants*, vol. 1, US DOE National Energy Technology Laboratory, DOE/NETL, Final Report 2010/1434, 2010.

44. G. Cortes, E. Tzimas and S. Peteves, *Technologies for Coal Based Hydrogen and Electricity Co-Production Power Plants with CO₂ Capture.* European Commission Joint Research Centre Institute for Energy, 2009.

45. Sukor, N. R., Shamsuddin, A. H., Mahlia, T. M. I., & Mat Isa, M. F. (2020). Techno-Economic Analysis of CO₂ Capture Technologies in Offshore Natural Gas Field: Implications to Carbon Capture and Storage in Malaysia. Processes, 8(3), 350.

46. Mari, V.; Kristin, J.; Rahul, A. Hydrogen production with CO₂ capture. *International Journal of Hydrogen Energy* 2016, 41, 4969–4992.

47. Zhang, C. L. Cost analysis and development suggestion for hydrogen production from coal and natural gas. *Pet. Process. Petrochem.* 2018, 49, 94–98.

48. Montenegro Camacho, Y. S.; Bensaid, S.; Piras, G.; Antonini, M.; Fino, D. Techno-economic analysis of green hydrogen production from biogas autothermal reforming. *Clean Technol. Environ. Policy* 2017, 19, 1437–1447.

49. Khojasteh Salkuyeh, Y.; Saville, B. A.; MacLean, H. L. Techno-economic analysis and life cycle assessment of hydrogen production from natural gas using current and emerging technologies. *International Journal of Hydrogen Energy* 2017, 42, 18894–18909.

50. Y. Khojasteh Salkuyeh, B. A. Saville and H. L. MacLean, "Techno-economic analysis and life cycle assessment of hydrogen production from natural gas using current and emerging technologies," *International Journal of Hydrogen Energy*, vol. 42, pp. 18894–18909, 2017.

51. I. P. Koronaki, L. Prentza and V. Papaefthimiou, "Modeling of CO₂ capture via chemical absorption processes: An extensive literature review," *Renewable and Sustainable Energy Reviews*, vol. 50, pp. 547–566, 2015.

52. A. Sieminski, *International Energy Outlook 2016.* U.S. Energy Information Administration (EIA), 2016.

53. R. M. Dell and D. A. J. Rand, "Energy storage – a key technology for global energy sustainability," *Journal of Power Sources*, vol. 100, pp. 2–17, 2001.

54. R. S. Haszeldine, "Carbon capture and storage: How green can black be?" *Science,* vol. 325, no. 5948, pp. 1647–1652, 2009.
55. J. Tollefson, "Hydrogen vehicles: Fuel of the future?" *Nature*, vol. 464, no. 7293, pp. 1262–1264, 2010.
56. A. Ajanovic and R. Haas, "Prospects and impediments for hydrogen and fuel cell vehicles in the transport sector," *International Journal of Hydrogen Energy*, vol. 1, 2020.
57. M. Miotti, J. Hofer and C. Bauer, "Integrated environmental and economic assessment of current and future fuel cell vehicles," *International Journal of Life Cycle Assessment*, vol. 22, pp. 94–110, 2017.
58. N. Casas, J. Schell, L. Joss and M. Mazzotti, "A parametric study of a PSA process for pre-combustion CO_2 capture," *Separation and Purification Technology*, vol. 104, pp. 183–192, 2013.
59. R. Guerrero-Lemus and J. M. Martínez-Duart, *Renewable Energies and CO_2 Cost Analysis, Environmental Impacts and Technological Trends*, 2012 ed. Springer, 2013.
60. M. M. Maroto-Valer, *Developments and Innovation in Carbon Dioxide (CO_2) Capture and Storage Technology*. CRC Press, 2010.
61. H. Yang, Z. Xu, M. Fan, R. Gupta, R. B. Slimane, A. E. Bland and I. Wright, "Progress in carbon dioxide separation and capture: A review," *Journal of Environmental Sciences*, vol. 20, no. 1, pp. 14–27, 2008.
62. B. Yildiz and M. Kazimi, "Efficiency of hydrogen production systems using alternative nuclear energy technologies," *International Journal of Hydrogen Energy*, vol. 31, no. 1, pp. 77–92, 2006.
63. K. Smith, C. J. Anderson, W. Tao, K. Endo, K. A. Mumford, S. E. Kentish, A. Qader, B. Hooper and G. W. Stevens, "Pre-combustion capture of CO_2 results from solvent absorption pilot plant trials using 30 wt% potassium carbonate and boric acid promoted potassium carbonate solvent," *International Journal of Greenhouse Gas Control*, vol. 10, pp. 64–73, 2012.
64. M. C. Romano, P. Chiesa and G. Lozza, "Pre-combustion CO_2 capture from natural gas power plants, with ATR and MDEA processes," *International Journal of Greenhouse Gas Control*, vol. 4, no. 5, pp. 785–797, 2010.
65. C. F. Martin, E. Stockel, R. Clowes, D. J. Adams, A. I. Cooper, J. J. Pis, F. Rubiera and C. Pevida, "Hypercrosslinked organic polymer networks as potential adsorbents for pre-combustion CO_2 capture," *Journal of Materials Chemistry*, vol. 21, no. 14, pp. 5475–5483, 2011.
66. P. Ahmadi, I. Dincer and M. A. Rosen, "Development and assessment of an integrated bio-mass-based multi-generation energy system," *Energy*, vol. 56, 155–156, 2013.
67. B. C. Bates, Z. W. Kundzewicz, S. Wu and J. P. Palutikof. *Climate Change and Water: Technical Paper of the Intergovernmental Panel on Climate Change*. IPCC Secretariat, 2008.
68. G. O. Garcia, P. Douglas, E. Croiset and L. Zheng, "Technoeconomic evaluation of IGCC power plants for CO_2 avoidance," *Energy Conversion and Management*, vol. 47, pp. 2250–2259, 2006.
69. T. F. Wall, "Combustion processes for carbon capture," *Proceedings of the Combustion Institute*, vol. 31, pp. 31–47, 2007.
70. A. Sieminski, *International Energy Outlook 2016*. U.S. Energy Information Administration (EIA), 2016.
71. J. M. Beer, "High efficiency electric power generation: The environmental role," *Progress in Energy and Combustion Science*, vol. 33, pp. 107–134, 2007.
72. M. Anheden and G. Svedberg, "Exergy analysis of chemical-looping combustion systems," *Energy Conversion and Management*, vol. 39, no. 16–18, pp. 1967–1980, 1998.
73. N. V. Gnanapragasam, B. V. Reddy and M. A. Rosen, "Status of research and recent advances on hydrogen production using coal, biomass and other solid fuels," *International Journal of Energy Research*, Special Issue, in press.

10 Improvements in Process Design, Simulation, and Control

Haslinda Zabiri, Faezah Isa, Tahir Sultan,
Muhammad Afif Asyraf Affian, Anmol Fatima, and
Abdulhalim Shah Maulud

CONTENTS

10.1 INTRODUCTION

Rising demand for energy and the struggle to alleviate climate change are two of the paramount factors receiving massive attention globally [1]. For decades, energy has remained a staple requirement for a society, but the use of traditional fossil fuel–based energy resources cause environmental issues such as greenhouse gas emissions, particulate matter, smog, and so forth. Most of the greenhouse gas emissions, such as CO_2, are highly associated with the industrial revolution and power generation [2]. Various technological options are available for CO_2 capture such as chemical absorption, adsorption, pressure, and temperature swing adsorption using different solid sorbents, membranes, and so forth. Among these, chemical absorption is generally accepted as the most established and well-developed technology for all pressure conditions [3]. While CO_2 emissions are obviously detrimental to the environment, CO_2 must also be removed in natural gas processing to comply with more specialized downstream pipeline specifications and end-user specifications. Additionally, the presence of CO_2 and other acid gas impurities reduces the life span of equipment and pipeline due to corrosion.

In all cases, the need for an energy-efficient process with appropriate strategies is evident, which can only be accomplished by implementing optimal process designs and operations for CO_2 capture through system modeling and process simulation, process control, and process intensification and design. The importance of such studies lies in the primary objectives of minimizing energy consumption and optimizing production. Indeed, the minimization of process errors and instabilities contributes to environmental benefits and economical gains.

10.2   MODELING AND SIMULATION

Chemical absorption is the most well known and widely used process in the CO_2 capture industry [4]. In such systems, vapor-liquid equilibrium (VLE) and thermodynamic properties are the two vital components that must be considered in process development [3]. In general, the thermodynamic models can be classified as semi-empirical (SE), excess Gibb's energy (G^E), equation of state/Gibbs energy (EoS/ G^E), and statistical thermodynamic models (ST) [4]. The classical thermodynamic approach for the effective design of CO_2 capture plants is widely divided into three models: SE, G^E, and EOS/ G^E. As an aside, equilibrium and rate-based modeling approaches with significant improvements are also developed to achieve the desired CO_2 capture rate in CO_2 absorption plants [3–4].

Various thermodynamic model developments have been reported in the literature. Most of the modeling and simulation works of any process related to CO_2 removal cannot escape from the thermodynamic development, especially for new solvent synthesis. An accurate thermodynamic modeling approach is essential for the utilization and optimization of new solvents to reduce the regeneration/stripping cost in CO_2 absorption/stripping plants. Recently, Mouhoubi et al. [5] focused on the development of thermodynamic modeling for CO_2 absorption in aqueous solutions of N,N-diethylethanolamine (DEEA), N-methyl-1,3-propane diamine (MAPA), and their mixtures using a G^E-based e-NRTL model in Aspen PLUS. The G^E-based electrolyte NRTL is utilized by regressing several key parameters such as pure vapor pressures, excess enthalpies, dielectric constants, physical solubilities of CO_2, and partial and total pressures from experimental data. The global DEEA-MAPA-H_2O-CO_2 thermodynamic system was established by correlating the two subsystems. The prediction of VLE and heat of CO_2 absorption were compared with experimental values that displayed good agreements between experimental and predicted results. Thus, it was deduced that the designed model could be used for the evaluation and optimization of the CO_2 absorption/stripping process using DEEA-MAPA mixtures at a large scale. Overall, the modeling and simulation efforts presented in this research are important for establishing the groundwork for estimation of physical properties and operating parameter optimization.

The thermodynamic behavior of CO_2 in concentrated DEEA solution can be described via reliable experimental data and a steady thermodynamic method. Luo et al. [6] compared an equation-based empirical model and an SE-based Kent Eisenberg (KE) model to analyze the VLE characteristics of a DEEA-CO_2-H_2O system in concentrated DEEA solutions. The absolute average relative deviations (AARDs) between VLE-based experimental data and predicted data for both models are calculated in terms of CO_2 solubility. The overall results concluded that the empirical model has shown better predictions for a wide range of DEEA concentrations with lower AARD values compared to the KE model. In particular, the KE model showed higher deviations, especially at higher concentrations (3–4 M). This study revealed the importance of appropriate prediction of the VLE data, whereby the authors suggested further research is required in analyzing Henry's law constant in DEEA solutions with high CO_2 concentration.

Optimization of CO_2 partial pressure is one of the main factors that influence the cost of CO_2 capture plants. The accuracy of a modeling approach allows for better evaluation of the CO_2 partial pressure, and thus more optimized regeneration energy costs in CO_2 absorption/stripping plants. Kalatjari et al. [7] studied four thermodynamic models to simulate the absorption process: eNRTL-Bishnoi, Gabrielsen's approach, ASPEN eNRTL default, and ASPEN-eNRTL Regressed-DRS techniques. In Gabrielsen's approach, a very simple predictive correlation was used to calculate the partial pressure of CO_2 over aqueous MEA by assuming ideal gas and liquid phases. On the other hand, eNRTL-Bishnoi defined the reference state for CO_2 as an infinite dilution in pure water. This allows the change in the activity of CO_2 to be modeled as a function of amine strength. This research aims to evaluate the most suitable thermodynamic model that has sufficient adjustable parameters to calculate the optimal CO_2 partial pressure gas feeds containing high and low CO_2 concentrations at low pressure over a large temperature range. The study concluded that Gabrielsen's modeling approach provided the most satisfactory results among the four models when compared with experimental data. The approach adopted in this chapter can be further extended to evaluate other solvents.

The appropriate solvent circulation rate is required for the desired CO_2 capture rate, as it is directly influenced by the CO_2 equilibrium solubility, which is a function of temperature and pressure. Xiao et al. [8] analyzed the activity-structure relationship at the molecular level to find the differences in equilibrium for CO_2 solubilities for 2-dimethylamino (DMEA), diethylethanolamine (DEEA), and 3-dimethyl-amino-1-propanol (3DMA1P) for potential usage in CO_2 capture. The

SE-based modified Kent and Eisenberg (M-KE) model was used to compare the VLE data of amine-H_2O-CO_2 systems and to predict equilibrium CO_2 solubility. Results indicated that the proposed M-KE model was able to correlate the relationships very well and was able to provide reasonable prediction accuracies for all amines. The overall results concluded that the DEEA solvent has the greatest potential for CO_2 capture with regards to equilibrium CO_2 solubility, the second-order reaction rate constant, and CO_2 absorption heat.

Accurate chemical and physical solubility data is necessary to properly model CO_2 absorption by amine solutions in Aspen Plus. Esmaeili *et al.* [9] utilized a model based on the G^E-based E-NRTL and the EoS-based perturbed-chain statistical associating fluid theory (PC-SAFT), for the calculation of CO_2 solubility in a piperazine (PZ) and methyldiethanolamine (MDEA) mixture. The methods were used to calculate the activity and fugacity coefficients, respectively. E-NRTL is well known for its versatility in terms of electrolyte property method for aqueous and mixed solvent systems, whereas PC-SAFT is capable of predicting vapor properties even at high pressures. The resulting simulation of the CO_2 capture system provides an adequate agreement between model estimates and experimental data in terms of CO_2 absorption efficiency and CO_2 loadings, especially in high-pressure conditions. Moreover, Asadi *et al.* [10] developed a robust thermodynamic model using G^E-based E-NRTL property method to measure the solubility of CO_2 in mixtures with different mass compositions (10–80 wt%) to correlate and predict the partial pressures of CO_2 against loadings. Then, a high-pressure quasi-static equilibrium method is utilized to evaluate the CO_2 solubility in the aqueous mixed solution of sulfolane (physical solvent) and ethanolamine (AEEA-chemical solvent). New equilibrium data were generated for CO_2 solubility in an aqueous blended solution with other several solvent compositions. The results displayed that the E-NRTL model demonstrated high capability for correlation and prediction for binary, ternary, and quaternary systems. Furthermore, it was observed that the accuracy of the model could be enhanced by normalizing the activity coefficient of CO_2 in the mixed solvent. This modeling approach with the hybrid solvent of sulfolane and AEEA can remove CO_2 effectively over wide ranges of operating conditions, especially for acid gas removal plants.

Similarly, in Shirazizadeh and Haghtalab [11], the experimental VLE data of a new mixed (chemical + physical) solvent system of MDEA-dimethyl sulfoxide (DMSO) were modeled by the G^E-based E-NRTL property method for correlation of the experimental solubility of CO_2. The E-NRTL activity coefficient model was used to model the VLE data so that the molecule-molecule, ion-pair molecule binary interaction parameters of the model could be obtained. However, the value of AAD between the experimental and calculated values is still on the higher side of 10.95% for the DMSO + MDEA + H_2O + CO_2 system. This suggests the E-NRTL model cannot be used for such mixed-solvent systems without appropriate modifications. Table 10.1 summarizes the recent developments in the field of thermodynamic modeling for carbon capture systems.

TABLE 10.1

Summary of Latest Developments in Modeling of CO_2 Processes

Objective/system	Model	References
To obtain VLE experimental data to be utilized in the DEEA-CO_2-H_2O system	Kent Eisenberg (KE), E-NRTL, and UNIQUAC	[6]
To analyze the activity-structure relationship at the molecular level to elucidate the differences in equilibrium for CO_2 solubilities of five tertiary amines	M-KE	[8]
To develop a thermodynamic model for CO_2 absorption and desorption calculations using gas feeds in an aqueous MEA solution	ENRTL-Bishnoi, Gabrielsen's approach, Aspen eNRTL default, and APEN-eNRTL regressed-DRS	[7]
To develop thermodynamic modeling of CO_2 absorption in aqueous solutions of DEEA, MAPA, and their mixtures DEEA-H_2O-CO_2 and MAPA-H_2O-CO_2 systems	E-NRTL	[5]
To develop thermodynamic modeling for CO_2 capture in an aqueous solution of PZ and MDEA	E-NRTL and PC-SAFT	[9]
To measure the solubility of CO_2 in an aqueous combination of AEEA solution using a high-pressure quasi-static equilibrium technique	E-NRTL	[10]
To develop a model of CO_2 solubility in binary and ternary solutions for dimethyl sulfoxide (DMSO), MDEA over wide ranges of temperature and compositions	E-NRTL/Pitzer-Debye-Huckel, NRTL	[11]

10.3 PROCESS CONTROL

Various carbon capture (CC) technologies have seen usage spanning across absorption, distillation, membrane separation, adsorption, biological separation, and separation by hydrates based on chemical looping combustion, pre-combustion, post-combustion, and oxyfuel combustion. The control strategies are the heart of all the aforementioned processes, as CC percentage and operational cost affect the process overall efficiency drastically. CO_2 removal through absorption using amine solvents is the most widely used technology due to its high capture efficiency. However, it has drawbacks of high energy consumption, high process nonlinearity in variable interactions, process/environmental constraints, and large computational times. Therefore, the current research related to process control of CC is focused on the following four major aspects:

1. *Mathematical model*: The performance and computational complexity of any control strategy depend heavily on the identified mathematical model

accuracy and structure. Therefore, most of the work is focused on optimizing the highly accurate models for the controller implementation.

2. *Model predictive control (MPC)*: MPC-based control strategies have the potential to control multiple-input multiple-output (MIMO) systems with variable interaction. Therefore, a huge emphasis on the MPC-based control schemes is observed.

3. *Multi-scenario*: Most of the carbon is emitted from power plants based on coal fire, and the related CC technology is a highly energy-intensive process. Hence, there is a significant focus on the design of multi-scenario control strategies including nonlinear MPC control schemes, which are flexible enough to work through a wide choice of power plant conditions to optimize the energy costs.

4. *Alternative CC technologies*: Conventional CC processes like solvent absorption and membrane separation always face the challenges of process nonlinearities and uncertainties. Therefore, alternative CC technologies are being utilized for the efficient control of CC processes.

10.3.1 MOTIVATION

The power sector is one of the major contributors to large point CO_2 emissions. For such processes, CO_2 capture with alkanolamine solvents is the most suggested technique for large-scale purposes, as CO_2 capture plants have effectively shown the attainability of this technology worldwide. However, it faces the challenges of having complicated MIMO systems, process/environmental constraints, large computational times, and so forth. Therefore, an efficient, well-designed, and flexible control strategy is required to overcome all these issues. Several control schemes are being applied to control the CO_2 capturing systems efficiently. Proportional (PI), proportional-integral-derivative (PID), and different algorithms of MPC control schemes are generally used to control CC plants. However, each controller has certain limitations. As most CC plants are MIMO systems, especially the post-combustion capture (PCC) systems, conventional PI and PID controllers are not recommended for such systems as they can only perform effectively for single-input single-output (SISO) systems. The MPC control scheme is a highly touted alternative for covering the drawbacks of the PID scheme. The MPC control scheme is suitable for dealing with MIMO systems, the variable interactions, and process uncertainties due to the state-space model structure used for implementation. However, MPC performance is also limited in certain systems where the fast-dynamic systems can induce a high computational burden. Nowadays, most research is focused on bettering the performance of MPC-based control strategies through different methods. In the end, a control strategy is primarily planned for a certain objective, whether to get a highly accurate mathematical model or through attempting different types of MPCs (classical, nonlinear, multi-scenario, etc.). The current work is focused to look into how such challenges are tackled, and how CC systems with advanced control strategies are further optimized.

10.3.2 Evaluation of Best Control Strategy through Comparison

Evaluation of the best control strategy for any plant is a major aspect of process control systems. This evaluation is based on two salient features: (1) comparison of different control strategies on the same plant model to calculate the particular performance parameters of the controller such as offsets, overshoot, settling time, integral square error (ISE), and so forth; and (2) utilization of coordinated control strategies using the multiple controllers for improved flexibility of the controller [12–14]. Tirapanichayakul *et al.* [15] designed a fluidized bed reactor using an adsorption-based CO_2 capture system in ANSYS FLUENT. The transfer function model of the system is identified using System Identification Toolbox in MATLAB, with the fact that solid circulation rate at inlet and gas velocity has a substantial impact on the CO_2 removal rate. The solids circulation rate is set as manipulated variable and CO_2 capture rate as a control variable, while inlet gas velocity, composition, and temperature are used as an external disturbance. The PID control scheme is implemented on the model for disturbance rejection. The controller can show a better response when the amount of disturbance is changed. Cristea *et al.* [16] also implemented a decentralized PI-based control scheme on a PCC plant to improve its flexibility. The performance objectives of the controller are CO_2 capture rate and energy performance, under disturbance rejection and setpoint tracking modes. Though the control scheme has shown satisfactory results, the overshoots in the control variables are still on the higher side. The results of the above studies showed that the response of the controller is slow, and there are still offsets in the setpoint values, indicating the need for a better control strategy to reduce such issues.

He *et al.* [12] implemented three types of advanced MPC control strategies on two different types of CC plants: (1) integrated gasification combined cycle (IGCC) power plant having a membrane reactor (MR) as a CC system (IGCC-MR); and (2) supercritical pulverized coal (SCPC) fired power plant attached to monoethanol-amine (MEA) based PCC plant (SCPC-MEA). The purpose of this research is to decide the best control strategy for optimization and minimization of operational and maintenance costs of power plants during cyclic operations. The performance of the PID controller, Dynamix Matrix Control (DMC) based linear MPC (LMPC) controller, literature-based nonlinear MPC (NMPC), and the proposed NMPC has been compared under multiple scenarios such as setpoint tracking, disturbance rejection, and power generation cycling trajectory. For the IGCC-MEA scenario, power demand is increased from 700 MW (nominal value) to 800 MW in the form of a step-change for setpoint tracking for case 1. In case 2, the carbon content of the coal slurry is decreased by 15% for disturbance rejection mode. The power generation (MW) and slurry flowrate (kg/s) is considered in terms of controller performance for both cases to calculate the ISE values. For the SCPC-MEA scenario, case 3 considers the power generation plant cycling trajectory similar to the IGCC-MEA case 1. Case 4 discusses the hybrid mode of disturbance rejection and setpoint tracking intending to maintain the CO_2 removal rate at 90%. The proposed NMPC strategy has shown better results than others in terms of ISE value, showing

73% improvement for IGCC-MR and 96% for SCPC-MEA compared to the DMC-based MPC strategy. Moreover, the proposed NMPC scheme can achieve the new setpoint quickly and rejects the disturbances faster than the other control schemes, which is an important target within the research of process control of CC. However, the proposed NMPC scheme response is tuned to be slightly aggressive, which may cause instrumentation problems in the plant. Proper sensitivity tuning of the controller parameters can overcome this challenge, which can enhance the controller performance significantly.

While conventional carbon capture technologies such as solvent absorption and membrane separation have greater carbon capture efficiency, the process nonlinearities and uncertainties involved may hinder the control strategy performance. Therefore, recent works are also focused on alternative carbon capture technologies such as centrifugal fluid separation (CFS), particularly supersonic gas separation (SGS), which is believed to be an efficient CC technology for offshore operations. Othman et al. [17] implemented PID and MPC control schemes on an SGS separator for process feed conditioning of natural gas before the CO_2 separation process. First, two types of PID-based control strategies are designed. The first control scheme consists of various controllers: pressure controller, liquid level controller, heat exchanger temperature controller, and so forth. In the second control strategy, the outlet stream heat of the compressor is utilized for the pre-heating of the feed stream during a disturbance model of low temperature. All the controllers are tuned for the PID control scheme. Then, the MPC control scheme is designed using the MPC algorithm in Aspen HYSYS for both control strategies, and the performance of PID and MPC controllers are compared under the disturbances in CO_2 composition, steam, pressure, and temperature. The ISE values are calculated for all the controlled variables under the disturbance rejection scenario. The results show that MPC has low ISE values in terms of pressure controller, while PID has four times better performance in terms of flow rate control compared to MPC. The results remain consistent for all control variables for both cases, however, MPC performance is on the lower end due to the reduced fitting percentage of the identified plant model, which is expected of the MPC control scheme. Table 10.2 presents a summary of recent control strategies used for carbon capture systems.

10.3.3 Selection of a Mathematical Model through Nonlinearity Analysis

To reduce the dynamic complexity of the mathematical model, nonlinearity analysis is a viable alternative option for controller implementation. Wu et al. [18] performed nonlinearity analysis through the identification of multiple state-space models and designed a multi-MPC (M-MPC) for the flexible operability of the PCC plant. The M-MPC scheme is developed for large changes in CO_2 removal rate according to the nonlinearity investigation of the plant which is performed using the gap-metric method. The changes in the flow rate of the flue gas are utilized as an unmeasured disturbance while implementing the M-MPC schemes for disturbance rejection. Then, a combined M-MPC scheme is proposed to design an algorithm that can

TABLE 10.2

Summary of Recent Control Strategies Used for Carbon Capture Systems

CO₂ capture technology	Control strategy	Software/tool	Basic purpose/findings	References
Membrane reactor (MR)	**PID**	Aspen Plus® Dynamics MATLAB®	Evaluation of best MPC control scheme based on ISE calculations.	[12]
Post-combustion capture (PCC) using an absorption system with MEA solvent	DMC based Linear **MPC** Proposed Literature-based **NMPC** Nonlinear MPC **(NMPC)**		The proposed NMPC scheme has 73% and 96% better performance in terms of ISE for IGCC-MR and SCPC-MEA plant, respectively.	
Fluidized bed reactor using adsorption	**PID**	ANSYS FLUENT® System Identification Toolbox in MATLAB®	Disturbance rejection. Required CO₂ capture rate is achieved under the changing external disturbances.	[15]
Supersonic gas separation (SGS)	**PID** **MPC**	Aspen HYSYS® System Identification Toolbox in MATLAB®	MPC performance is better than PID in one case, while PID performance is four times better than MPC in the second case.	[17]
Solvent-based carbon capture	Multi MPC **(MMPC)**	gPROMS® MATLAB®	MPC has the best performance in terms of fast and close tracking of CO₂ capture rate.	[18] [19]
Absorption/stripping system using MEA solvent	Classical **MPC**	Aspen PLUS® Dynamics MPC Toolbox in Simulink	MPC performs efficiently when the carbon capture system faces the challenges of setpoint tracking and disturbance rejection under varying plant conditions.	
Post-combustion carbon capture plant	**MFAC** **GPC**	gPROMS®	The flexibility and stability of the control strategy can be easily achieved by online parameter stabilization.	[20]
Amine absorption (MEA)	Nominal **NMPC** multi-scenario **NMPC**	Python®	Nominal NMPC has less computational time and it increases as the number of scenarios increase in each assumed plant.	[21]
An MEA-based natural gas carbon capture integrated with PCC plant	**MPC** **PID** (From Literature)	GT Pro Modelica (TPL Library)	MPC has superior performance than PID in terms of quick response, settling time, overshoot, and offsets.	[22]

effectively reject the disturbances for a wide range of operating points, which can alter the whole M-MPC scheme. A membership function combines all the designed local MPCs and evaluates the best control moves which are implemented on the CO_2 capture plant. The proposed control strategy is explained in the schematic diagram as illustrated in Figure 10.1.

Nonlinearity analysis is performed by identifying different linear state-space models using subspace identification method under two scenarios: (1) 50%, 60%, 70%, 80%, 90%, and 95% CO_2 capture rate conditions, and (2) 0.07 kg/s, 0.10 kg/s, 0.1 kg/s, and 0.15 kg/s of flue gas flowrate. The gap metric values for all the models are calculated between the range of 0 to 1. The models having large gap metric values with smaller nonlinearities are picked, and an integrated multi-model system is designed using the multi-models and multi-predictive controllers. This approach enhances the controller performance. This multi-model MPC control scheme performance is evaluated under the ramp change of 0.14%/30 sec in the CO_2 capture rate. The results show that this MMPC strategy has the best performance in terms of its fast and close tracking of CO_2 capture rate. However, the computational time and system complexity may increase as well due to multiple model involvement.

Ra *et al.* [22] implemented the 5×5 (five manipulated variables (MVs) and five control variables (CVs) MPC control strategy on a natural gas combined-cycles

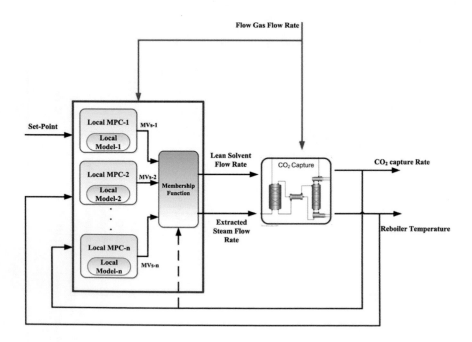

FIGURE 10.1 Schematic diagram of MMPC control scheme for the PCC process [18].

plant integrated with a PCC plant. The nonlinear behavior of the models is reduced by designing a network of local linear state-space models; the MPC algorithm then utilizes this network as a quadratic program for the computation. The major objective of this control strategy is to minimize the deviation of control variables from setpoints. A step change is introduced in the power demand, while the response of net mechanical power, superheating temperature, reheating temperature, capture ratio, and reboiler temperature is observed in the MPC. Furthermore, the MPC results are compared with PID-based results in the literature. The results have shown that power demand has achieved the new setpoint quickly and all other CVs have achieved the nominal values again after little disturbance with zero offsets. However, PID results have a very high amount of overshoot especially for capture ratio and reboiler temperature. Moreover, offsets present in the PID also makes MPC a better option for such integrated complex capture systems. However, the response of MPC is still slow compared to more advanced MPC control strategies, which can be improved by the appropriate tuning of objective function weights.

In CC process control, simulation and modeling are the key steps, as the performance and the dynamic complexity of model-based control strategy depends significantly on the fitting percentage and dynamic structure of the identified mathematical model. A work by Wu et al. [18] is focused on the optimization of highly accurate models for controller implementation through nonlinearity analysis for the identification of the best state-space model for controller implementation. A state-space model is an appropriate option for the MPC control scheme as it considers the real dynamics of the CC plant. Sultan et al. [19] designed a 2×2 classical MPC control strategy for the CC plant through a second-order linear state-space model. The controller is implemented for setpoint tracking mode under two cases: (1) ±5% step change in CO_2 composition for setpoint tracking mode, and (2) ±15% step change in stripper temperature for disturbance rejection scenario. The controller can achieve the setpoint quickly in 1.2 seconds without offsets and overshoot. For the disturbance rejection scenario, the controller has rejected the disturbances in 7.5 seconds. The controller performance is extraordinary due to the accuracy of the state-space model. It is difficult to get the high-percentage fitting for state-space models during the identification process. However, the percentage fittings for this case are 68.42% and 63.07% for CO_2 composition and stripper temperature, respectively. Hence, the MPC controller can work adequately for this model. However, the controller response is very aggressive, which may damage the sensitive parts of the plant, especially the valves.

Nonlinear autoregressive methods are also widely used for model identification which can exhibit comprehensive and deep knowledge of the plant. The correlation among input and output variables is described through a nonlinear autoregressive exogenous (NARX) model by Akinola et al. [23], which is obtained from various models due to the multiple-input single-output (MISO) nature of the system. Then the most accurate model in terms of process dynamics is selected, utilizing the forward regression orthogonal least squares (FROLS) algorithm. The results verified that the identified model, with its capability in predicting the real dynamics of the CC plant, is believed to be the most feasible model for controller implementation for CC.

10.3.4 MODEL-FREE APPROACH

Moreover, model-free control approaches have also been utilized for PCC systems to avoid the model complexities and nonlinearities. Li *et al.* [20] designed a model-free adaptive control (MFAC) algorithm for a post-combustion carbon capture plant that uses only the measured input-output data of the plant. The performance of the model-free controller and model-based (Generalized Predictive Controller and PID) controller is compared in terms of CO_2 capture level. The advantage of the MFAC is that it can be tuned online through the process plant's input-output data instead of defining the model parameters in offline mode through system identification. Therefore, online tuning parameter determination with the MFAC becomes a flexible control strategy that can be implemented easily. Though the response of the model-free approach is fast and satisfactory, it is clearly limited in emulating the true plant behavior, as an accurate mathematical model of complex CC systems exhibits the true dynamics more effectively.

10.3.5 IMPLEMENTATION OF NONLINEAR MPC SCHEMES

The ubiquity of process nonlinearities always presents challenges in control, in the form of uncertainties. Patron *et al.* [21] implemented two types of MPC controllers on a nonlinear dynamic mechanistic model: a nominal NMPC, and multi-scenario NMPC. The Pyomo environment is used for the implementation of both controllers, which is an optimization tool in Python. Nominal NMPC refers to the controller with all known uncertainties. Multi-scenario NMPC refers to the controller with different scenarios, having multiple uncertain parameters based on the size of uncertainty/disturbance or the number of scenarios. A dynamic identified model is utilized for the implementation of NMPC, termed as a partial differential-algebraic system of equations (PDAEs); which consists of ordinary differential equations (ODEs), partial differential equations (PDEs), and algebraic equations (AEs). The performance of a single scenario, nominal NMPC, and the multi-scenario NMPC is compared in terms of computational time and offsets in CO_2 capture rate. Three types of disturbances are introduced in the absorber column: (1) effect of uncertainty/disturbance size, (2) number of realizations/scenarios, and (3) variations in flue gas flow rate. The results show that the nominal NMPC has less computational time, as it considers only one scenario. However, the computational time increases as the number of scenarios increase in each assumed plant.

In another NMPC-based control study, Akinola *et al.* [24] designed an NMPC control scheme for a PCC process using a NARX input model identified by the FROLS method. The performance of NMPC is compared with LMPC under two scenarios. In the first case, both the controllers are set to flue gas flow rate changes with a ramping rate of 0.42%/min and 0.67%/min. In the second case, flue gas flowrate variation is the same as in the first case. However, the CO_2 capture level (CV) is varied, with ramping rates of 0.79%/min, 0.94%/min, and 2.4%/min, and the stripper temperature

(CV) is changed with a ramping rate of 0.092%/min. The results show the NMPC has better performance than LMPC in terms of disturbance rejection by 55.3% and 92.74% for CO_2 capture level and reboiler temperature, respectively. Similarly, in the second case, NMPC has a superior performance of 17.86% and 18.67% for CO_2 removal level and reboiler temperature, respectively. The better performance of the NMPC is attributed to the nonlinear model identified using the FROLS method, as it counts the nonlinearity behavior of the model with high accuracy. The NMPC performance may degrade in large and quick ramp changes using the NARX model with a high quantity of uncertainties.

10.4   PROCESS INTENSIFICATION

10.4.1   PROCESS INTENSIFICATION (PI) ON POST-COMBUSTION CO_2 CAPTURE (PCC) SYSTEM

Process intensification (PI) has received much attention from both researchers and engineers due to its applicability in optimizing much of the conventional technology used in the industry. The term 'process intensification' constantly changes in definition due to rapid changes within the research field, brought about by the inpouring of new knowledge and technology. Reay [25] identified PI as any development in the chemical engineering sector that gives smaller, better, more energy-efficient, and better optimized systems than the conventional ones. Another definition set by the European Processsor Initiative [26] suggests that PI is not limited to the miniaturization of unit operations but includes innovative principles in process design, a substantial reduction in capital and operating expenditure, improvements in process safety, and so forth. The advancement of chemical engineering processes can be credited to the success of several PI technologies such as the invention of reactive distillation, a combination of the reactor and distillation column. It is widely known as the earliest intensified system ever created in the chemical engineering field. It has been reported that implementations of reactive distillation have seen more than 150 commercial petrochemical applications [27]. Static mixers are also broadly recognized as an intensified operation. The ability of a static mixer to improve the agitation of fluid mixture with no moving parts not only increases the efficiency of the process but also reduces the overall capital costs to the facility using it. Adoption of these PI technologies could potentially improve the CO_2 capture industries drastically with an adequate amount of research and development. PI can be divided into two areas, which are known as intensified equipment and intensified methods.

10.4.2   INTENSIFIED EQUIPMENT

Intensified equipment is the resulting product of a successfully implemented process intensification strategy. The intensified equipment can be categorized into two types, which are reactor and nonreactor forms. Some of the equipment are rotating packed bed (RPB), static mixer, microchannel reactor, and so forth. An RPB is a device comprising a rotor, inside of which consists of a mechanical shaft and large packing area that can enhance the mass transfer performance of the fluids. By taking advantage of the centrifugal forces, the gas-liquid interaction could be improved compared to certain conventional packed bed designs [28]. For instance, Yu *et al.* [29] studied the use of alkanolamine solutions mixed with either piperazine or diethylenetriamine (DETA) for CO_2 capture using an RPB. It was found that the height of the transfer unit (HTU) is far better than the conventional packed bed, largely due to the superior mass transfer which provided a higher CO_2 capture efficiency. In addition to that, Lin *et al.* [30], used cross-flow RPB in the absorption of CO_2. It is suggested that the gas has relatively low resistance to the gas flow, rendering the system unaffected by the centrifugal force. An in-depth review of RPB is explained in the section below.

10.4.3 INTENSIFIED METHODS

Intensified methods are often related to the configuration of equipment in a system, involving the combination of multiple processes into a single unit operation. A widely recognized and used example of such intensified method is reactive distillation, which is basically a combination of a reactor and distillation column into a single unit operation. In the petrochemical industry, it is used to produce MTBE, one of the most valuable products manufactured by the industry. Moreover, the intensified method was reported to save from as low as 10% up to a staggering 80% in terms of capital costs and energy consumption within the petrochemical industry [27]. This result shows the potential that process intensification carries in improving the efficiency of the current conventional distillation system used for CO_2 capture.

10.4.4 LEADING PI TECHNOLOGIES FOR THE PCC INDUSTRY

This section reviews some of the more advanced PI technologies adopted for PCC processes. Most of the intensified processes are implemented within the absorber, stripper, and heat exchanger units, which are identified to carry a large influence on CO_2 capture performance, energy consumption, and capital costs. A summary of the leading intensified equipment for the PCC industry is shown in Table 10.3.

10.4.4.1 Rotating Packed Bed (RPB) Absorber

The RPB absorber has become widely known in the field of PCC as a better alternative to the conventional absorber. The standard absorber consists of large, packed columns designed to facilitate the transfer of CO_2 from one specific mixture to another. However, with the advent of an RPB-based absorber, the size can be reduced due to increased absorption rate and efficiency by taking advantage of the inherent centrifugal force produced by the equipment. Oko *et al.* [31] conducted a study on using RPB with high concentrations of MEA solvent to improve the CO_2 capture while reducing volume footprint. Highly concentrated MEA solutions (>30 wt%) are used concerning the RPB, due to the smaller packing volume compared to the conventional absorber.

A higher concentration of MEA solutions is required to enhance the driving force and the mass transfer performance of the system [32], which carries the trade-off of higher corrosiveness and viscosity in the concentrated MEA solution. The highly

TABLE 10.3

Summary of the Leading Intensified Equipment for the PCC Industry

Intensified equipment	Description	Advantages	Limitation	References
RPB absorber	RPB absorber leverages its small packing volume and high centrifugal forces to use concentrated MEA solution (>30 wt%) for the absorption of CO_2.	Several studies showed its feasibility in the CO_2 capture industry. It also possesses smaller packing volume and centrifugal forces that help in improving the mass transfer performance.	No known drawbacks for CO_2 capture were found up to date.	[28, 31–32]
Printed circuit heat exchanger (PCHE)	A PCHE is classified as a plate-fin type heat exchanger which is created based on is based on two vital chemical processes, known as diffusion bonding and chemical etching.	PCHE units are 5 times smaller and lighter than the conventional shell and tube heat exchanger, having the same pressure drop and energy duty.	No known limitations up to date.	[35–36]
RPB stripper	RPB absorber its small packing volume and high centrifugal forces to use concentrated MEA solution (>30 wt%) for the absorption of CO_2.	RPB as a stripper not only hugely reduces the required stripper volume but also utilizes less regeneration energy as compared to the conventional stripper.	The packing column has been intensified, but the reboiler size still the same.	[28, 37, 40]

corrosive nature of the MEA solution could shorten the life span of the equipment, and the high viscosity results in the reduction of wetting potential of the packing, reducing the overall mass transfer performance [31]. These perceived drawbacks made concentrated MEA solutions widely inappropriate in conventional packed bed absorbers for CO_2 capture [31]. However, since the RPB can generate ample centrifugal forces, a highly viscous MEA solution becomes a viable choice. Using a stainless steel material for the RPB, the inherent corrosion issue is also easily rectified.

In terms of energy usage, the reboiler is known to consume the highest amount of energy. An RPB-based PCC process has estimated the reboiler duty using the Oexmann [33] approach, accounting for 2940 MJ/ton CO_2 for 80 wt% MEA solvent compared to about 4000 MJ/ton CO_2 for the typical 30 wt% MEA solvent [34]. Combining the compact size of RPB with the more concentrated MEA solvent, the technical and economical feasibility of using this equipment in the PCC process becomes more apparent.

10.4.4.2 Printed Circuit Heat Exchanger (PCHE)

A printed circuit heat exchanger (PCHE) is classified as a plate-fin type heat exchanger. The invention of this equipment is based on two vital chemical processes: diffusion bonding and chemical etching. The contacting fluids will be transferring heat from the channels created by chemical etching. Subsequently, the diffusion bonding attaches all the layers of the etched plates to form a PCHE.

As a result of the diffusion bonding, PCHE is more compact in terms of size and weight. According to Li et al. [35], the volume of a PCHE unit is five times smaller and lighter than the conventional shell and tube heat exchanger while having the same pressure drop and energy duty. It was reported that the configurations of the heat exchanger could further reduce the pressure drop in the unit. Kim et al. made the comparison between a zig-zag channel and air-foil fin PCHE to evaluate the heat transfer performance and the pressure drop of both different configurations. It was reported that the pressure drop in the air-foil fin is one-twentieth of the zig-zag PCHE. In addition to that, the PCHE effectiveness was more than 98%, and it can operate at maximum permissible pressure and temperature of 600 bars and 800 °C [36]. The air-foil fin PCHE was decidedly preferred due to the better heat transfer performance and operability at a higher range of pressure and temperature.

10.4.4.3 RPB Stripper

Studies have been made on RPB strippers to evaluate the effectiveness of the intensified process in reducing cost and energy duty by Jassim et al. [37]. The range of parameters studied was for the regeneration runs of 30 wt%, 54 wt%, and 60 wt% MEA solutions, with mass flow rates between 0.2 to 0.6 kg/s. The RPB stripper column has the following configuration:

- Internal diameter: 0.156 m
- External diameter: 0.398 m
- Axial height: 0.025 m
- Packed bed material: stainless steel
- Specific surface area: 2132 m^2/m^3.

A comparison has been observed for both conventional and RPB strippers at similar performance [37]. It was stated that the intensified RPB stripper height has been effectively reduced by a factor of 8.4 and the diameter has decreased by a factor of 11.3. This is considered a breakthrough, considering both strippers have similar performances in terms of energy consumption and CO_2 removal rate. The reduced size of the intensified stripper compared to the conventional stripper renders it an economically feasible option where saving space is an objective.

Furthermore, with the conventional stripper, the energy used to capture CO_2 from a coal-fired power plant was reported to be between 3240 and 4200 MJ/ton CO_2 [38]. Most of the energy consumption comes from reboiler duty, which is used to regenerate the solvent. Therefore, Cheng et al. [39] introduced a back pressure regulator (BPR) into the stripper to regenerate the rich solvent at a temperature of higher than 100 °C under pressures of more than the atmospheric pressure. In both setups, the usage of RPB as a stripper similarly not only reduced the required stripper volume by a large amount but also utilizes less regeneration energy as compared to the conventional absorber [28].

10.5 PROCESS DESIGN

To design an effective system, the selection of CC technology and the process units play a vital role, with the designs being further refined in the modeling stage. When designing any process, it is very important to ensure that the flow rates, yields, and product purities are well described. From there, the development of simple but reliable process models is necessary to describe the behavior relationships in the data that is consistent with the actual behavior of the CO_2 capture plant. The detailed modeling of the processes has many challenges, including lack of process knowledge, selection of optimal parameters from largely available options, and so forth. Hence, researchers have endeavored to accomplish optimal integration of CC units through deeper fundamental investigations of process modeling, simulation, and optimization.

Oh et al. [41] designed an integrated CC plant with a coal-based power plant to reduce the efficiency penalty through heat integration. Such integrated processes can affect the overall efficiency of the power plant, which was observed through the techno-economic analysis of the developed process models. The process model included a pulverized coal boiler and steam cycle. To fulfill the industrial design specifications, an MEA-based PCC CO_2 capture plant is first designed on a pilot scale. The CO_2 capture plant is also optimized through the exploitation of system interactions to enhance the overall power generation efficiency of the integrated process. The findings have shown that the proposed process integrated design of the method leads to substantial improvements in energy efficiency. The developed process models were later used to test the effect of a redesigned steam cycle for an integrated overall system.

The power demand or load varies significantly during the operation of such integrated systems. Therefore, Oh et al. [41] evaluated the economic efficiency of the integrated process under changing load conditions. The minimization of regeneration energy without sacrificing CC performance is still a challenging aspect in CO_2

capture systems. To find a more economically integrated system, a CO_2 capture plant under part-load condition is modeled with both power plants: the NGCC plant and the coal-based power plant. A case study was then done to verify the productivity of the suggested design and optimization framework, which concluded to improved overall efficiency.

A PCC CO_2 capture plant integrated with a 600 MW NGCC plant is designed by Dutta et al. [42]. The purpose of the integrated design is to optimize the efficiency loss through plant operability. Two different plants having low-efficiency penalties were selected for further analysis. The plant design structure is highly tied to design constraints based on the absorber design and operability. Two absorber configurations were evaluated under two scenarios: (1) the flow rate of flue gas at full load, and (2) the average time in load conditions to capture the plant flexibility. To confirm their operability, dynamic simulations of the designed plants were conducted under the existing control structure to confirm their operability. Compared to the other two configurations, the cycle with the overhead condenser was found to be the optimum.

An additional case was also investigated [42] for discovering techniques in improving turnaround productivity through thermodynamic analysis, primarily on energy and exergy examinations. The findings of the investigations were utilized to assess the operating conditions of a liquid nitrogen (LN_2) based energy storage structure, with an open Rankine cycle discharging system. The productivity of an open Rankine cycle power plant is substantially upgraded by utilizing a reheating cycle. The strategy involves several reheat and expansion stages to fix refrigeration issues with the cyclic exhaust stream and turbines working at high pressures. The ideal cycle conditions for the lab-scale LN2 Energy Storage System (LESS), however, are successfully found and proposed within the work.

Streb et al. [43] designed and optimized a new vacuum pressure swing adsorption (VPSA) process cycle for co-production of high purity, high-recovery CO_2 and H_2 from a multi-component feed stream. Four separate VSPA cycles that could purify CO_2 up to 95% while co-producing H_2 with the same specifications as were defined in the study. The characteristics of the cycles included a purge under vacuum, with part of the H_2 product and H_2-rich outflow recycled during the initial part of blowdown. The performances of the cycles were evaluated, and it was concluded that the most promising cycle was with a compressor and light purge step, to attain very high product purities.

Sharma et al. [44] proposed an 8-bed, 14-stage intensified VPSA unit to remove CO_2 from a highly partial-pressured stream for a miniaturized scale process for refining methane with steam. The method utilized a solid adsorbent to keep the cycle duration small. The energy inefficiency in the CO_2 removal process was decreased by controlling the blowdown and pressure while maintaining the CO_2 purity and recovery conditions. The loss is seen at a minimum when CO_2 recovery and purities were fixed to approximately 90% and 95%, respectively. However, it was noticed that the VPSA process design could be further improved if the PSA unit could be updated to consider the decrease in general flow rate and higher H_2 composition in the feed.

Dubois and Thomas [45] designed different CO_2 capture plant structures, including lean/rich vapor compression (L/RVC), solvent split flow (SSF), and rich solvent recycle (RSR), which were all applied to a flue gas coming from a cement plant utilizing three distinct solvents, specifically MEA, MDEA, and PZ mix. For every design and solvent, different parametric investigations were performed to define the optimal operating conditions, which minimized the solvent regeneration energy. The total thermodynamic work and utility expenses were additionally investigated. The outcomes showed that LVC and RVC managed to save the most expenses while also lessening the condenser cooling energy. MDEA was also found to be the solvent of choice, among the three studied.

Joss [46] developed a model of the temperature swing adsorption (TSA) process to facilitate the design of TSA processes. An equilibrium-based short model for a zeolite-adsorbate system was developed to gain a better understanding of the design of TSA processes. A parametric analysis of operating conditions was performed on purity, recovery, specific thermal energy consumption, and productivity. Based on the developed model, the cycle configurations are designed and evaluated for their

TABLE 10.4
Summary of Latest Developments in the Design of CO_2 Processes

Objective	Strategy	Reference
To minimize the net efficiency penalty	Integrated design of supercritical coal-fired power plant	[41]
To analyze economical strategy for the CO_2 capture plant operation under varying operating loads	A systematic design structure with optimization	[41]
To design a post-combustion CO_2 capture plant as a trade-off between operability and mitigation of the efficiency penalty	Two modified plant configurations with a low-efficiency penalty	[42]
To improve turnaround efficiency	Thermodynamic studies based on energy and exergy analyses	[42]
To co-produce high-purity, high-recovery CO_2, and H_2 from a multi-component feed stream	A new vacuum pressure swing adsorption (VPSA) process cycle	[43]
To capture CO_2 from a high partial pressure stream in a small-scale steam methane-reforming process	An enhanced 8-bed, 14-step vacuum pressure swing adsorption unit	[44]
To compare different configurations for process design	Rich solvent recycle (RSR), solvent split flow (SSF), lean/rich vapor compression (L/RVC)	[45]
To facilitate the process design of TSA processes	An equilibrium-based short model for a pertinent adsorbent	[46]
To observe the performance of a biomass steam gasification to produce a high-quality product gas stream	An integrated scheme of torrefaction, gasification, and CO_2 capture	[47]

use with zeolite in terms of CO_2 capture and storage. The cycle configuration was observed to have a significant impact on process efficiency. Consequently, the results indicated that the CO_2 separation from flue gases could be carried out using TSA.

Bach *et al.* [47] developed a combined process of torrefaction, gasification, and CO_2 capture to study their overall efficiencies within a steam gasification process that utilizes biomass as fuel, equipped with a pre- and post-treatment plant to produce high-quality gas. The double fluidized bed method is chosen as the primary gasification system. The biomass feed is upgraded through torrefaction, and CO_2 is removed from the product gas stream using an MEA-based CO_2 capture process. The quality of the product gas stream produced from this integrated process and the operational efficiency are documented and compared with their reference case, based on the gasification of unprocessed biomass. Table 10.4 presents a summary of the latest developments in the design of CO_2 processes.

REFERENCES

1. K. M. S. Salvinder *et al.*, "An overview on control strategies for CO_2 capture using absorption/stripping system," *Chemical Engineering Research and Design*, vol. 147, pp. 319–337, 2019, doi:10.1016/j.cherd.2019.04.034.
2. K. M. S. Salvinder, H. Zabiri, F. Isa, S. A. Taqvi, M. A. H. Roslan and A. M. Shariff, "Dynamic modelling, simulation and basic control of CO_2 absorption based on high pressure pilot plant for natural gas treatment," *International Journal of Greenhouse Gas Control*, vol. 70, pp. 164–177, 2018, doi:10.1016/j.ijggc.2017.12.014.
3. H. Suleman, A. S. Maulud and Z. Man, "Review and selection criteria of classical thermodynamic models for acid gas absorption in aqueous alkanolamines," *Reviews in Chemical Engineering*, vol. 31, no. 6, pp. 599–639, 2015.
4. F. Isa, H. Suleman, H. Zabiri, A. S. Maulud, M. Ramasamy, L. D. Tufa and A. M. Shariff, "An overview on CO_2 removal via absorption: Effect of elevated pressures in counter-current packed column," *Journal of Natural Gas Science and Engineering*, vol. 33, pp. 666–677, 2016.
5. S. Mouhoubi, L. Dubois, P. Loldrup Fosbøl, G. De Weireld and D. Thomas, "Thermodynamic modeling of CO_2 absorption in aqueous solutions of N,N-diethylethanolamine (DEEA) and N-methyl-1,3-propanediamine (MAPA) and their mixtures for carbon capture process simulation," *Chemical Engineering Research and Design*, vol. 158, pp. 46–63, 2020, doi:10.1016/j.cherd.2020.02.029.
6. X. Luo *et al.*, "Experiments and modeling of vapor-liquid equilibrium data in DEEA-CO_2-H_2O system," *International Journal of Greenhouse Gas Control*, vol. 53, pp. 160–168, 2016, doi:10.1016/j.ijggc.2016.07.038.
7. H. R. Kalatjari, A. Haghtalab, M. R. J. Nasr and A. Heydarinasab, "Experimental, simulation and thermodynamic modeling of an acid gas removal pilot plant for CO_2 capturing by mono-ethanol amine solution," *Journal of Natural Gas Science and Engineering*, vol. 72, 2019, doi:10.1016/j.jngse.2019.103001.
8. M. Xiao, H. Liu, H. Gao and Z. Liang, "CO_2 absorption with aqueous tertiary amine solutions: Equilibrium solubility and thermodynamic modeling," *The Journal of Chemical Thermodynamics*, vol. 122, pp. 170–182, 2018, doi:10.1016/j.jct.2018.03.020.
9. A. Esmaeili, Z. Liu, Y. Xiang, J. Yun and L. Shao, "Modeling and validation of carbon dioxide absorption in aqueous solution of piperazine + methyldiethanolamine by PC-SAFT and E-NRTL models in a packed bed pilot plant: Study of kinetics and thermodynamics," *Process Safety and Environmental Protection*, vol. 141, pp. 95–109, 2020, doi:10.1016/j.psep.2020.05.010.

10. E. Asadi, A. Haghtalab and H. A. Shirazizadeh, "High-pressure measurement and thermodynamic modeling of the carbon dioxide solubility in the aqueous 2-((2-aminoethyl)-amino)-ethanol + sulfolane system at different temperatures," *Journal of Molecular Liquids*, vol. 314, 2020, doi:10.1016/j.molliq.2020.113650.

11. H. A. Shirazizadeh and A. Haghtalab, "Measurement and modeling of CO_2 solubility in binary aqueous DMSO and MDEA and their ternary mixtures at different temperatures and compositions," *Fluid Phase Equilibria*, vol. 528, 2021, doi:10.1016/j.fluid.2020.112845.

12. X. He and F. V. Lima, "Development and implementation of advanced control strategies for power plant cycling with carbon capture," *Computers & Chemical Engineering*, vol. 121, pp. 497–509, 2019, doi:10.1016/j.compchemeng.2018.11.004.

13. X. Wu, M. Wang and K. Y. Lee, "Flexible operation of supercritical coal-fired power plant integrated with solvent-based CO_2 capture through collaborative predictive control," *Energy*, vol. 206, 2020, doi:10.1016/j.energy.2020.118105.

14. X. Wu, M. Wang, J. Shen, Y. Li, A. Lawal and K. Y. Lee, "Reinforced coordinated control of coal-fired power plant retrofitted with solvent based CO_2 capture using model predictive controls," *Applied Energy*, vol. 238, pp. 495–515, 2019, doi:10.1016/j.apenergy.2019.01.082.

15. C. Tirapanichayakul, B. Chalermsinsuwan and P. Piumsomboon, "Dynamic model and control system of carbon dioxide capture process in fluidized bed using computational fluid dynamics," *Energy Reports*, vol. 6, pp. 52–59, 2020, doi:10.1016/j.egyr.2019.11.041.

16. V.-M. Cristea, M. I. Burca, F. M. Ilea and A.-M. Cormos, "Efficient decentralized control of the post combustion CO_2 capture plant for flexible operation against influent flue gas disturbances," *Energy*, vol. 205, 2020, doi:10.1016/j.energy.2020.117960.

17. N. A. Othman, L. D. Tufa, H. Zabiri, A. A.-M. M. Jalil and K. Rostani, "Enhancing the Supersonic Gas Separation operating envelope through process control strategies of the feed conditioning plant for offshore CO_2 removal from natural gas," *International Journal of Greenhouse Gas Control*, vol. 94, 2020, doi:10.1016/j.ijggc.2019.102928.

18. X. Wu, J. Shen, Y. Li, M. Wang and A. Lawal, "Flexible operation of post-combustion solvent-based carbon capture for coal-fired power plants using multi-model predictive control: A simulation study," *Fuel*, vol. 220, pp. 931–941, 2018, doi:10.1016/j.fuel.2018.02.061.

19. T. Sultan, H. Zabiri, S. Ali Ammar Taqvi and M. Shahbaz, "Plant-wide MPC control scheme for CO_2 absorption/stripping system," *Materials Today: Proceedings*, 2020, doi:10.1016/j.matpr.2020.11.467.

20. Z. Li, Z. Ding, M. Wang and E. Oko, "Model-free adaptive control for MEA-based post-combustion carbon capture processes," *Fuel*, vol. 224, pp. 637–643, 2018, doi:10.1016/j.fuel.2018.03.096.

21. G. D. Patron and L. Ricardez-Sandoval, "A robust nonlinear model predictive controller for a post-combustion CO_2 capture absorber unit," *Fuel*, vol. 265, 2020, doi:10.1016/j.fuel.2019.116932.

22. J. Rúa, M. Hillestad and L. O. Nord, "Model predictive control for combined cycles integrated with CO_2 capture plants," *Computers & Chemical Engineering*, 2020, doi:10.1016/j.compchemeng.2020.107217.

23. T. E. Akinola, E. Oko, Y. Gu, H.-L. Wei and M. Wang, "Non-linear system identification of solvent-based post-combustion CO_2 capture process," *Fuel*, vol. 239, pp. 1213–1223, 2019, doi:10.1016/j.fuel.2018.11.097.

24. T. E. Akinola, E. Oko, X. Wu, K. Ma and M. Wang, "Nonlinear model predictive control (NMPC) of the solvent-based post-combustion CO_2 capture process," *Energy*, vol. 213, 2020, doi:10.1016/j.energy.2020.118840.

25. D. Reay, "The role of process intensification in cutting greenhouse gas emissions," *Applied Thermal Engineering*, vol. 28, no. 16, pp. 2011–2019, 2008, doi:10.1016/j.applthermaleng.2008.01.004.

26. E. P. I. Roadmap, "European roadmap for process intensification," 2007. [Online]. Available: https://efce.info/efce_media/-p-531.pdf.

27. J. Harmsen, "Process intensification in the petrochemicals industry: Drivers and hurdles for commercial implementation," *Chemical Engineering and Processing: Process Intensification*, vol. 49, no. 1, pp. 70–73, 2010, doi:10.1016/j.cep.2009.11.009.

28. M. Wang, A. S. Joel, C. Ramshaw, D. Eimer and N. M. Musa, "Process intensification for post-combustion CO_2 capture with chemical absorption: A critical review," *Applied Energy*, vol. 158, pp. 275–291, 2015, doi:10.1016/j.apenergy.2015.08.083.

29. C.-H. Yu, H.-H. Cheng and C.-S. Tan, "CO_2 capture by alkanolamine solutions containing diethylenetriamine and piperazine in a rotating packed bed," *International Journal of Greenhouse Gas Control*, vol. 9, pp. 136–147, 2012, doi:10.1016/j.ijggc.2012.03.015.

30. C.-C. Lin, Y.-H. Lin and C.-S. Tan, "Evaluation of alkanolamine solutions for carbon dioxide removal in cross-flow rotating packed beds," *Journal of Hazardous Materials*, vol. 175, no. 1, pp. 344–351, 2010, doi:10.1016/j.jhazmat.2009.10.009.

31. E. Oko, C. Ramshaw and M. Wang, "Study of intercooling for rotating packed bed absorbers in intensified solvent-based CO_2 capture process," *Applied Energy*, vol. 223, pp. 302–316, 2018, doi:10.1016/j.apenergy.2018.04.057.

32. J. Ying and D. A. Eimer, "Determination and measurements of mass transfer kinetics of CO_2 in concentrated aqueous monoethanolamine solutions by a stirred cell," *Industrial & Engineering Chemistry Research*, vol. 52, no. 7, pp. 2548–2559, 2013, doi:10.1021/ie303450u.

33. J. Oexmann, A. Kather, S. Linnenberg, U. Liebenthal, *Post-Combustion CO_2 Capture: Energetic Evaluation of Chemical Absorption Processes in Coal-Fired Steam Power Plants*. Cuvillier Verlag, 2011.

34. A. Cousins, A. Cottrell, A. Lawson, S. Huang and P. H. M. Feron, "Model verification and evaluation of the rich-split process modification at an Australian-based post combustion CO_2 capture pilot plant," *Greenhouse Gases: Science and Technology*, vol. 2, no. 5, pp. 329–345, 2012, doi:10.1002/ghg.1295.

35. Q. Li, G. Flamant, X. Yuan, P. Neveu and L. Luo, "Compact heat exchangers: A review and future applications for a new generation of high temperature solar receivers," *Renewable and Sustainable Energy Reviews*, vol. 15, no. 9, pp. 4855–4875, 2011, doi:10.1016/j.rser.2011.07.066.

36. D. E. Kim, M. H. Kim, J. E. Cha and S. O. Kim, "Numerical investigation on thermal-hydraulic performance of new printed circuit heat exchanger model," *Nuclear Engineering and Design*, vol. 238, no. 12, pp. 3269–3276, 2008, doi:10.1016/j.nucengdes.2008.08.002.

37. M. S. Jassim, G. Rochelle, D. Eimer and C. Ramshaw, "Carbon dioxide absorption and desorption in aqueous monoethanolamine solutions in a rotating packed bed," *Industrial & Engineering Chemistry Research*, vol. 46, no. 9, pp. 2823–2833, 2007, doi:10.1021/ie051104r.

38. International Energy Agency, "Prospects for CO_2 capture and storage," in *Energy Technology Analysis*, 2004. [Online]. Available: http://ccs-info.org/onewebmedia/iea_oecd_ccs_prospects.pdf

39. H. H. Cheng, C. C. Lai and C. S. Tan, "Thermal regeneration of alkanolamine solutions in a rotating packed bed," *International Journal of Greenhouse Gas Control*, vol. 16, pp. 206–216, 2013, doi:10.1016/j.ijggc.2013.03.022.

40. H. H. Cheng and C. S. Tan, "Removal of CO_2 from indoor air by alkanolamine in a rotating packed bed," *Separation and Purification Technology*, vol. 82, pp. 156–166, 2011, doi:10.1016/j.seppur.2011.09.004.

41. S. Y. Oh, S. Yun and J. K. Kim, "Process integration and design for maximizing energy efficiency of a coal-fired power plant integrated with amine-based CO_2 capture process," *Applied Energy*, vol. 216, pp. 311–322, 2018, doi:10.1016/j.apenergy.2018.02.100.

42. R. Dutta, P. Ghosh and K. Chowdhury, "Process configuration of liquid-nitrogen energy storage system (LESS) for maximum turnaround efficiency," *Cryogenics*, vol. 88, pp. 132–142, 2017, doi:10.1016/j.cryogenics.2017.10.003.

43. A. Streb, M. Hefti, M. Gazzani and M. Mazzotti, "Novel adsorption process for co-production of hydrogen and CO_2 from a multicomponent stream," *Industrial & Engineering Chemistry Research*, vol. 58, no. 37, pp. 17489–17506, 2019, doi:10.1021/acs.iecr.9b02817.

44. I. Sharma, D. Friedrich, T. Golden and S. Brandani, "Monolithic adsorbent-based rapid-cycle vacuum pressure swing adsorption process for carbon capture from small-scale steam methane reforming," *Industrial & Engineering Chemistry Research*, vol. 59, no. 15, pp. 7109–7120, 2020, doi:10.1021/acs.iecr.9b05337.

45. L. Dubois and D. Thomas, "Comparison of various configurations of the absorption-regeneration process using different solvents for the post-combustion CO_2 capture applied to cement plant flue gases," *International Journal of Greenhouse Gas Control*, vol. 69, pp. 20–35, 2018, doi:10.1016/j.ijggc.2017.12.004.

46. L. Joss, "Optimal design of temperature swing adsorption processes for the capture of CO_2 from flue gases," 2016. [Online]. Available: https://www.semantic-scholar.org/paper/Optimal-design-of-temperature-swing-adsorption-for-Joss/a660a9172bc7a68cadadd201d942ddf1a2c0bb62.

47. Q. V. Bach, H. R. Gye, D. Song and C. J. Lee, "High quality product gas from biomass steam gasification combined with torrefaction and carbon dioxide capture processes," *International Journal of Hydrogen Energy*, vol. 44, no. 28, pp. 14387–14394, 2019, doi:10.1016/j.ijhydene.2018.11.237.

11 Special Case Studies in Sustainable Carbon Capture

Jasper A. Ros, Juliana G.M-S. Monteiro,
Earl L.V. Goetheer, Randi Neerup, Wentao Gong,
Sai Hema Bhavya Vinjarapu, Jens Kristian Jørsboe,
Sebastian Nis Bay Villadsen, and
Philip Loldrup Fosbøl

CONTENTS

DOI: 10.1201/9781003162780-11

11.1   CASE STUDY 1: CO_2 CAPTURE IN THE WASTE-TO-ENERGY SECTOR

Waste incineration is used to reduce the volume of nonrecyclable waste and capture or destroy hazardous substances, as compared to landfilling of waste. Additionally, incineration enables the recovery of the energy released by the oxidation of the organic waste, to generate high pressure steam, which is often used in a combined heat and power (CHP) system. Generally, the generated electricity is transferred to the grid, and the heat is used for district heating. The gross efficiency (usable energy vs. total energy input) of the CHP system is much higher than systems with electricity generation only.

Waste-to-energy (WtE) plants are generally one or two orders of magnitude smaller than full-scale power plants. The average WtE plant in Europe has an incineration capacity of ca. 200 kton/year [1]. In 2019, AVR (Duiven, The Netherlands) had commissioned the first commercial full-scale CO_2 capture plant in the WtE sector. The captured CO_2 is liquefied and transported for direct use in the horticulture sector. The capture plant of AVR is shown in Figure 11.1.

The main advantage of implementing CO_2 capture in the WtE sector is that a large part of the waste (ca. 50–60%) is biogenic in nature. The same percentage of CO_2 has a biogenic origin. Negative emissions can be achieved when capturing and storing the majority of the CO_2 in the flue gas. Additionally, the business model for the WtE plant is generally based on the incineration of the waste. Therefore, energy (in the form of low-pressure steam) is often available at relatively low prices. This is especially relevant in the summer when the need for district heating is low or nonexistent. To illustrate the implications of CO_2 capture in the WtE sector, this chapter discusses three case studies. The three case studies are based on different scales, representative for the WtE sector: 60 kton/year (Norwegian average), 200 kton/year (European average), and 500 kton/year (Dutch average). The case study of the 200 kton/year WtE plant was developed and reported in the ALIGN-CCUS project (www.alignccus.eu/). Both the first-generation MEA solvent and the second-generation CESAR1 solvent (a mix of AMP and Piperazine) have been evaluated in that project. It was found that MEA is the preferred solvent when energy/heat is relatively cheap. Therefore, the case studies described here regard MEA (30 wt%) as

FIGURE 11.1 The commercial AVR CO_2 capture plant in Duiven, The Netherlands.

the solvent of choice. A detailed description of the capture plant can be found in the ALIGN-CCUS work [2].

11.1.1 Reference WtE and CO_2 Capture Plant Description

The simplified reference WtE plant used in this study is shown in Figure 11.2. The energy content of the waste is estimated at 11.2 MJ/kg (wet basis, 20 wt% moisture). Simple mass and energy balance calculations are used to calculate the flue gas composition and temperature before entering the waste heat recovery unit (WHRU), where superheated high-pressure steam is produced at 40 bars and 400 °C. After the WHRU, the flue gas continues to the flue gas pre-treatment and finally to the CO_2 capture plant. The produced superheated steam is used for electricity generation, and the low-pressure steam (at 3 bar) is used for district heating and/or CO_2 capture. The main results of the WtE plant modeling without CO_2 capture can be found in Table 11.1.

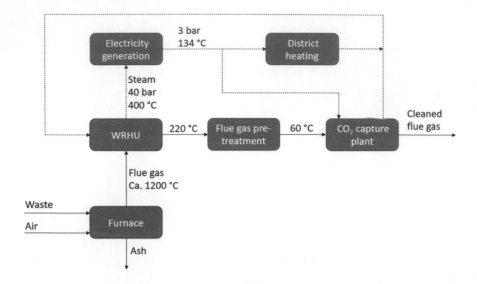

FIGURE 11.2 A schematic overview of a simplified WtE plant. Flue gas streams are represented by a continuous line, while the steam cycle is represented by a dotted line.

Figure 11.3 shows a schematic representation of the CO_2 capture plant simulated in this study. The simulation tool used in this work is ProTreat. The MEA model in ProTreat has been validated against VLE data [3] and pilot plant operation [2].

The CO_2 capture plant is modeled assuming a 70% flooding parameter in all columns and is further optimized by varying the solvent flow rate to minimize the reboiler duty. The main results are shown in Table 11.2. Approximately 47.9% of the steam for district heating and 4.4% of the electricity generated are used in the CO_2 capture plant (regardless of scale). Note that liquefaction is not included in this study, which would increase the electricity demand.

TABLE 11.1
Main Results from the WtE Plant Modeling

Parameter	Units	60 kton/ year	200 kton/ year	500 kton/ year
Electricity generation	MWe	3.01	10.24	25.74
District heating generation	MWth	13.89	47.22	118.65
Flue gas flow rate	kg/hr	57,000	190,000	475,000
CO_2 concentration in flue gas	vol% (wet)	8.94	8.94	8.94

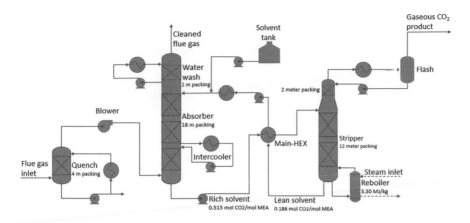

FIGURE 11.3 Schematic representation of the CO_2 capture plant used in this study, including modeling parameters that are constant for all three cases.

TABLE 11.2

Main Operational Results of the CO_2 Capture Plant

Parameter	Units	60 kton/ year	200 kton/ year	500 kton/ year
CO_2 capture rate	%	90	90	90
Reboiler duty	MWth	6.66	22.2	55.5
Electricity demand	kWe	133	443	1109
Solvent flow rate	ton/hr	100,500	335,000	837,500

11.1.2 TECHNO-ECONOMIC ANALYSIS

The techno-economic analysis has been performed using the Aspen Capital Cost Estimator V10 (ACCE). The total direct cost of material is taken from ACCE for all equipment, together with the Engineering and Procurement costs. The cost methodology shown in Table 11.3 is followed.

The total cost of CO_2 capture for the three cases is estimated at 30 to 55 €/ton CO_2, depending on the scale of the plant. The division of OPEX and CAPEX is shown in Figure 11.4. The higher cost for the small scale can be fully accounted to the higher specific CAPEX (economy of scale) and fixed OPEX (as the same amount of operators and technologist costs are expected for the different scales). At the larger scales, the variable OPEX becomes the cost-dominating factor.

A breakdown of the specific variable OPEX costs is shown in Figure 11.5 (specific costs per ton of CO_2 are identical for all three cases). The heat costs dominate the variable OPEX, as is common for post-combustion CO_2 capture processes. In this

TABLE 11.3

Main Assumptions for the Techno-Economic Analysis

Parameter	Units	Value
Cost year (Europe)	–	2019
Discount factor	%	8
Depreciation of plant	Years	15
Plant availability	%	90
Project contingencies	%	30
Maintenance costs	% of TPC	2.5
Labor percentage of maintenance	% of maintenance	40
Operators costs (six operators)	k€/year	360
Technologist costs	k€/year	100
Insurance costs	% of TPC	2
Administrative and overhead labor costs	% of total labor costs	30
Heat costs	€/GJ	4
Electricity costs	€/kWh	0.1
Cooling water costs*	€/m³	0.3

Cooling water makeup is estimated at 1 cubic meter/GJ cooling.
Abbreviation: TPC = Total process cost.

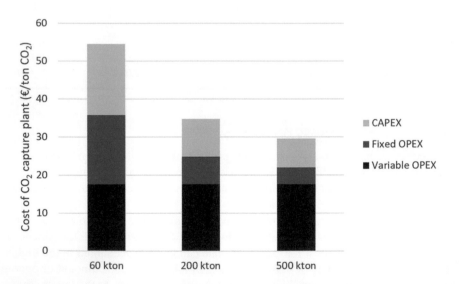

FIGURE 11.4   The cost of CO_2 capture for the three sizes of WtE plants.

study, the assumption is made that heat costs are 4 €/GJ. In reality, this value is variable and could be lower in the WtE context, especially when the demand for district heating is low. This could decrease the total cost of CO_2 capture further below 30 €/ton for large-scale plants.

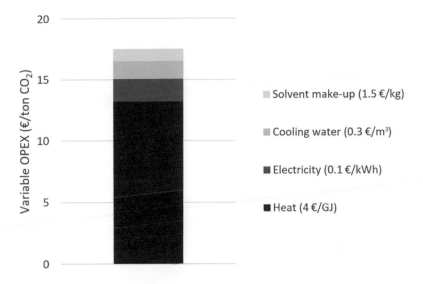

FIGURE 11.5   The variable OPEX costs.

In all three cases, the columns (absorber, stripper, and quench) dominate the equipment costs (60%–65%), followed by the heat exchangers, including the reboiler (15%–20%), as shown in Figure 11.6. The other equipment combined accounts for approximately 20%–25% of the costs. It is also shown that the columns follow the economies of scale effect to a lower extent than the other units, as the percentage of column costs increases with scale.

11.1.3   Case Study 1: Concluding Notes

This chapter evaluates the cost of CO_2 capture in the WtE sector, focusing on the effect of the plant scale. The results indicate that the cost of CO_2 capture for a 60 kton/year scale plant (Norwegian average) is 55 €/ton CO_2, which is significantly higher than a 500 kton/year scale plant (Dutch average) evaluated at 30 €/ton CO_2. Considering that most of the CO_2 in the WtE flue gases is biogenic, CCS in this sector offers a very cost-competitive negative emissions option. Additionally, for smaller WtE plants where energy is relatively cheap, cost optimization could be achieved by optimizing the design of the plant and reducing the CAPEX rather than minimizing the energy demand of the plant.

11.2   CASE STUDY 2: CO₂ CAPTURE ON LNG-FUELED VESSELS

11.2.1   Introduction to Carbon Capture and Storage in the Maritime Sector

The maritime sector has set a goal to reduce carbon emissions by at least 50% by 2050 [4]. To this extent, a lot of research is conducted towards deploying zero-emission

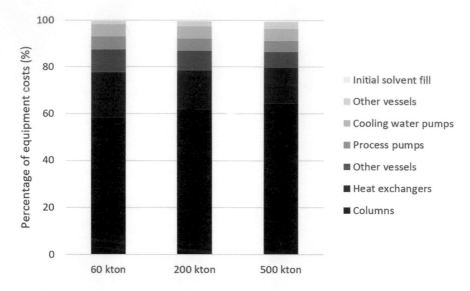

FIGURE 11.6   The impact of different equipment on the total cost.

fuels (ZEFs). However, these fuels are still at a relatively low TRL, and costs are high. Carbon capture and storage on existing or newly built vessels could play a large role in decarbonizing the maritime sector before 2050. At the moment of writing, a limited number of papers are available that discuss the conceptual designs of CO_2 capture plants on board LNG- and diesel-fueled vessels [5–7]. Still, no piloting or demonstration efforts have been reported so far. Ship-based carbon capture (SBCC) plants contain a CO_2 capture, liquefaction, and temporary on-board storage section. A techno-economic analysis for SBCC with 30 wt% MEA capture solvent is performed for a hypothetical LNG-fueled ship in this study. Recommendations for successful large-scale implementation of the technology are given.

11.2.2   Reference Vessel and CO_2 Capture Plant Description

The case study discusses a hypothetical vessel with an LNG-fueled electric propulsion (single) engine with a maximum continuous rating (MCR) of 9.2 MW [8]. The operational profile of the considered ship is shown in Figure 11.7, assuming that 20% of the total time the ship is in the harbor with the main engine off. The total single voyage time (for which the CO_2 storage tanks are designed) is assumed at 14 days.

The synergy of implementing CO_2 capture on an LNG-fueled ship comes from the heat integration between the exhaust gas and the CO_2 capture plant and the heat integration between the LNG evaporation and CO_2 liquefaction, as shown in Figure 11.8. This drastically reduces the OPEX of such systems, only electricity is

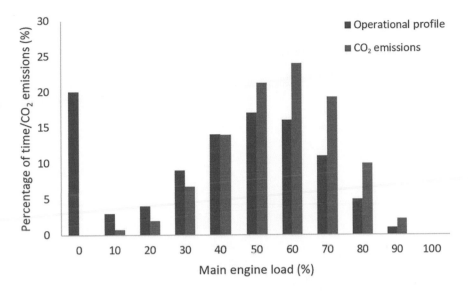

FIGURE 11.7 Operational profile and corresponding CO_2 emissions for the reference vessel.

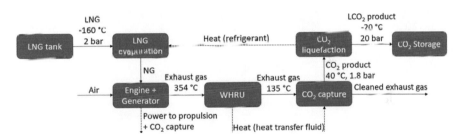

FIGURE 11.8 Simplified schematic overview of an LNG-fueled vessel integrated with a CO_2 capture plant.

needed as a utility (cooling water is cheap at sea), and the total cost will be CAPEX dominated.

The CO_2 capture plant is designed at 75% engine load of the main engine. The main results from the reference vessel (which serves as input for the SBCC design) can be found in Table 11.4. The electricity demand of the CO_2 capture plant is estimated (before modeling) at 2.5% of the vessel's power generation, which is added to the propulsion system power demand.

Schematic representation of the CO_2 capture and liquefaction plant used for this study can be found in Figure 11.9. The main restriction for the capture plant design is the maximum height of equipment, assumed to be limited to 15 meters. This leads to a maximum packing height of 7 meters for each column. The water wash is

TABLE 11.4

Main Results from the Reference Vessel

Parameter	Units	Value
Propulsion power demand	kW	6900
SBCC power demand (estimation)	kW	175
Total main engine power demand	kW	7075
WHRU heat recovery	kWth	3163
Cooling capacity of LNG	kWth	232.1
Flue gas flow rate	kg/hr	47592
CO_2 concentration in flue gas	vol% (wet)	4.17

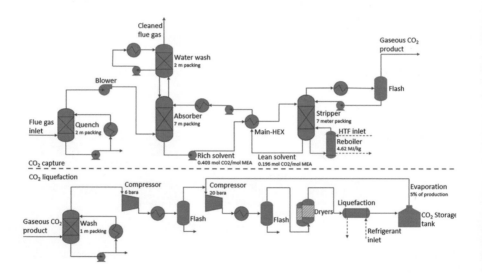

FIGURE 11.9 CO_2 capture, liquefaction, and storage process for the SBCC case study.

designed as a separate column instead of being placed on top of the absorber. For the heat transfer fluid (HTF) between the flue gas and the capture plant, oil is assumed, while for the refrigerant to transfer the heat from the CO_2 to the LNG, ammonia is assumed. The CO_2 capture plant is modeled with ProTreat, while the CO_2 liquefaction plant is modeled with Aspen.

The main results from the capture, liquefaction, and storage plant are shown in Table 11.5. The CO_2 capture rate can be limited either to the heat available in the flue gas or the cooling capacity of the LNG for the CO_2 liquefaction. In this case, both values are similar (and thus limiting), and a capture rate of 80.2% is achieved. To simplify the analysis in this study, it is assumed that all parameters are constant at

TABLE 11.5
Main Results of the CO_2 Capture and Liquefaction Plant

Parameter	Units	75% engine load (design)
CO_2 capture percentage	%	80.2
CO_2 capture flow rate	kg/hr	2467
Solvent flow rate	kg/hr	55,000
Hot-oil flow rate	kg/hr	230,000
Ammonia flow rate	kg/hr	4000
Reboiler duty	kWth	3163
Total electricity demand of the process	kWe	201.6
Total cooling duty of plant	kWth	4939
CO_2 liquefaction duty	kWth	232.1

TABLE 11.6
Main Cost Assumptions for the SBCC TEA

Parameter	Units	Value
Cost year (Europe)	–	2019
Discount factor	%	8
Depreciation of plant	Years	15
Labor percentage of maintenance	% of maintenance	40
Operators costs	k€/year	0
Technologist costs	k€/year	100
Insurance costs	% of TPC	2
Administrative and overhead labor costs	% of total labor	30
LNG costs	€/ton	400
Solvent costs	€/ton	1500

Abbreviation: TPC = Total process cost.

different engine loads, and 80.2% of the CO_2 in the exhaust gas at a specific engine load can be liquefied. For engine loads higher than 75%, is it assumed that part of the exhaust gas is vented so that no flooding occurs in the quench and absorber columns.

11.2.3 TECHNO-ECONOMIC ANALYSIS

The techno-economic analysis is performed using the Aspen Capital Cost Estimator V10 (ACCE). The main assumptions for this study can be found in Tables 11.6 and 11.7.

The main results for the base case (case 1) can be found in Figure 11.10. The figure shows that the calculated total cost of CO_2 capture is high (at 168 €/ton CO_2)

TABLE 11.7

Specific Assumptions per Case Study

Parameter	Units	Case 1	Case 2	Case 3	Case 4
Average engine load	%	40	75	40	75
CO_2 Capture rate	ton/yr	11675	22387	11675	22387
EPC costs	–	Aspen	Aspen	10% of Aspen	10% of Aspen
Process contingency	%	30	30	10	10
Maintenance costs	% of TPC	2.5	2.5	1	1
Installation factor	–	Aspen	Aspen	50% of material	50% of material

Note: Case 1 is the base case; Case 2 discusses the constant high-engine load of the engine; Case 3 discusses the standardization of the SBCC concept; and Case 4 is a combination of Cases 2 and 3.

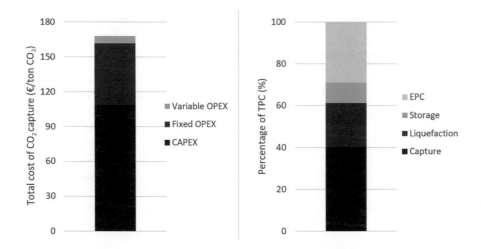

FIGURE 11.10 Division of CAPEX into capture, liquefaction, storage, and EPC costs (left);
division of CAPEX, fixed OPEX, and variable OPEX of the total CO_2 capture costs (right).

and is fully CAPEX dominated (as fixed OPEX is mostly a function of CAPEX).
The CAPEX division shows that the capture plant accounts for 40%, the liquefaction
plant for 20%, the storage tanks for 10%, and EPC for 30% of the total capture costs.

A lower specific CAPEX (€ per unit mass of CO_2) could lead to a drastic reduction of the total CO_2 capture costs. This can be achieved by increasing the CO_2 capture flow rate for a given plant size and/or lowering the plant costs. Increasing the
total CO_2 capture flow rate without changing the equipment design can be achieved
by avoiding running the engine at low engine load (see Table 11.5). To illustrate
this, case 2 is defined in which the engine constantly operates at 75% engine load,
as opposed to the average engine load based on the operational profile shown in
Figure 11.7, which approximately doubles the CO_2 capture rate with the same installation. Standardization of CO_2 capture plants for SBCC is proposed to lower the plant

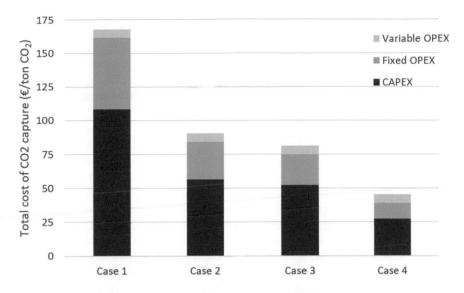

FIGURE 11.11 The results of the four case studies considered in this study.

costs, with a drastic reduction of equipment and EPC costs (case 3). Case 4 combines both strategies: running the engine at a high engine load and standardizing the CO_2 capture plant. The assumptions per case are given in Table 11.7, and the resulting CO_2 capture costs are found in Figure 11.11.

Figure 11.11 also suggests that the total capture costs of SBCC can be drastically reduced when considering the strategies proposed in this work. The total capture costs can drop to anywhere between 45 and 80 €/ton CO_2, depending on the exact operational profile of the ship. Moreover, sailing more often at higher engine loads will have additional economic and environmental benefits (i.e., lowering CH_4 slip).

11.2.4 CASE STUDY 2: CONCLUDING NOTES

In this study, CO_2 capture on LNG-fueled vessels is evaluated. Because of the high synergy between the LNG-fueled vessel and the CO_2 capture plant, OPEX is reduced drastically compared to studies considering CO_2 capture cases in power and industry. This means that for SBCC on LNG-fueled vessels, the CO_2 capture costs are fully determined by CAPEX. In the design proposed in the current study, a height restriction was imposed on the equipment, leading to a specific reboiler duty (SDR) of 4.62 MJ/kg CO_2, which is considerably above the optimal range obtained using 30 wt% MEA in gas-fired power plants, with similar CO_2 content in the flue gas. This illustrates clearly that the solvent optimization for SBCC should not be centered on the SRD, and attention should be given to mass transfer rates. For the reference case investigated in this study, the capture costs are determined at 168 €/ton CO_2. To lower these costs, avoiding operating the engine at low engine load and standardization of the equipment should be considered, which could drop the total cost of CO_2 capture to anywhere between 45 and 80 €/ton CO_2.

11.3   CASE STUDY 3: CO_2 CAPTURE FOR BIOGAS UPGRADING

Biogas is an attractive energy resource to replace fossil fuel. Biogas is produced by the biological degradation of organic matter by microorganisms under anaerobic conditions. The anaerobic digestion (AD) of biomass from agriculture, industry, and sewage sludge results in a sustainable energy resource. The raw biogas can be used to generate electricity and heat for the local producer of the biogas. Upgraded biogas can be distributed to the natural gas grid or used for transportation fuels [9–11].

The biological decomposition of organic matter is complex due to the interaction of highly specialized microorganisms [12–13]. All kinds of biomass contain carbohydrates, fats, and proteins as main components. The production of biogas can be divided into four steps: (1) hydrolysis, (2) acidogenesis, (3) acetogenesis, and (4) methanation. The production of biogas is illustrated in Figure 11.12. In the hydrolysis step, molecules of high molecular weight (e.g., polysaccharides, proteins, and fat polymers) are broken down to monomers. This allows the microorganisms to access and digest the organic material. In the acidogenesis, approximately 20% of the monomers are converted into CO_2 and H_2, and 50% of the monomers are converted to acetic acid [14]. The remaining is broken down to volatile fatty acids (VFA), and other components are produced, such as NH_3, H_2, H_2S, and siloxanes. Methanogens can use acetic acid, H_2, and CO_2. Due to higher molecular weight, VFA and other components produced in step 2 are digested in acetogenesis (step 3) before they can be used in the final process. In the final step, methanogens convert the intermediate products into CH_4, CO_2, and water. Seventy percent of methane production comes from the degradation of acetic acids [14]. The remaining 30% is produced from CO_2 and H_2 [14].

The chemical composition of the raw biogas varies with the source of biomass, which typically is manure, wastewater sludge, and industrial waste. Furthermore, biogas composition is also highly dependent on the pH in the reactor [15]. The chemical composition of biogas from various biomass is listed in Table 11.8. The main components of raw biogas are CH_4 and CO_2, but it also contains impurities such as water, O_2, H_2, and volatile organic components [16].

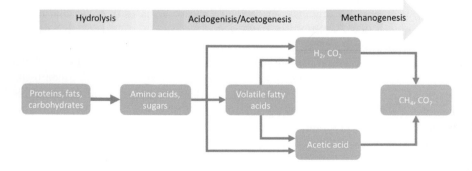

FIGURE 11.12   Production of CH_4 and CO_2 from biomass [4, 14].

TABLE 11.8

Chemical Composition of Biogas Produced from Different Sources [13, 16–20]

Component	Units	Landfill	AD	Sewage sludge	Impact
CH_4	vol%	35–65	53–70	50–70	–
CO_2	vol%	15–50	30–47	20–50	Low calorific value
					Corrosion (wet gas)
H_2O	vol%	0–5			Corrosion
H_2	vol%	0–3	0	0–5	–
O_2	vol%	0–5	0		Explosive mixtures
N_2	vol%	5–40	0–0.2		Low calorific value
H_2S	ppm	0–100	0–10,000	0–1	Toxic
					Corrosion
					Damages catalysts
					Emissions of H_2S, SO_2, SO_3
NH_3	ppm	0–5	<100		Corrosion
					Emissions
Cl^-	mg/m^3	20–200	0–5		Corrosion
Siloxanes	mg/m^3	0–50		0–41	Formation of SiO_2 and quartz

The impact of CO_2 and other impurities on the biogas quality is stated in Table 11.8. Siloxanes are typically present in landfill and sewage sludge gas and originate from household products such as shampoo, soaps, oils, and detergents. They deposit on valves, cylinders, and in the combustion chamber and cause blockage by forming a silica layer. H_2S is formed during the biological reduction of amino acids and peptides. The presence of H_2S in the biogas leads to corrosion of most metals. NH_3 forms from the hydrolysis of materials containing proteins [15]. When combusted, NO_x is formed [17]. NH_3 also leads to corrosion of steel material. Biogas is often used directly on-site or distributed to external customers through separate pipelines [17]. The upgrading of biogas enables biogas to be distributed through natural gas grids [17] by removing most of the CO_2 and other harmful impurities such as H_2S, O_2, NH_3, and siloxanes [16].

11.3.1 Biogas Production

Since the 10th century BC, biogas has been used for heating water in Assyria and AD of biomass has been applied in ancient China [10, 21]. Actual documentation on the digestion of biomass has been known since the mid-19th century, when digesters were built and constructed in New Zealand and India [10]. Cameron et al. were granted a US patent [22] in 1899 to invent a continuous flow septic tank using sewage bacteria, which did improve the liquidation and purification process of sewage. The produced gas was used to fuel street lamps [10, 23].

Biogas plants for wastewater have been constructed and used worldwide since the 1920s to supply heat [14] and energy for cooking and lighting [21]. In the 1950s, advances in biogas processes were impeded by the economic benefits of fossil fuels. However, the oil crisis in the mid-1970s initiated research into new types of biogas plants [14].

The number of operating biogas plants in Europe increased by 51% in two years, reaching 18,493 with 729 upgrading plants in 2019 [24]. In 2019, the European biogas plants produced approximately 167 TWh, and biomethane production reached 26 TWh. However, the implementation of new installed biogas plants has leveled off due to political decisions during the past six years [25]. Germany has been leading in biogas development for decades and contributes to 62% of the European biogas plants [26].

In 2018, the number of upgrading plants increased by approximately 13% despite a stall in installing new biogas plants [27]. Eighteen (18) European countries produce biomethane, led by Germany and France with 232 and 131 plants, respectively [24, 27]. The feedstock used for production mainly comes from energy crops, agricultural residues, manure, and sewage sludge [27].

A few European countries have developed their national standards for gas grid injections of upgraded biogas, as presented in Table 11.9. Table 11.9 gives the minimum limits according to the Wobbe index, along with the limits of allowed impurities. Germany and France allow for two types of gas to be injected into the grid: (1) gas for unlimited injection (H-gas) and (2) limited injection (L-gas) [17]. The H- and L-gas are natural gas with high and low heating values, respectively. Sweden uses a standard for biogas utilized as vehicle fuel [28] and for injection into the gas grid. France has a stricter limit towards oxygen in the biogas compared to other European countries [28]. Denmark and Austria have a limit of total sulfur <10 mg S/Nm^3, whereas Germany and France have a limit of 30 mg S/Nm^3.

Biogas production in Denmark is increasing, and in 2020, the annual production was 15 pJ [33]. Most of the produced biogas is used for electricity. Still, it is estimated that it will be upgraded to biomethane and distributed to the natural gas grid in the long term. Figure 11.13 shows one of the oldest biogas plants [34], Hashøj, in Denmark. It is also the first plant in Denmark, implementing upgrading technology.

Most of the produced biogas is based on manure, and the plants are located near the farms [35]. However, biogas is also produced from wastewater treatment plants [29, 33, 35]. In 2020, Denmark had 186 biogas plants [33] and 46 upgrading plants [33].

The total emission of CO_2, based on the European biogas plants, was 200 megatons in 2020. The European Biogas Association (EBA) estimated a potential production of 46 TWh of biogas, leading to 560 megatons of CO_2 emissions by 2030, corresponding to more than 10% of the European CO_2 emission [22]. The CO_2 emitted from these plants all have a negative emission potential, which could help to reduce the impact of climate change.

11.3.2 BIOGAS UPGRADING

Biogas leaving the digesters consists of different kinds of impurities, as mentioned in the previous section. The major impurity is carbon dioxide with traces of water vapor, hydrogen sulfide, siloxanes, nitrogen, oxygen, and other particulates. The

TABLE 11.9
Some European Countries' Standard Requirements for Grid Injection [28–32]

Compounds	Unit	Denmark	France	Germany	Sweden	Switzerland	UK	Austria	The Netherlands
Wobbe index	MJ/Nm³	50.8	H: 49.1–56.5 L: 43.2–46.8	H: 46.1–56.5 L: 37.8–46.8	45.4–48.6	47.9–56.5	47.2–51.4	47.9–56.5	43.46–44.4
Calorific value	MJ/Nm³	39.6–46.0	H: 38.5–46.1 L: 34.2–47.8	30.2–47.2	39.6–43.2	38.5–47.2	-	38.5–46.1	31.6–38.7
CH_4	vol%	97.3	>95	98.3	>97	>96	78–100	>96	>85
CO_2	vol%	<2.5	<2.5	<6	<3	<6	<3	<3	<6
O_2	vol%	<0.5	<0.01	<3	<1	<0.5	<1	<0.5	<3
Sulphur (total)	mg S/Nm³	<10	<30	<30	<23	<30	<50	<10	<45
H_2S	mg/m³	<5	<5	<5	<10	<5	-	<5	<5
NH_3	mg/m³	<3	<3	<3	<20	-	-	-	<3

FIGURE 11.13   Hashøj biogas plant, Denmark.

Photo: **Philip Loldrup Fosbøl.**

concentrations of these impurities are dependent on the raw material used for biogas production. Some of these components are corrosive and need to be removed before removing CO_2 so that the upgrading equipment is not subjected to mechanical wear.

11.3.2.1 Removal of Hydrogen Sulfide

H_2S is formed during the microbiological reduction of sulfur-containing compounds in the raw organic material. H_2S is a poisonous gas, and dissolved H_2S can cause corrosion. The concentration of H_2S depends on the organic matter used and can be reduced either by precipitation in the digester liquid, or by treating the gas before carbon capture [28]. There are several technologies available for the removal of H_2S from the biogas mixture [28].

In some of the technologies used for CO_2 capture, H_2S reacts with the sorbents, causing issues with their regeneration. To prevent such issues, it is critical to remove any H_2S present in the biogas mixture before separating CO_2 and CH_4. Some of the methods used to remove H_2S are adsorption on activated carbon, chemical absorption, and biological treatment [28].

11.3.2.2 Removal of Water

Biogas is typically saturated with water vapor when it leaves the digester. The water vapor can condensate in the pipelines, and the presence of sulfur oxides leads to corrosion. At low temperatures, water is likely to freeze in the pipes causing operational difficulties. Therefore, this water vapor is removed before further processing of biogas. Adsorption using sorbents like charcoal, SiO_2, molecular sieves, or absorption using glycol or hygroscopic salts are some of the methods usually employed [28]. Another approach is to vary the conditions to condensate the water vapor and separate it from the biogas [28].

11.3.2.3 Removal of Other Impurities

Components like N_2, O_2, ammonia, siloxanes, and particulates are also present in traces in biogas. As biogas is produced by anaerobic digestion, O_2 and N_2 appear only if the air is present in the digester. They can be removed by adsorption, using molecular sieves, charcoal, or activated carbon [28]. These components are difficult to separate from CH_4, and hence their presence should be avoided. The concentration of ammonia depends on the organic material and is removed when the gas is dried or upgraded [28]. Siloxanes can be separated either by absorption, adsorption, or by cooling the gas. Mechanical filters can remove particulates present in biogas [28].

11.3.3 Biogas Upgrading Implementing Carbon Capture Technologies

Currently, four biogas upgrading technologies exist on a commercial basis. These are (1) water scrubbing, (2) chemical scrubbing, (3) membrane, and (4) pressure swing absorption (PSA). Cryogenic and organic physical scrubbing are other biogas upgrading technologies, but they are still developing [15]. Figure 11.14 outlines the progress in the upgrading plants and the technology within the Task 37 countries. The IEA Bioenergy Task 37 is an international working group who addresses the entire biogas production chain [36]. The number of upgrading plants have been increasing significantly since 2007, and the use of water scrubbing and chemical scrubbing has more than doubled from 2007 to 2018.

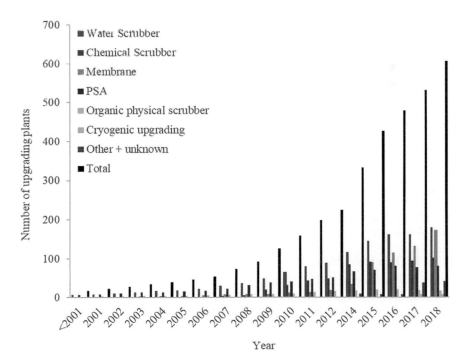

FIGURE 11.14 The number of upgrading plants and upgrading technologies within the Task 37 countries [37].

The following subsections discuss the several full-scale technologies available for the removal of CO_2 from biogas.

11.3.3.1 Water scrubbing

Water scrubbing is an absorption-based method that exploits the fact that CH_4 has a significantly lower solubility in water than CO_2. Water scrubbing achieves a CH_4 purity of 80%–99%. The purity of CH_4 depends on the concentration of N_2 and O_2 in the biogas, as these components cannot be separated from CH_4. Raw gas compression and circulation of the processing water contribute to most of the process's energy requirements [38]. Ease of operation, adjustable capacities, high efficiency (>97%), and low CH_4 losses (<2%) are some of the advantages of this process [16]. High investment and operation costs, possible foaming, and biological contamination are some disadvantages of this technology [33].

11.3.3.2 Cryogenic Separation

Differences in the condensing temperatures of CH_4 and CO_2 can be used for separating them. This is achieved by condensation and distillation. This process needs to compress the raw gas to pressures as high as 200 bars, leading to high-energy consumption and accounting for 5%–10% of the biomethane produced. Cryogenic separation is advantageous in producing liquid and high-purity biomethane with less than 1% loss of CH_4. Additionally, a high-purity (98%) CO_2 stream is also produced [38]. It is also environmentally friendly [33]. However, it requires pre-treatment of the raw gas. High investment and operational costs and high-energy requirements are some of the disadvantages of this process [33].

11.3.3.3 Physical Absorption

Physical absorption is based on the same principle as water scrubbing. Instead of water, organic solvents like methanol and dimethyl ethers of polyethene glycol (DMPEG) are used to absorb CO_2. Therefore, physical absorption also has limitations like water scrubbing, where N_2 and O_2 cannot be separated from CH_4. However, the solubility of CO_2 is higher in organic solvents than in water, which can reduce the size of the absorber. The energy consumption of physical absorption is equivalent to that of water scrubbing [38]. Physical scrubbing has a high efficiency resulting in a purity of CH_4 greater than 97%, with low CH_4 losses. Some of the drawbacks of this technology include high investment and operation costs, difficult operation, and incomplete regeneration of the solvent [16].

11.3.3.4 Chemical Absorption

In chemical absorption, the solvents used chemically react with the solute gas. Chemical absorption is favorable over physical absorption at low CO_2 concentrations. Amines are widely used as solvents for this process [38]. The chemical reaction between amines and CO_2 is strongly selective, making methane losses less than <0.1% [28]. There are some liquid losses due to evaporation that need to be replaced. The liquid, in which carbon dioxide is chemically bound, is regenerated by heating. High temperatures are required for solvent regeneration, leading to high-energy consumption. Chemical absorption can produce a very pure stream of CH_4 (>99%), and the CO_2 absorption capacity of the solvents is much higher compared to physical solvents and water [16]. Degradation of the solvents, corrosion of the equipment, and high-energy consumption for solvent regeneration are some of the disadvantages of this process [16].

11.3.3.5 Pressure Swing Adsorption

This technology is based on the selective absorption of gas molecules on solid surfaces according to their molecular sizes. Pressure swing adsorption (PSA) technology can separate CH_4 from CO_2, N_2, and O_2, as CH_4 is a larger molecule. The purity of CH_4 is usually around 96%–98% after upgrading with PSA, and the losses are about 2%–4% [38]. PSA is advantageous due to this high purity of CH_4. Moreover, the process does not involve liquid chemicals [33], has low energy consumption, and is tolerant to impurities [16]. Nevertheless, it is disadvantageous in terms of high operation and investment costs, extensive process control requirements, and CH_4 losses [16].

11.3.3.6 Membrane Separation

The natural gas industry has implemented gas permeation membranes to remove impurities from natural gas. The application of membranes can thus be extended to biogas upgrading [39]. In membrane separation, CO_2 and H_2S move through the membrane to the permeate side, and CH_4 stays at the inlet side [38]. Some of the CH_4 molecules can pass through the membrane, making it difficult to obtain a high purity of CH_4 without simultaneously incurring large losses of CH_4 [38]. Deng and Hagg reported that a CO_2-selective polyvinyl amine/polyvinyl alcohol blend membrane could optimally deliver a 98% pure biomethane with a recovery of 99% [40]. Membrane technology is advantageous, as it is an easy process with high-energy efficiency and low costs [38]. The operation is simple and the process is highly reliable [16]. Some of the shortcomings of this technology include low membrane selectivity, CH_4 losses, and the requirement of several stages to obtain high purity [16]. The advantages and disadvantages of all the technologies are summarized in Table 11.10.

11.3.4 UTILIZATION OF CO_2

The CO_2 obtained from the upgraded biogas can potentially be used as feedstock and upgraded using a methanation reaction known as the Sabatier reaction. The Sabatier reaction was initially used to remove traces of CO_2 from the feed gas for ammonia synthesis. This technology can be transferred to the case of biogas upgrading as well. The mixture of CO_2 and CH_4 can be used as the feed for CO_2 methanation. The reaction mechanism is as follows [43]:

$$CO_2 + 4H_2 \rightarrow CH_4 + 2H_2O$$

Methanation of CO_2 is an exothermic reaction, and the CH_4 present in biogas is used as a diluent to control the temperature to prevent a runaway reaction. However, the reaction rate at lower temperatures is very slow. Therefore, the current focus on this technology is concerning catalyst development to enhance the reaction rate at lower

TABLE 11.10

Comparison of the Advantages and Disadvantages of the Different Processes

Process	Advantages	Disadvantages
Water scrubbing	• Simple process • Low methane loss • Low operation and maintenance costs	• High water and energy requirements • Scope for biological contamination
Chemical scrubbing	• High CO_2 absorption per unit volume • Low methane losses • High purities • High rate of absorption	• High energy needed to produce steam • Pre-treatment of the raw gas required • High-energy consumption for solvent regeneration • Corrosion of the equipment
Pressure swing adsorption	• Dry process • High gas quality • Low methane losses • Low energy requirements	• Complex process • Extensive process control • High operation and investment costs • An off-gas with 15%–20% of methane is produced, requiring further treatment [42]
Cryogenic Separation	• Environmentally friendly • Produces liquid biomethane • Low methane losses.	• High investment and operational costs • Pre-treatment of gas is required. • High-energy requirements for condensation and cooling
Membrane separation	• Simple process • Environmentally friendly • Low consumption of energy • Low cost	• Low membrane selectivity • Pre-treatment of gas required • Low purity of CH_4 produced

Source: Adnan *et al.* [41].

temperatures while maintaining high stability in an operating window of approximately 500 °C [43]. Nano-sized nickel particles are deemed to be most promising in this aspect. Through this technology, biogas can be upgraded to the natural gas grid quality without the removal of CO_2.

Another method of upgrading the CO_2 and CH_4 in biogas is through syngas conversion to organic substances such as methanol. This process is achieved in two stages: syngas production and synthesis of methanol. Syngas is obtained through a combination of steam reforming and dry reforming, which are as follows [44]:

Steam reforming: $$CH_4 + H_2O \rightarrow CO + 3H_2$$

Dry reforming: $$CH_4 + CO_2 \rightarrow 2CO + 2H_2$$

Conversion to methanol requires additional H_2 to obtain the correct stoichiometric ratio between CO and H_2, which is 1:2 [44].

$$CO + 2H_2 \rightarrow CH_3OH$$

This is an alternative way to methanization, which allows for energy storage. The additional H_2 required can potentially be obtained from renewable sources of energy like wind or solar, which enable green electrolysis of water. These are some ways of upgrading both CO_2 and CH_4 in biogas to store energy instead of removing CO_2.

11.3.5 CASE STUDY 3: CONCLUDING NOTES

In the last decades, biogas and biomethane production have increased worldwide due to rising awareness of the need for CO_2 emissions reduction. Additionally, the conversion of biogas into biomethane is increasing in Europe to use it for the national gas grid and as transportation fuels. The main technologies for CO_2 removal currently used are scrubbing using either water or a chemical as solvent, and membranes. Today, most of the CO_2 is emitted into the atmosphere, but new technologies open up using CO_2 as a building block to obtain chemicals and fuels. It can also be stored in geological formations. By 2030, the biogas sector has a potential for 560 megatons of negative CO_2 emissions.

11.4 CASE STUDY 4: APPLICATION OF CARBON CAPTURE IN THE CEMENT AND IRON/STEEL SECTOR

11.4.1 CARBON CAPTURE FOR THE CEMENT SECTOR

The estimated global cement production between 2014 and 2019 has remained at an average of 4.1 Gt/yr, with each ton of cement releasing 0.54 tons of carbon dioxide despite many process improvements (base case: year 2018). These CO_2 emissions from cement production constitute 3% of global greenhouse gas emissions [45]. Half of these emissions come from fuel consumption, while the rest are from the process itself, as calcium carbonate (limestone) calcines to calcium oxide (lime), releasing carbon dioxide as a by-product [46]. This reaction is endothermic and occurs in the kiln operating at near 1000°C.

The carbon capture technologies in the cement sector include post-combustion capture, oxy-combustion capture, and calcium looping. The post-combustion capture in the cement sector is similar to the one in coal-fired power generation, except that the former has a slightly higher content of carbon dioxide. The technology is preferred over oxy-combustion due to the ease of retrofit in existing plants, without much changes in the original setup. Recently, calcium looping has been tested. Plaza *et al.* (2020) compared carbon capture technologies in the cement industry using data from recent CCS projects (e.g., Norway's full chain CCS project, the Norcem CO_2 capture project, and the CEMCAP project). They concluded that chemical absorption using a liquid solvent is preferable, as it is the most mature technology to date and can be retrofitted to the existing cement facilities [47].

11.4.1.1 Case study: SkyMine® Beneficial CO_2 Use Project

The SkyMine® Beneficial CO_2 Use Project is the largest CCUS project in the cement sector to date. Started in 2010, the objective of this project is to use the SkyMine®

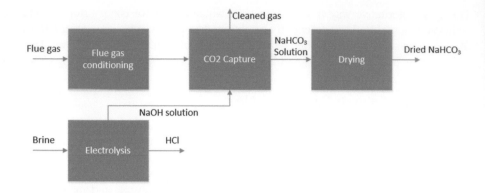

FIGURE 11.15 Simplified process flow diagram of SkyMine® process at Capitol Aggregates' San Antonio cement plant, Texas.

Source: **Adapted from [48].**

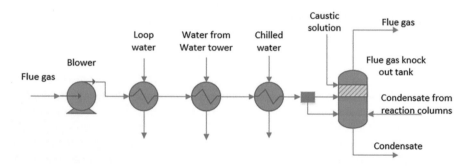

FIGURE 11.16 Simplified process flow diagram of flue gas conditioning processes.

Source: **Adapted from [49].**

technology for pilot-scale carbon capture at Capitol Aggregates' San Antonio cement plant located in Texas (US). The pilot plant is designed to capture 75,000 tons of CO_2 per year from a slipstream of exhaust gas from the coal-fired cement plant [48]. In the SkyMine® process, as presented in Figure 11.15, CO_2 from the flue gas is removed with a NaOH solution. The NaOH is generated using electrolysis of NaCl brine in a membrane cell. The by-products of this process, $NaHCO_3$ and HCl, are sellable products.

Figure 11.16 shows the four process steps of the SkyMine® process: flue gas conditioning, CO_2 absorption, low energy dewatering, and electrochemical production. The flue gas from the kiln contains mainly N_2 (48.89 vol%), CO_2 (29.58 vol%), H_2O (11.20 vol%), O_2 (10.22 vol%), and a small number of impurities such as SO_2, NO_2, NO, CO, and Hg [49].

In the flue gas conditioning step as shown in Figure 11.16, the flue gas goes through a blower, which ensures that enough flue gas is available in the CO_2 absorber columns. The gas is then cooled from 129 °C to 40 °C with three coolers. In the first cooler, flue gas is cooled from 129 °C to 77 °C using water in a closed loop. The heat recovered by the loop water is used to heat the gas entering the bicarbonate dryer. The second cooler utilizes cooling water from a water tower. The temperature of flue gas after this cooler is reduced to 49 °C. The final cooler uses chilled water to reduce the flue gas temperature to 40 °C. After cooling, the condensates and flue gas are separated. The flue gas goes through a knock-out tank, where any remaining condensate is removed. In the knock-out tank, the flue gas passes a dilute caustic spray, which reacts with SO_x and NO_x and removes them from the gas as condensate. The condensates from the knock-out tank, coolers, and reaction columns are collected at the bottom of the knock-out tank. Finally, the impurities in the condensates (e.g., lead, mercury) are removed with carbon filters [49].

Figure 11.17 shows the CO_2 absorption step. The flue gas is separated into four streams, two of which are sent to two parallel carbonate columns. In these columns, flue gas flows upwards through a 20% NaOH solution and rapidly produces a sodium carbonate solution. The sodium carbonate solutions are pumped into a tank, where

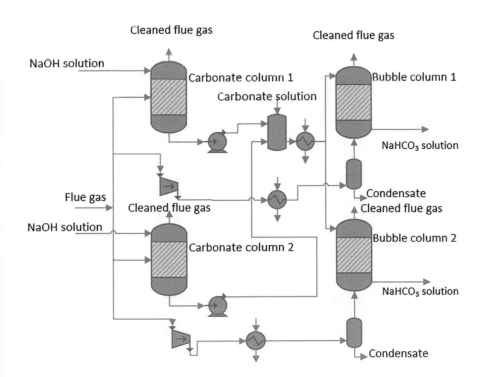

FIGURE 11.17   Simplified process flow diagram of CO_2 absorption process in the SkyMine®
process.

Source: **Adapted from [49].**

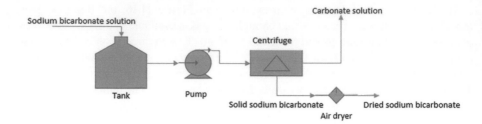

FIGURE 11.18   Simplified process flow diagram of the dewatering process in the SkyMine®
process.

Source: **Adapted from [49].**

they are mixed with a recycled carbonation solution stream, which comes from the
dewatering process. The remaining two flue gas streams are compressed, cooled,
dehydrated, and fed into two bubble columns. In these two columns, CO_2 reacts
with sodium carbonate, produced from the carbonate columns, to form sodium
bicarbonate.

Figure 11.18 illustrates the dewatering process. The sodium bicarbonate solution
is initially pumped to a centrifuge where solid sodium bicarbonate and carbonate
solution are separated. The carbonate solution is concentrated and recycled, whereas
the solid sodium bicarbonate is dried in an air dryer.

Hydrogen and chlorine gas produced from the electrolysis is converted into HCl
solution. The power consumption of the plant is 2.7 MW. However, the electrical
power required for the electrolysis process is 18.42 MW. Hence, the total electrical
power consumption of the SkyMine® process is 21.12 MW. This corresponds to the
specific energy consumption of 8 GJ/t captured CO_2 [48].

11.4.2   CARBON CAPTURE IN THE IRON AND STEEL INDUSTRY

The estimated CO_2 emission from the iron and steel sector contributes to 7.2% of the
global greenhouse gas emission. On average, the CO_2 intensity of steel production in
2018 is 1.8 t CO_2/t crude steel [50]. Similar to cement production, the CO_2 emissions
in the iron and steel sector are from fuel combustion and the reduction reaction of
iron oxide.

$$Fe_2O_3(s) + 3CO(g) \rightarrow 2Fe(s) + 2CO_2(g)$$

In a blast furnace, where this reaction occurs, the temperature is usually around 1000
°C and coal is used as the primary fuel. Carbon monoxide, which acts as the reducing
gas in this reaction, is typically produced from either coal or natural gas.

Most carbon capture technologies applicable to the iron/steel sector are in the
research and demonstration phase. The EU ULCOS project has investigated 80 various options for reducing the CO_2 emissions from blast furnaces. The study recommended the use of CCS technologies for CO_2 emissions reduction. The Japanese

COURSE 50 is another ongoing research program using amine-based carbon capture for the steel industry. Other projects include the Swedish STEPWISE project, the Australian CO_2BTP project, and the Korean POSCO program [51].

11.4.2.1 Case study: DMX™ Demonstration in Dunkirk (3D) project

European-funded DMX™ Demonstration in Dunkirk (3D) project is one of the largest ongoing CCS demonstration projects in the steelmaking industry. The project aims to demonstrate the efficiency of the DMX™ carbon capture technology on an industrial-scale pilot plant and to implement the operational CCS units on ArcelorMittal's blast furnace in Dunkirk by 2025. The captured CO_2 in this project will be stored in the North Sea reservoir developed by other projects like the Northern Light.

The DMX™ technology, developed by IFP Energies Nouvelles (IFPEN), is based on using a solvent that exhibits a liquid-liquid separation. This is shown in Figure 11.19.

Figure 11.19 shows that when CO_2 is absorbed in the DMX™ solvent, the solvent separates into two liquid phases: a water-rich phase with very high carbon dioxide loading and an amine-rich phase with low carbon dioxide loading. Because CO_2 is only absorbed in one of the phases, the CO_2 lean phase does not need to go to the stripper for regeneration. The flow rate to the stripper and the energy consumption of the regeneration is hence reduced. The DMX™ process is very similar to a conventional MEA process, as shown in Figure 11.20. The capture system consists of an absorber, where CO_2 is absorbed into the solvent; a decanter, where the CO_2-lean phase and the CO_2-rich phase are separated; and a stripper, where CO_2-rich amine is regenerated with heat.

The advantages of DMX™ compared to the first-generation amine-based processes are as follows [53]:

- High cyclic solvent capacity (0.1–1 mol CO_2/mol solvent)
- Low heat required for regeneration (<2.3 GJ/t CO_2 for 90% capture rate)
- High CO_2 purity

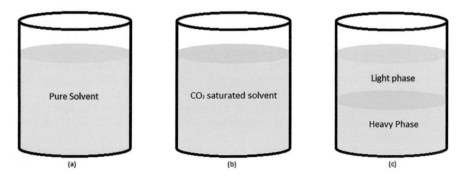

FIGURE 11.19   DMX™ solvent principle.

Source: Adapted from [52].

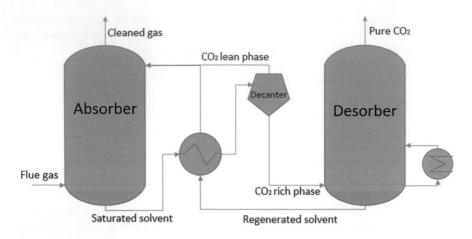

FIGURE 11.20   Process flow diagram of the IFP Energies Nouvelles DMX™ process.

Source: Adapted from [53].

- No degradation of the solvent (up to 160 °C), allowing the regeneration step to carry out at a higher temperature and pressure
- Producing CO_2 at higher pressure (6–7 bars)
- No corrosion of the solvent allowing the use of carbon steel material for the construction of future industrial units
- The first techno-economic evaluations have shown an OPEX reduction and CO_2 capture cost up to 30% compared to the first-generation processes using amine-based solvents.

REFERENCES

1. F. Neuwahl, G. Cusano, J. Gómez Benavides, S. Holbrook and S. Roudier, "Best available techniques (BAT) reference document for waste incineration: Industrial emissions directive 2010/75/EU," *Luxembourg*, 2019, doi:10.2760/761437.
2. S. Garcia, M. V. D. Spek, H. Weir, A. Saleh, C. Charalambous, E. Sanchez-Fernandez, J. Ros, E. Skylogianni, J. Monteiro, H. M. Kvamsdal, R. Skagestad, G. Haugen and N. H. Eldrup, "Guidelines and Cost-drivers of capture plants operating with advanced solvents (ALIGN-CCUS D1.4.3)," 2021. [Online]. Available: https://www.alignccus.eu/sites/default/files/[WEBSITE]%20ALIGN-CCUS%20D1.4.3%20Guidelines%20and%20Cost-drivers%20of%20capture%20plants%20operating%20with%20advanced%20solvents_1.pdf.
3. H. M. F. Paul, C. Ashleigh, J. Kaiqi, Z. Rongrong and G. Monica, "An update of the benchmark post-combustion CO_2 capture technology," *Fuel*, vol. 273, p. 117776, 2020, doi:10.1016/j.fuel.2020.117776.
4. International Maritime Organization, "Adoption of the initial IMO strategy on reduction of GHG emissions from ships and existing imo activity related to reducing GHG emissions in the shipping sector," p. 27, 2018. [Online]. Available: https://www.imo.org/en/MediaCentre/HotTopics/Pages/Reducing-greenhouse-gas-emissions-from-ships.aspx#:~:text=The%20initial%20GHG%20strategy%20envisages,that%20total%20annual%20GHG%20emissions.

5. M. Feenstra, J. Monteiro, J. T. van den Akker, M. R. M. Abu-Zahra, E. Gilling and E. Goetheer, "Ship-based carbon capture on-board of diesel or LNG-fuelled ships," *The International Journal of Greenhouse Gas Control*, vol. 85, pp. 1–10, 2019, doi:10.1016/j. ijggc.2019.03.008.

6. J. Monteiro, "CO$_2$ASTS – carbon capture, storage and transfer in shipping, a technical and economic feasibility study: Public concise report," 2020. [Online]. Available: https://www.mariko-leer.de/wp-content/uploads/2020/06/200513-CO2ASTS-Public-Concise-Report.pdf.

7. X. Luo and M. Wang, "Study of solvent-based carbon capture for cargo ships through process modelling and simulation," *Applied Energy*, vol. 195, pp. 402–413, 2017, doi:10.1016/j.apenergy.2017.03.027.

8. MAN, *Project Guide – Marine Four-Stroke Dual Fuel Engine Compliant with IMO Tier III (MAN 51/60DF)*. MAN, 2018.

9. M. Arshad, M. A. Zia and F. A. Shah, *An Overview of Biofuel*. Springer, 2017, doi:10.1007/978-3-319-66408-8.

10. T. Bond and M. R. Templeton, "History and future of domestic biogas plants in the developing world," *Energy for Sustainable Development*, vol. 15, no. 4, pp. 347–354, doi:10.1016/j.esd.2011.09.003.

11. R. Muoz, L. Meier, I. Diaz and D. Jeison, "A review on the state-of-the-art of physical/ chemical and biological technologies for biogas upgrading," *Reviews in Environmental Science and Biotechnology*, vol. 14, no. 4, pp. 727–759, doi:10.1007/s11157-015-9379-1.

12. P. Weiland, "Biogas production: Current state and perspectives," *Applied Microbiology and Biotechnology*, vol. 85, no. 4, pp. 849–860, doi:10.1007/s00253-009-2246-7.

13. J. Kwany and W. Balcerzak, "Production logistics and participation of biogas in obtaining primary energy in Poland," *Energy & Environmental Science*, vol. 28, no. 4, pp. 425–436, doi:10.1177/0958305X17695277.

14. P. J. Jørgensen, *Biogas green energy. Process, design, energy supply, and environment*. Faculty of Agricultural Sciences, Aarhus University, 2009, p. 36.

15. I. Angelidaki *et al.*, "Biogas upgrading and utilization: Current status and perspectives," *Biotechnology Advances*, vol. 36, no. 2, pp. 452–466, doi:10.1016/j. biotechadv.2018.01.011.

16. E. Ryckebosch, M. Drouillon and H. Vervaeren, "Techniques for transformation of biogas to biomethane," *Biomass and Bioenergy*, vol. 35, no. 5, pp. 1633–1645, doi:10.1016/j. biombioe.2011.02.033.

17. M. Persson, O. Jonsson and A. Wellinger, "Biogas upgrading to vehicle fuel standards and grid," 2006. [Online]. Available: https://www.sciencedirect.com/science/article/pii/ S2351978916302451.

18. A. Jaffrin, N. Bentounes, A. M. Joan and S. Makhlouf, "Landfill biogas for heating greenhouses and providing carbon dioxide supplement for plant growth," *Biosystems Engineering*, vol. 86, no. 1, pp. 113–123, doi:10.1016/S1537-5110(03)00110-7.

19. M. Ajhar, M. Travesset, S. Yüce and T. Melin, "Siloxane removal from landfill and digester gas – a technology overview," *Bioresource Technology*, vol. 101, no. 9, pp. 2913–2923, doi:10.1016/j.biortech.2009.12.018.

20. G. Soreanu *et al.*, "Approaches concerning siloxane removal from biogas – a review," *Canadian Biosystems Engineering / Le Genie des biosystems au Canada*, vol. 53, pp. 8.1–8.18.

21. P. J. He, "Anaerobic digestion: An intriguing long history in China," *Waste Management*, vol. 30, no. 4, pp. 549–550, doi:10.1016/j.wasman.2010.01.002.

22. D. Cameron, F. J. Commin and A. J. Martin, "Process of and apparatus for treating sewage." [Online]. Available: https://patents.google.com/patent/US634423A/en.

23. C. Merdinger, "The development of modern sewerage: Part II," *Military Engineering*, vol. 45, pp. 123–127. [Online]. Available: https://www.jstor.org/stable/44561586.

24. "EBA Statistical Report 2020," 2021. [Online]. Available: www.europeanbiogas.eu/eba-statistical-report-2020/.

25. EBA, "European Biogas Association Annual Report 2019," 2019. [Online]. Available: www.europeanbiogas.eu/wp-content/uploads/2020/01/EBA-AR-2019-digital-version.pdf.

26. M. Torrijos, "State of development of biogas production in Europe," *Procedia Environmental Science*, vol. 35, pp. 881–889, doi:10.1016/j.proenv.2016.07.043.

27. EBA/GIE, "Biomethane Map 2020," p. 1, 2020. [Online]. Available: https://www.europeanbiogas.eu/wp-content/uploads/2021/01/EBA_StatisticalReport2020_abridged.pdf.

28. A. Petersson, J. B. Holm-Nielsen, and D. Baxter, "Biogas upgrading technologies – developments and innovations." [Online]. Available: https://www.ieabioenergy.com/wp-content/uploads/2009/10/upgrading_rz_low_final.pdf.

29. EnviDan A/S, "Intelligent udnyttelse af kulstof og energi på renseanlaeg," 2013. [Online]. Available: https://naturstyrelsen.dk/media/138744/iceu-del-1-endelig-version.pdf.

30. L. Maggioni and C. Pieroni, "Report on the Biomethane Injection into National Gas Grid," [Online]. Available: www.isaac-project.it/wp-content/uploads/2017/07/D5.2-Report-on-the-biomethane-injection-into-national-gas-grid.pdf

31. "Work package: WP06," 2009. [Online]. Available: https://businessdocbox.com/Logistics/109413727-Work-package-final-report.html.

32. F. Graf and U. Klaas, "State of biogas injection to the gas grid in Germany," *International Gas Union World Gas Conference Papers*, vol. 5, pp. 3867–3879, 2009.

33. "Biogas in Denmark," 2021. [Online]. Available: https://ens.dk/en/our-responsibilities/bioenergy/biogas-denmark

34. "Hashøj biogas," 2021. [Online]. Available: http://www.hashoejbiogas.dk/.

35. J. Bjerg and A. M. Fredenslund, "Biogas i danske kommuner Afprøvede løsninger," 2014. [Online]. Available: https://orbit.dtu.dk/en/publications/biogas-i-danske-kommuner-afpr%C3%B8vede-l%C3%B8sninger.

36. "Task 37: Energy from biogas," 2020. [Online]. Available: https://task37.ieabioenergy.com/about-task-37.html

37. "Task 37: Upgrading plant list 2019," 2019. [Online]. Available: https://task37.ieabioenergy.com/plant-list.html

38. Q. Sun, H. Li, J. Yan, L. Liu, Z. Yu and X. Yu, "Selection of appropriate biogas upgrading technology-a review of biogas cleaning, upgrading and utilisation," *Renewable and Sustainable Energy Reviews*, vol. 51, pp. 521–532, doi:10.1016/j.rser.2015.06.029.

39. F. M. Baena-Moreno, E. le Saché, L. Pastor-Pérez and T. R. Reina, "Membrane-based technologies for biogas upgrading: A review," *Environmental Chemistry Letters*, vol. 18, no. 5, pp. 1649–1658, 2020, doi:10.1007/s10311-020-01036-3.

40. L. Deng and M. B. Hägg, "Techno-economic evaluation of biogas upgrading process using CO_2 facilitated transport membrane," *The International Journal of Greenhouse Gas Control*, vol. 4, no. 4, pp. 638–646, doi:10.1016/j.ijggc.2009.12.013.

41. A. I. Adnan, M. Y. Ong, S. Nomanbhay, K. W. Chew and P. L. Show, "Technologies for biogas upgrading to biomethane: A review," *Bioengineering*, vol. 6, no. 4, pp. 1–23, doi:10.3390/bioengineering6040092.

42. R. Augelletti, M. Conti and M. C. Annesini, "Pressure swing adsorption for biogas upgrading. A new process configuration for the separation of biomethane and carbon dioxide," *The Journal of Cleaner Production*, vol. 140, pp. 1390–1398, doi:10.1016/j.jclepro.2016.10.013.

43. K. Stangeland, D. Kalai, H. Li and Z. Yu, "CO_2 methanation: The effect of catalysts and reaction conditions," *Energy Procedia*, vol. 105, no. 1876, pp. 2022–2027, doi:10.1016/j.egypro.2017.03.577.

44. S. N. B. Villadsen, P. L. Fosbøl, I. Angelidaki, J. M. Woodley, L. P. Nielsen and P. Møller, "The potential of biogas; the solution to energy storage," *ChemSusChem*, vol. 12, no. 10, pp. 2147–2153, May 2019, doi:10.1002/cssc.201900100.
45. L. Peter, T. Vass, H. Mandová and A. Gouy, "Cement – analysis IEA," *International Energy Agency 2020*. [Online]. Available: www.iea.org/reports/cement
46. E. Worrell, N. Martin and L. Price, "Potentials for energy efficiency improvement in the US cement industry," *Energy*, vol. 25, no. 12, pp. 1189–1214, 2000.
47. M. G. Plaza, S. Martínez and F. Rubiera, "CO_2 capture, use, and storage in the cement industry: State of the art and expectations," *Energies*, vol. 13, no. 21, p. 5692, 2020.
48. J. Walters, "Recovery act: SkyMine® beneficial CO_2 use project," *Skyonic Corporation 2016*. [Online]. Available: www.osti.gov/servlets/purl/1241314.
49. J. Jones, C. Barton, M. Clayton, A. Yablonsky and D. Legere, "SkyMine® carbon mineralization pilot project," *Skyonic Corporation 2011*. [Online]. Available: www.osti.gov/servlets/purl/1027801.
50. "Steel's contribution to a low carbon future and climate resilient societies – worldsteel position paper," W. S. Association, Ed. 2020.
51. S. Jahanshahi, J. G. Mathieson and H. Reimink, "Low emission steelmaking," *Journal of Sustainable Metallurgy*, vol. 2, no. 3, pp. 185–190, 2016.
52. A. Hamidian, R. Bonnart, M. Lacroix and V. Moreau, "A pilot plant in Dunkirk for DMX process demonstration," 15th Greenhouse Gas Control Technologies Conference, Abu Dhabi, Mar. 15–18, 2021, vol. 1, SSRN.
53. L. Raynal et al., "The DMX™ process: An original solution for lowering the cost of post-combustion carbon capture," *Energy Procedia*, vol. 4, pp. 779–786, 2011.

12 Modeling the Socio-Economic Impacts of Carbon Capture and Storage Deployment: Current Practices and Pathways Forward

Judy Jingwei Xie, Piera Patrizio, and Niall Mac Dowell

CONTENTS

12.1 INTRODUCTION

The first evidence of anthropogenic climate change was observed almost a century ago [1]. Historically, the contribution to global emissions has been spatially and temporally unequal [2]. Consequently, the historical responsibility for climate change, resulting contemporary impacts, [3], and the socio-economic consequences of the mitigation thereof also present different regional patterns making it essential to discuss and negotiate the equitable distribution of emission-reduction targets. This is manifestly a

DOI: 10.1201/9781003162780-12

323

nontrivial task, with the global negotiation of our collective commitments to mitigating climate change having commenced several decades ago with the advent of international organizations such as the Intergovernmental Panel on Climate Change (IPCC) in 1988. The United Nations Framework Convention on Climate Change (UNFCCC) has also put forth international treaties such as the Paris Agreement. The European Union especially has been a leading figure in its progressive climate policy, evident in the European Green Deal's pledge towards carbon neutrality in 2050. Following the UK, nations including Sweden, Scotland, France, Denmark, and New Zealand have also made official commitments to net-zero by 2050.

Historically, climate scientists and economists have used integrated assessment models (IAMs) to explore the feasibility of achieving climate targets and setting the agenda for global environmental negotiations. Emission stabilization pathways arising from these models highlight the key role of carbon dioxide removal (CDR) technologies to keep the 1.5 °C temperature target within reach [4]. Under different transition scenarios, the removal capacity requirement of CCS may range from 150 to 1200 $GtCO_2$ by the end of the century [5]. However, the least-cost approach underpinning these pathways and the lack of technological detail incorporated in IAMs provide limited information on the role of these technologies in national energy systems. The unique mitigation service (providing low carbon base load electricity) and removal service (capturing CO_2 at point sources [6]) provided by CCS are often masked in IAMs. As the research and development for the downstream utilization of CO_2 continue and the market expands [7], the timely deployment of CCS would require modeling techniques beyond least-cost approaches.

Moreover, as frequently argued by energy transition scholars [8], the transition towards a net-zero economy will imply profound structural changes for national and regional economies. The 'transition policies' adopted by nations include a portfolio of energy, industrial, climate, and trade policies required to enable the transition to a net-zero paradigm. These changes and upheaval brought by the decarbonization of national energy systems will be most evident in the labor markets [9–10], as carbon-intensive industries and their associated labor forces become less competitive and decline, and low carbon activity growth and their associated employment increases [11]. These activities will not be evenly distributed, as different regions have different levels of industrial and economic activity. Hence, the socio-economic aftermath of decarbonization will depend on the regional portfolio of physical, human, social, and intangible capitals.

Translating net-zero emissions commitments into effective energy decarbonization pathways requires further understanding of the intertwined dimensions of the transition. Therefore, the social, economic, and political drivers of energy system transition can be identified and translated into meaningful metrics for energy modelers and analysts. Local and regional areas will be an important unit of analysis to identify technically feasible, financially viable, and socially equitable energy transitions [12]. Despite the difficulty in accounting for the socio-economic factors than the pure cost-based metrics [13], it is believed that they are necessary factors to move climate action forward now.

The following sections are structured as follows. Section 12.2 will discuss the relevant socio-economic drivers within the different stages of CCS deployment. As quantitative modeling techniques will play a major role in designing effective energy decarbonization and CCS deployment strategies, the remaining sections will explore how to include socio-economic drivers into quantitative models. Section 12.3 will

evaluate existing quantitative modeling approaches for energy system decarbonization, followed by their respective examples. Section 12.4 will discuss the trade-offs
of these models and their flexibility to accommodate the socio-economic dimension.
Section 12.5 will present an outlook and future path forward for more socially equitable policymaking for the CCS deployment.

12.2 SOCIO-ECONOMIC DRIVERS OF ENERGY TRANSITION

In recent history, societal engagement has tangibly impacted the policymaking and the
deployment of sustainability-focused infrastructure projects. This needs to reflect the
demand-side behaviors of CCS deployment, which has not yet been reflected in most
literature [14]. Coal workers in the United States resisting the defunding of coal power
plants and the yellow vests in France protesting the high fuel taxes are examples of the
lack of public trust and policymakers' disconnection with the people's essential needs
[15]. The lack of engagement with local stakeholders in the early stage of the CCS demonstration plant in Barendrecht, The Netherlands, also caused massive project delays
[16]. A framework of 'socio-energy system design' [10] can help to account for the
diverse social dimension of energy policies, including local communities, which may
benefit from or be harmed by said policies. The design of these policies can impact the
structure of a local economy and spark political conflicts in governance.

The interplay of the social, technical, and economical dimension of CCS deployment is presented in Figure 12.1. The modeling techniques suitable for tackling these
dimensions are explained in more detail in the next section. Currently, there is no

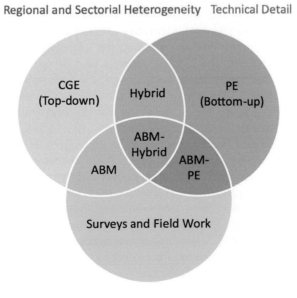

FIGURE 12.1   Socio- and techno-economic impacts of CCS deployment in the context of
research methods.

one-size-fits-all approach in accounting for socio-economic issues; thus, these issues should be adequately studied and accounted for under case-by-case circumstances. This is not to serve as an exhaustive list of all the socio-economic factors involved in the decarbonization transition but to remind modelers of the complex issues involved within and outside of their models.

12.2.1 Sustainable Development and Economic Growth Indicators

The United Nations Sustainable Development Goals (UN SDGs) is a blueprint set in 2015 by the UN General Assembly, demonstrating the intertwining relationship between sustainable growth and development. Importantly, the indicators embedded within the SDG framework [17] could be mapped out to broader potential implications of deploying CDR concepts at scale. Nations were encouraged to adopt the global SDG indicators along with their individually set national indicators. Although nations have supported the SDG framework, the set of quantifiable indicators provided by the UN has not yet been comprehensively reviewed at the time of writing.

While there are increasing efforts in the energy modeling community to quantify the impact of CDR on natural resources such as land, biomass, and water use [4, 18–19], the potential effects of deploying and upscaling these technologies across other SDGs are scarce, notably regarding social, institutional, or policy dimensions. Consideration of regional socio-economic impacts such as distributional effects (SDG 10), employment creation (SDG 8), and economic growth (SDG 9) of CDR deployment is scarce in the literature, despite indications that regional, cultural, socio-economic, and political differences and their influence on policy design could strongly affect progress towards sustainable development.

Although the UN SDG framework is widely referred to as a standard for sustainable development in international negotiations, the aggregate nature of these indicators hinders the ability to trace status, progress, and cross-national comparison. Within this context, the Organisation for Economic Co-Operation and Development (OECD) has developed a set of Green Growth Indicators [20] that can be adopted at the national level to assess the progress towards decarbonization and economic growth. Some relevant indicators presented in Table 12.1 include the percentage of research and development (R&D) expenditure in green growth and gross domestic product (GDP) per unit of CO_2 emitted. While these indicators do not account for regional-specific circumstances and need improvement when comparing progress in developing countries [21], they represent a useful set of metrics to address the socio-economic impacts of national decarbonization strategies. For example, Iceland was heavily penalized in the sustainability ranking by the "proportional of land area covered by forests" indicator, despite its renewable heavy energy infrastructure. All relevant SDGs and OECD indicators for evaluating the socio-economic impacts of CCS adoption are summarized in Table 12.1.

TABLE 12.1

Sustainable Development and Economic Growth Indicators Relevant to CCS Deployment

Framework	Goal	Example Indicator
SDG Framework [17]	7 – Affordable and Clean Energy	The proportion of population with primary reliance on clean fuels and technology
	8 – Decent Work and Economic Growth	The unemployment rate, by sex, age, and persons with disabilities
	9 – Industry, Innovation and Infrastructure	CO_2 emission per unit of value-added, proportion of medium and high-tech industry value added in total value-added
	10 – Reduced Inequalities	Labor share of GDP
	12 – Responsible Consumption and Production	Amount of fossil fuel subsidies per unit of GDP (production and consumption)
	13 – Climate Action	Total greenhouse gas emissions per year
	15 – Life on Land	Progress towards sustainable forest management
OECD Green Growth Indicators [20]	*The socio-economic characteristic of growth*	
	Productivity and trade	Relative importance of trade: (exports + imports)/GDP
	Labor markets	Labor force participation
	The environmental and resource productivity of the economy	
	1 – CO_2 productivity	GDP per unit of energy-related CO_2 emitted
	2 – Energy productivity	GDP per unit of total primary energy supply (TPES) XE "Total Primary Energy Supply (TPES)"
	The environmental dimension of quality of life	
	14 – Environmentally induced health problems and related costs	Population exposure to air pollution and the related health risks
	15 – Exposure to natural or industrial risks and related economic losses	N/A
	Economic opportunities and policy responses	
	17 – Research and development expenditure of importance to green growth	Environmental technology (% of total R&D)
	18 – Patent of importance to green growth	Environment-related patents (% of a country's patent families worldwide)
	20 – Production of environmental goods and services XE "environmental goods and services" (EGS)	Gross value added in the EGS sector (% of GDP), employment in the EGS sector

12.2.2 REGIONAL SPECIALIZATION

In addition to the equitable distribution of climate targets, much of the regionally initiated commitment towards low carbon transition has been found contingent on the respective areas' political will and socio-economic resources. As demonstrated by Jewell *et al.*, nations committed to the Powering Past Coal Alliance have also shown inherently small dependence on and existing natural resources of coal [22]. Within the European Union, an evaluation of the job creation and GVA potential of decarbonization in the UK, Spain, and Poland [12] showed that Poland would face much higher GVA losses from eliminating coal in its electricity generation than the others. Due to its land-use requirement and potential competition with agricultural needs, bioenergy with carbon capture and storage (BECCS) especially presents region-specific challenges through its deployment [23]. Furthermore, the technological availability of advanced CCS methods such as DACCS can impact the regional capability for deployment. Hence the dependency of regional resources, availability of geological constraints, and the structure of the local economy should be fully accounted for during the planning and deployment of CCS technologies.

12.2.3 SECTORIAL HETEROGENEITY

Besides region-specific drivers such as the geopolitical setting and natural resources involved, the heterogeneity of national economic structure, including national industrial strength and labor market composition, plays a major role in the successful implementation of energy transition strategies. The potential socio-economic gains associated with low carbon technologies will depend on how economies can rely on domestic industries and expertise for their implementation. In addition, the commercialization of low carbon technologies and the diffusion of particular CDR practices will bring unequal socio-economic impacts across different sectors of the economy. In quantifying the employment effects of the decarbonization of the US coal sector, Patrizio *et al.* had shown that the deployment of BECCS would mitigate significant job losses in the mining industry. In contrast, job creation would be predominantly observed in the forestry and transport sectors [24].

The structural changes through the decarbonization of our society will also lead to skill- and innovation-biased transitions in the job market [25]. As more renewable energy technology gets deployed, the long-term improvements in efficiency and automation could lead to the displacement of manual labor activities. The re-skilling and re-training of labor forces from high carbon intensity sectors towards those of low carbon intensity present a nontrivial challenge for policymakers.

12.2.4 PUBLIC ACCEPTANCE

The local deployment of CCS technologies is dependent on both technological advancement and the social processes in the adoption [26]. To further prevent project delays in CCS demonstration plants [16], public perception should be carefully analyzed before the implementation. The deployment of CCS technologies could introduce perceived unequal safety risks to local communities. Through interviews, researchers have found that the public in both the US and the UK have low prior knowledge and understanding of CCS technologies. The participants also believed that CCS would allow for the continued dominance of the fossil fuels industry and distract meaningful investments in other sustainable energy solutions [27]. Seigo *et al.* [26] developed a technology acceptance framework that can be used to survey the public perception and guide educational outreach. As a positive case for public perception, Swiss DACCS start-up Climeworks built out a successful carbon offset subscription campaign to crowdsource funds for further development. It demonstrates the public's enthusiasm for a decarbonization future with CCS technologies. As the public becomes more interested and committed to a societal green transition, the fossil fuels industry is gradually losing its social license to operate. It is instrumental to communicate and frame the importance of CCS as part of the sustainable energy portfolio towards net-zero.

12.3   MODELING APPROACHES

Aggregate climate models such as IAMs are predominantly used by international organizations like the IPCC for setting the global agenda for climate action. Together with a consistent modeling approach, the IAMs modeling community has been developing the concept of 'shared socioeconomic pathways' (SSP) to provide a range of modeling outcomes with different levels of socio-economic development and climate change mitigation [28]. While attempting to describe plausible trajectories of future societal development, the SSPs are necessarily based on a set hypothesis about which societal elements are key determinants to climate change mitigation.

The following subsections present a review of the main approaches available for modeling the energy transition. In particular, the subsections focus on energy models detailing CCS within their framework by presenting their applications in specific case studies. The potential quantification of the socio-economic impacts associated with CCS deployment within these models will also be explored. The modeling structure and the intrinsic assumptions can determine the outcome of the models [29–30]; thus, they should be defined based on the problems they are used to tackle. Table 12.2 summarizes all the models discussed, including their definitions of CCS technologies and socio-economic considerations.

12.3.1   Computational General Equilibrium or Top-Down Models

Computational general equilibrium (CGE) models adopt a top-down approach to representing the interaction of the economy and the environment. Rooted in the equilibrium of supply and demand in the interconnected markets of the entire economy [47], they are commonly used for policy analysis through the comparison of different equilibriums caused by policy 'shocks' [48]. However, they often lack technological details, especially in the description of power production and CDR technologies [49].

Few exceptions are represented by the top-down DEMETER-CCS and ENTICE-BR models. Gerlagh *et al.* adopted DEMETER-CCS [31] to compare the economic effects of a portfolio of energy policy actions, including energy efficiency improvements, fuel switch, and post-combustion CCS deployment. A recycling carbon tax to fund renewables deployment was found to be the most cost-effective pathway. The explicit

TABLE 12.2

Summary of the Models Evaluated and Their Definitions of CCS Technology and Socio-Economic Factors

Type	Publication	Model	CCS	Socio-economic factors
CGE	Gerlagh 2006 [31]	DEMETER-CCS	Post-combustion	N/A
	Grimaud 2011 [32]	ENTICE-BR	IEA 2006 and IPCC 2005 report	N/A
	Küster 2007 [33]	CGE/MPSGE	N/A	Skill-based labor impacts
	Golosov 2014 [34]	DSGE	N/A	Economic welfare[*]
	Li 2017 [35]	Dynamic recursive CGE	Post-combustion deployed after 2020	GDP,[*] sectorial output
PE	Koljonen 2009 [36]	Global ESAP TIAM	Fossil fuel electricity generation with CCS	N/A
	Realmonte 2019 [37]	TIAM-Grantham	DACCS, BECCS, afforestation, traditional CCS	N/A
	Ioakimidis 2012 [38]	TIMES	CCS on coal and lignite plants	N/A
	Heuberger 2017 [39]	ESO	CCGT-CCS, Coal-CCS, BECCS	N/A
	Fajardy 2018 [23]	MONET	BECCS	N/A
Hybrid	McFarland 2004 [40]	MIT EPPA with bottom-up detail	Gas-CCS, coal-CCS	Economic welfare[*]
	Kumbaroğlu 2003 [41]	SCREEN with bottom-up detail	N/A	GDP,[*] employment
	Fujimori 2014 [42]	AIM/CGE	BECCS, NG-CCS, coal-CCS, oil-CCS	GDP[*]
	Sue Wing 2008 [43]	SAM	N/A	Economic welfare[*]
	Patrizio 2020 [12]	ESO-JEDI	CCGT-CCS, Coal-CCS, BECCS	GVA, jobs
	Patrizio 2018 [24]	BeWhere-US and JEDI	CCS and BECCS retrofitted coal plants	Job creation
ABM	Gerst 2013 [44]	ENGAGE	N/A	GDP,[*] decision rules of households and firms
	Chappin 2009 [45]	ABM	Coal-CCS	Decision rules of power companies
	Rai 2015 [46]	ABM	N/A	Decision rules of households, social networks

[*] *Although GDP and economic welfare have been included as a social accounting metric in existing models, they do not demonstrate the complex landscape of socio-economic factors.*

representation of CCS technologies is also featured in the decentralized, top-down ENTICE-BR model, which has been adopted to explore the value of carbon tax mechanism and R&D incentives to foster the development of CCS [32]. The modeling outcome showed that the optimal time path for carbon tax followed a reverse U-shape and that the policy instruments complemented each other.

Throughout the landscape of CGE models, there is a marked paucity of consideration of the socio-economic implications of energy technology deployment. While quantifying the economic impacts associated with certain energy policy strategies (expressed in the form of GDP variations from a benchmarking value), the current lack of regional and sectorial details of CGE models prevents the adequate representation of the key drivers identified in Section 12.2. Hence, further developments in addressing the regional and sectorial factors would be an important research direction for CGE models.

One of the criticisms of traditional general equilibrium models is their inflexibility in accounting for the dynamic changes in the market. Dynamic stochastic general equilibrium (DSGE) models are a new modeling technique that uses the equilibrium theory but allow for incorporating future shocks and endogenous technology change in the formulation [50]. The CGE/MPSGE model [33] used a dynamic recursive approach to disaggregate energy policy impact on different labor skill sets in disaggregated sectors in the economy. They demonstrated that renewable energy subsidies may not automatically lead to emissions reduction. Even if emissions reduction was achieved, it might lead to a contraction in employment due to investments in less labor-intensive renewables. The use of CCS, however, was not considered in this model. Golosov et al. also used a DSGE formulation to determine the optimal global carbon tax and arrived at higher optimums than the Nordhaus model [34]. Coal was found to be the main driver of welfare loss instead of oil. Although the authors did not include CCS technologies in the formulation, they argue that it could be included if capture and storage cost is less than $60 per ton of carbon. Li et al. used a recursive dynamic CGE analysis on the impact of electric vehicles (EV) and CCS deployment on China's low carbon economy transition. Their model assumed to only adopt CCS to fossil-fueled power plants after 2020 to account for its technological availability. They have found a basic-level promotion of EV and CCS could reduce GDP loss and achieve emission reduction [35].

12.3.2 PARTIAL EQUILIBRIUM OR BOTTOM-UP MODELS

In contrast to the CGE models discussed previously, PE models are typically bottom-up and disaggregated in nature [48], as they contain more technological detail of the selected sector. Their technological focus allows for important features, such as predicting learning curves of emerging technologies, which cannot be easily reflected in traditional CGE models. They also allow for more elaborate incorporation of emerging technologies such as CCS. Unlike CGE models that account for the entire economy, PE models take all other things being equal, assuming the rest of the sectors to be static.

The global TIAM TIMES energy system model [36] showed an increase in demand for wind, bioenergy, and CCS, where the capacity for CCS storage was under high uncertainty. Realmonte *et al.* investigated the deployment of CCS in the electricity system by comparing a TIAM-Grantham model and a hybrid WITCH model. They concluded that DACCS could significantly reduce mitigation costs as part of a decarbonization portfolio instead of a substitute for any NETs [37]. An interesting case study on the regional assets shaping the energy transition was evaluated through the Integrated MARKAL/EFOM System (TIMES) model [38]. Due to Greece's dependence on lignite in electricity generation, the implementation of CCS in the Greek energy system to already licensed lignite power plants were shown to help transition the electricity system to imported electricity and renewables such as wind. The Electricity Systems Optimization (ESO) framework [39] has included CCS technologies in the mix of future energy portfolio and have expanded its capacities into technology learning. Through an evaluation of the systems value of a suite of electricity generating technologies including CCS-equipped power plants, the incorporation of CCS, on-shore wind, and grid-level energy storage in the UK has shown actual savings in the cost of the system. Due to the technological focus of PE models, the incorporation of socio-economic factors can be difficult on its own. The ESO model formulation combined with an input-output socio-economic accounting will be discussed in the next section.

Some researchers have also explored NETs as a standalone sector to evaluate the regional factors, such as the value of a bioeconomy for the case of BECCS. Through a value chain optimization process of the upstream resource use, the MONET framework [18] found that BECCS could lead to either positive or negative emissions. The sustainability assessment of BECCS should also integrate regional specificity, especially regarding land use calculations. Despite the transport cost, imported biomass can be economical in meeting regional targets when the local production factors are not ideal [23]. Additionally, the MONET model used a set of key levers for improved sustainability of BECCS, which can be traced under the UN SDG framework. These key levers include exploiting alternative biomass processing options (SDG 12) and measuring and limiting the impacts of direct and indirect land-use change (SDG 15).

12.3.3 AGENT-BASED MODELS

Owing to the complexity of the system transition involved in the decarbonization of our society, optimization-based modeling techniques are becoming less capable of capturing the nuisance in actors' interactions. Instead of focusing on the equilibrium of a system, agent-based models (ABMs) identify the system of adaptive agent behaviors and generate future scenarios based on these decision-making metrics [51]. Compared to traditional optimization-based models used to control an energy system, ABMs provide a unique insight in projecting the future outcome due to actors' adaptive behaviors [52]. Rather than being a direct substitute to the approaches presented above, they can provide a complementary perspective of the decarbonization transition. ABMs are especially useful in presenting the heterogeneity of agents, less-than-rational decision making, and their interactive behaviors [51]. In addition to these benefits, ABMs can also be operated in a spatial environment [53] and represent the regional specificity of CCS deployment. They can be an interesting modeling technique for the application of CCS technology deployment due to various public education campaigns. The application of ABM can be applied to various levels of granularity and range from the global negotiation level to the deployment of specific technologies [54]. However, compared to other research fields, the use of ABM in the energy system has been scarce [55], especially in the case of CCS deployment.

For example, the ENGAGE model [44] uses a 'Putnam two-level game' concept, which captured the interactive decision-making of international negotiation and domestic economic decisions of households and firms. Through the policy experiments conducted on the model, carbon tax revenue on carbon-free technology R&D is the most effective pathway to increase real GDP. Chappin *et al.* explored the impact of the carbon emission trading (CET) system in the Dutch electricity system based on the decision rules of power companies and observed relatively minimal effect in the first few decades [45]. Rai *et al.* modeled the household adoption of residential photovoltaics in Austin, Texas, using ABM based on demographic, attitudinal, social network, and environmental variables. They found that solar rebate programs need to provide sufficiently high incentives to impact the adoption in low-income households, and that changes in rebate level impact existing adopters due to their social networks [46].

12.3.4 Hybrid Models

Top-down models provide a comprehensive view of the economy. Bottom-up models have technical detail but little representation of the economic processes. Recognizing the value of both economics and engineering views, energy modelers have proposed various hybrid modeling efforts, aiming to bridge the gap between these perspectives. Although these hybrid models can demonstrate meaningful scenarios that neither model alone can show, they are still rooted in economic equilibrium theory, thus they still hold the same flaws for neglecting market imperfections.

Although written before the significant reduction of natural gas prices, McFarland *et al.* incorporated bottom-up engineering detail on CCS technologies in the top-down MIT Emissions Prediction and Policy Analysis (EPPA) model. They argued that CCS can lead to lower welfare loss and should not be used in only the electricity generation system. The timing and penetration of CCS technologies were also dependent on the formulation of climate policy [40]. Kumbaroğlu *et al.* developed a hybrid model including bottom-up details in a dynamic CGE model to evaluate climate policies in the Swiss energy system. CCS technologies were not included in the evaluation. A linear increase in carbon tax on fossil fuels was found to have no significant impact on national employment and GDP [41]. Fujimori *et al.* used a recursive dynamic global CGE model with bottom-up technology details in the energy sector compared to the aggregated method. Despite demonstrating similar energy demand, formulation with bottom-up information suggested more BECCS technologies be used for the mitigation scenario of reaching a global emission target of 19 Gt-CO_2 eq/year in 2050 [42].

The linkage of complex CGE formulations with bottom-up engineering models is a nontrivial task. Sue Wing developed a simpler social accounting matrix (SAM) to define the macroeconomic framework and has shown its promising capabilities in applying the US electric power system [43]. As one of the earlier developers of hybrid models, Sue Wing demonstrated difficulty calibrating economic and engineering data and recommends more consistent data collections in the future. Patrizio *et al.* have also added a socio-economic accounting feature using a JEDI (Jobs and Economic Development Impact) framework to the aforementioned ESO model. They found vastly different socio-economic outcomes (i.e., jobs and value-added) of decarbonization across Europe, especially in the losses in coal-dependent Poland [12]. Similarly, the BeWhere-US model and JEDI formulation demonstrated that retrofitting US coal plants with BECCS would retain and create more jobs than the baseline scenario of coal retirement to natural gas [24].

12.4 MODELING TRADE-OFFS

As presented in Table 12.3, each model type evaluated in this chapter has its respective advantages and limitations. The selection of an appropriate model type depends on the problem statement and the assumptions involved. While the CGE formulation can reflect the economy the best, ABMs can include social decision metrics. PE models generally maintain more technical details, especially in CCS technologies.

The CGE method can be used for high-level negotiation and planning of emission mitigation targets as they provide tracible information about the entire economy.

TABLE 12.3

Advantages, Limitations, and the Potential Incorporation of Socio-Economic Dimensions of Each Model Type

Type	Advantages	Limitations	Socio-economic potentials
CGE	Tracible; comprehensive view of the whole economy	Lack technological details	Sustainable development and economic growth indicators
PE	Technological details of a certain sector	Lack sectorial interactions	N/A
Hybrid	Both economic and engineering perspectives	Integration challenges	Sustainable development and economic growth indicators, regional specialization, sectorial heterogeneity
ABM	Realistic agent decision making rules; capable of solving various problem scopes	Potential scarcity of existing calibration data (initialization issues)	Public acceptance, regional specialization, sectorial heterogeneity

Critics of CGE models have argued that their aggregation ignores the heterogeneity of the stakeholders and that their equilibrium assumptions neglect the imperfect foresight of the market. The inclusion of CCS technologies has also been rare due to the aggregation of the technological details in CGE models. Developments in dynamic stochastic CGE (DSGE) models could improve possibilities for emerging technologies like CCS. However, the lack of technological detail may not be as detrimental when evaluating optimal emission reduction targets at a high level. CGE has explored economic indicators such as GDP, economic welfare, and skill-based labor impacts. Further developments in the regional and sectoral details can provide crucial information on the deployment of CCS.

PE models, on the other hand, provide a higher level of technical detail. They can incorporate methods such as learning curves and consider technologies under early-stage development, such as CCS. However, the common assumption of ceteris paribus may be unrealistic when using the model to reflect system-level and socio-economic impacts. Although they cannot represent socio-economic factors independently, PE models can be coupled with input-output tables such as JEDI or CGE formulations to gain the best of both worlds.

ABMs use the agents' decision-making process instead of a mathematical optimization formulation to provide a unique model structure and perspective. While the equilibrium formulations described before tackle the supply side of the policy-making, the agent-based technique could also cover the demand side reaction to the policies. In addition to being effective in understanding the local deployment of new technologies, ABMs can also evaluate the game theory behind international negotiations. As a rather new modeling technique, especially in the application of the energy system, the decision-making process of stakeholders is not as easily defined and understood as the thermodynamics of a powerplant. Existing databases used for calibration also may not be readily available. However, ABMs coupled with existing public perception research can provide new insights that the previously mentioned model frameworks could not accomplish. ABMs are structurally ideal for reflecting the social dimension of the energy system in addition to the economic metrics defined in CGE models.

Existing research in hybrid models can provide both top-down and bottom-up perspectives, allowing for both the inclusion of CCS technologies and the evaluation of socio-economic impacts. However, linking PE and CGE models is not trivial and would require higher computational power [56]. The calibration process of a hybrid model also requires both economic and engineering databases to be uniform, which is difficult due to separate data collection and maintenance [43]. The combination of PE models and input-output tables can be more achievable than a CGE formulation. Since hybrid models continue to use the same equilibrium assumptions as CGE and PE models, they still neglect the imperfect foresight and heterogeneity of stakeholders. Recent developments in DSGE, however, can improve the dynamic behaviors in hybrid models. These models can also be designed to include sustainable development indicators, regional specializations, and sectoral heterogeneity without sacrificing the technology details of CCS deployment. In addition to top-down and bottom-up hybridization, future research could expand to a feedback interaction with

ABM formulations to reflect the dynamic structure of the economy. A recent case study on UK residential heating investment decisions has shown promising results [57] and could extend to the application of CCS deployment.

12.5   OUTLOOK AND PATH FORWARD

Understanding the urgency of transitioning and especially CCS deployment, policymakers should move to immediate implementation at the local level. The necessary deployment of existing CCS technologies has historically faced numerous deployment challenges. The authors have found a scarcity in socio-economic considerations in the IAMs widely used by policymakers. Besides the equitable distribution of emission reduction targets, sustainable development indicators (Table 12.1) should be used to add more dimensions to the traditionally cost-based evaluations. For example, the hybridization of top-down and bottom-up models can dynamically include socio- and techno-economic impacts of CCS deployment. Combining a PE formulation and an input-output table could be a starting point for exploring these complex impacts. While factors like public perception can be tackled by local stakeholder engagement, they could also be included in ABMs to reflect agent decision-making behaviors beyond cost reduction.

The decarbonization of our society will present complex structural changes in our economy and communities. Modelers should develop further research in understanding the sectoral and skill-based impacts in the labor market and the social implications in the regions and communities involved. Recent progress in dynamic general equilibrium and the hybridization of bottom-up engineering models with ABM have demonstrated promising pathways to understanding the uncertainty and human dimensions of the transition. The deployment of CCS should be evaluated on a region-specific basis to show a realistic representation of the physical, social, political, and natural factors. To ensure a sustained structural transition of our economies, the socio-economic impacts of the decarbonization pathways should be carefully studied using the appropriate quantitative method discussed in this chapter.

REFERENCES

1. G. S. Callendar, "The artificial production of carbon dioxide and its influence on temperature," *Quarterly Journal of the Royal Meteorological Society*, pp. 223–240, 1938.
2. E. Neumayer, "In defence of historical accountability for greenhouse gas emissions," *Ecological Economics*, vol. 33, pp. 185–192, 2000.
3. J. C. Ciscar and L. Szab, "The integration of PESETA sectoral economic impacts into the GEM-E3 Europe model: Methodology and results," *Climatic Change*, vol. 112, pp. 127–142, 2012.
4. S. Fuss *et al.*, "Negative emissions – part 2: Costs, potentials and side effects," *Environmental Research Letters*, vol. 13, no. 6, 2018.
5. J. Rogelj *et al.*, "Scenarios towards limiting global mean temperature increase below 1.5°C," *Nature Climate Change*, vol. 8, no. 4, pp. 325–332, 2018.
6. D. W. Keith, G. Holmes, D. St. Angelo and K. Heidel, "A process for capturing CO_2 from the atmosphere," *Joule*, vol. 2, no. 8, pp. 1573–1594, 2018.
7. C. Hepburn *et al.*, "The technological and economic prospects for CO_2 utilization and removal," *Nature*, vol. 575, no. 7781, pp. 87–97, 2019.

8. J. Markard, "The next phase of the energy transition and its implications for research and policy," *Nature Energy*, vol. 3, pp. 628–633, 2018.

9. R. Hanna, Y. Xu and D. G. Victor, "After COVID-19, green investment must deliver jobs to get political traction," *Nature*, vol. 582, pp. 178–180, 2020.

10. C. A. Miller, J. Richter and J. O'Leary, "Socio-energy systems design: A policy framework for energy transitions," *Energy Research and Social Science*, vol. 6, pp. 29–40, 2015.

11. OECD, *Monitoring the Transition to a Low-Carbon Economy*. OECD, 2015.

12. P. Patrizio, Y. W. Pratama and N. M. Dowell, "Socially equitable energy system transitions," *Joule*, vol. 4, no. 8, pp. 1700–1713, 2020.

13. C. Böhringer and A. Löschel, "Computable general equilibrium models for sustainability impact assessment: Status quo and prospects," *Ecological Economics*, vol. 60, no. 1, pp. 49–64, 2006.

14. G. F. Nemet *et al.*, "Negative emissions – part 3: Innovation and upscaling," *Environmental Research Letters*, vol. 13, no. 6, 2018.

15. M. Martin and M. Islar, "The 'end of the world' vs. the 'end of the month': Understanding social resistance to sustainability transition agendas, a lesson from the Yellow Vests in France," *Sustainability Science*, vol. 16, 2020.

16. S. Brunsting, M. D. Best-Waldhober, C. F. J. Y. Feenstra and T. Mikunda, "Stakeholder participation practices and onshore CCS: Lessons from the Dutch CCS Case Barendrecht," *Energy Procedia*, vol. 4, pp. 6376–6383, 2011.

17. United Nations, "Global indicator framework for the sustainable development goals and targets of the 2030 agenda for sustainable development," 2020. [Online]. Available: https://unstats.un.org/sdgs/indicators/Global Indicator Framework after 2019 refinement_Eng.pdf%0Ahttps://unstats.un.org/sdgs/indicators/Global Indicator Framework_A.RES.71.313 Annex.pdf.

18. M. A. Fajardy, "Can BECCS deliver sustainable and resource efficient negative emissions?" *Energy & Environmental Science*, vol. 10, pp. 1389–1426, 2017.

19. P. Smith *et al.*, "Land-management options for greenhouse gas removal and their impacts on ecosystem services and the sustainable development goals," *Annual Review of Environment and Resources*, vol. 44, pp. 255–286, 2019.

20. OECD, *Green Growth Indicators 2017*. OECD, 2017.

21. S. E. Kim, H. Kim and Y. Chae, "A new approach to measuring green growth: Application to the OECD and Korea," *Futures*, vol. 63, pp. 37–48, 2014.

22. J. Jewell, V. Vinichenko, L. Nacke and A. Cherp, "Prospects for powering past coal," *Nature Climate Change*, vol. 9, pp. 592–597, 2019.

23. M. Fajardy and S. A. Chiquier, "Investigating the BECCS resource nexus: Delivering sustainable negative emissions," *Energy & Environmental Science*, vol. 11, pp. 3408–3430, 2018.

24. P. Patrizio *et al.*, "Reducing US coal emissions can boost employment reducing US coal emissions can boost employment," *Joule*, vol. 2, no. 12, pp. 2633–2648, 2018.

25. S. Fankhauser, F. Sehlleier and N. Stern, "Climate change, innovation and jobs," *Climate Policy*, vol. 8, pp. 421–429, 2008.

26. S. L. O. Seigo, S. Dohle and M. Siegrist, "Public perception of carbon capture and storage (CCS): A review," *Renewable and Sustainable Energy Reviews*, vol. 38, pp. 848–863, 2014.

27. E. Cox, E. Spence and N. Pidgeon, "Public perceptions of carbon dioxide removal in the United States and the United Kingdom," *Nature Climate Change*, vol. 10, pp. 744–749, 2020.

28. K. Riahi *et al.*, "The shared socioeconomic pathways and their energy, land use, and greenhouse gas emissions implications: An overview," *Global Environmental Change*, vol. 42, pp. 153–168, 2017.

29. E. A. Stanton, F. Ackerman and S. Kartha, "Inside the integrated assessment models: Four issues in climate economics," *Climate and Development*, vol. 1, no. 2, pp. 166–184, 2009.

30. P. Söderholm, "Modeling the economic costs of climate policy: An overview," *American Journal of Climate Change*, vol. 1, pp. 14–32, 2012.

31. R. Gerlagh and B. van der Zwaan, "Options and instruments for a deep cut in CO_2 emissions: Carbon dioxide capture or renewables, taxes or subsidies?" *The Energy Journal*, vol. 27, no. 3, pp. 25–48, 2006. Published by International Association for Energy Economics Stable.

32. A. Grimaud, G. Lafforgue and Magné, "Climate change mitigation options and directed technical change: A decentralized equilibrium analysis," *Resource and Energy Economics*, vol. 33, no. 4, pp. 938–962, 2011.

33. R. Küster, I. Ellersdorfer and U. Fahl, "A CGE-analysis of energy policies considering labor market imperfections and technology specifications," pp. 1–32, 2007. [Online]. Available: https://papers.ssrn.com/sol3/papers.cfm?abstract_id=960725.

34. M. Golosov, J. Hassler, P. Krusell and A. Tsyvinski, "Optimal taxes on fossil fuel in general equilibrium," *Econometrica*, vol. 82, no. 1, pp. 41–88, 2014.

35. W. Li, Z. Jia and H. Zhang, "The impact of electric vehicles and CCS in the context of emission trading scheme in China: A CGE-based analysis," *Energy*, vol. 119, no. 2017, pp. 800–816, 2017.

36. T. Koljonen, M. Flyktman and Lehtil, "The role of CCS and renewables in tackling climate change," *Energy Procedia*, vol. 1, pp. 4323–4330, 2009.

37. G. Realmonte *et al.*, "An inter-model assessment of the role of direct air capture in deep mitigation pathways," *Nature Communications*, vol. 10, pp. 1–12, 2019.

38. C. Ioakimidis, N. Koukouzas, A. Chatzimichali, S. Casimiro and A. Macarulla, "Energy policy scenarios of CCS implementation in the Greek electricity sector," *Energy Procedia*, vol. 23, pp. 354–359, 2012.

39. C. F. Heuberger, I. Staffell and N. a. Shah, "A systems approach to quantifying the value of power generation and energy storage technologies in future electricity networks," *Computers and Chemical Engineering*, vol. 107, pp. 247–256, 2017.

40. J. R. McFarland, J. M. Reilly and H. J. Herzog, "Representing energy technologies in top-down economic models using bottom-up information," *Energy Economics*, vol. 26, no. 4, pp. 685–707, 2004.

41. G. Kumbaroğlu and R. Madlener, "Energy and climate policy analysis with the hybrid bottom-up computable general equilibrium model SCREEN: The case of the Swiss CO_2 act," *Annals of Operations Research*, vol. 121, pp. 181–203, 2003.

42. S. Fujimori, T. Masui and Y. Matsuoka, "Development of a global computable general equilibrium model coupled with detailed energy end-use technology," *Applied Energy*, vol. 128, pp. 296–306, 2014.

43. I. Sue Wing, "The synthesis of bottom-up and top-down approaches to climate policy modeling: Electric power technology detail in a social accounting framework," *Energy Economics*, vol. 30, no. 2, pp. 547–573, 2008.

44. M. D. Gerst *et al.*, "Agent-based modeling of climate policy: An introduction to the ENGAGE multi-level model framework," *Environmental Modelling & Software*, vol. 44, pp. 62–75, 2013.

45. E. J. L. Chappin and G. P. J. Dijkema, "On the impact of CO_2 emission-trading on power generation emissions," *Technological Forecasting and Social Change*, vol. 76, no. 3, pp. 358–370, 2009.

46. V. Rai and S. A. Robinson, "Agent-based modeling of energy technology adoption: Empirical integration of social, behavioral, economic, and environmental factors," *Environmental Modelling & Software*, vol. 70, pp. 163–177, 2015.

47. I. Sue Wing, "Computable general equilibrium models for the analysis of economy-environment interactions," *Research Tools in Natural Resource and Environmental Economics*, pp. 255–305, 2011.
48. A. Nikas, H. Doukas, and A. Papandreou, "A detailed overview and consistent classification of climate-economy models," *Understanding Risks and Uncertainties in Energy and Climate Policy*, pp. 1–54, 2019.
49. C. Böhringer, "The synthesis of bottom-up and top-down in energy policy modeling," *Energy Economics*, vol. 20, no. 3, pp. 233–248, 1998.
50. J. D. Farmer, C. Hepburn, P. Mealy and A. Teytelboym, "A third wave in the economics of climate change," *Environmental and Resource Economics*, vol. 62, no. 2, pp. 329–357, 2015.
51. A. Hoekstra, M. Steinbuch and G. Verbong, "Creating agent-based energy transition management models that can uncover profitable pathways to climate change mitigation," *Complexity*, vol. 2017, 2017.
52. T. Ma and Y. Nakamori, "Modeling technological change in energy systems – from optimization to agent-based modeling," *Energy*, vol. 34, no. 7, pp. 873–879, 2009.
53. T. Filatova, P. H. Verburg, D. C. Parker and C. A. Stannard, "Spatial agent-based models for socio-ecological systems: Challenges and prospects," *Environmental Modelling & Software*, vol. 45, pp. 1–7, 2013.
54. E. J. L. Chappin and G. P. J. Dijkema, "Agent-based modeling of energy infrastructure transitions," *International Journal of Critical Infrastructures*, vol. 6, no. 6, 106-–130, 2010.
55. C. S. E. Bale, L. Varga and T. J. Foxon, "Energy and complexity: New ways forward," *Applied Energy*, vol. 138, pp. 150–159, 2015.
56. P. I. Helgesen and A. Tomasgard, "From linking to integration of energy system models and computational general equilibrium models – Effects on equilibria and convergence," *Energy*, vol. 159, pp. 1218–1233, 2018.
57. J. Sachs, Y. Meng, S. Giarola and A. Hawkes, "An agent-based model for energy investment decisions in the residential sector," *Energy*, vol. 172, pp. 752–768, 2019.

13 Emerging Technologies for Sustainable Carbon Capture

Mariam Ameen, Dzeti Farhah Mohshim,
Rabia Sharif, Rizwan Nasir, and Hilmi Mukhtar

CONTENTS

13.1 SUSTAINABLE DEVELOPMENT FRAMEWORK

The study of sustainable development tries to analyze the benefits and trade-offs involved in pursuing the many goals of environmental conservation, social equality, economic growth, and poverty eradication. A diversity of approaches has emerged from analyzing climate change and related challenges in keeping with this complexity [1]. The United Nations Sustainable Development Goals (SDG) are the blueprint to achieve a sustainable future for all. They address a full scale of global challenges, including inequality, poverty, climate change, peace and justice, and environmental degradation. In research, industries, and development related to energies, one could address all goals. However, most of them are more selective towards SDG 6 (clean water), SDG 7 (clean energy), SDG 12 (responsible consumption and production), and SDG 13 (climate action) [2–3].

DOI: 10.1201/9781003162780-13

The fundamental goal of the future is to evade the dangerous anthropogenic interference (DAI) with the environment and climate system [4]. The average global temperature has already increased by approximately 1.0 °C over pre-industrial levels due to human activities in climatic change. There is strong confidence that it will reach 1.5 °C between 2030 and 2052 if current rates continue [5]. The Paris Agreement has an objective of "holding the increase in the global average temperature to well below 2 °C above pre-industrial levels and preferably limit it to 1.5 °C above pre-industrial levels" [4]. To reach this goal, the new emissions must be rapidly reduced to maintain the balance between emissions and removal by the sink. The rapid evolution of technologies and innovation is an excellent example of low-carbon energy projects being guided to market readiness now [6]. All new plants in the industrial sector must use cutting-edge low-carbon technologies, and agriculture must follow suggested sustainable practices. Some examples of practices are as follows [7]:

- Sustain renewables growth
- Develop 1.5 °C vision for aviation and shipping
- No new coal plants
- New industrial installations low carbon after 2020
- Last fossil fuels car sold before 2035
- Zero deforestation by 2020
- Best practice in agriculture
- New buildings zero emissions from 2020
- Renovate 3%–5% of buildings per year.

Therefore, it is much needed to safely implement these practices in the next decade to achieve the expected sustainability by mid-century.

13.2 OFFSHOOT ADVANCEMENTS IN CARBON CAPTURE TECHNOLOGIES

13.2.1 CO_2 CAPTURE USING SUSTAINABLE AND BIOLOGICAL MATERIALS VIA BIO-ORGANIC LIQUIDS

Climate change due to the excessive increase of CO_2 emissions into the atmosphere during the last few decades is mainly due to industrialization and the global transportation sector. Fossil fuel–powered industries like coal-fired plants, petrochemicals, chemical industry, cement, and transportation are major sources of greenhouse gas emissions like CO_2 [8]. Since the last decade, the issue has been raised when global Earth temperature increased and drastic climate change was observed. Researchers have introduced several scientific methods to reduce carbon emission and CO_2 capture techniques [20]. History shows many options: solvent adsorption, membrane cryogenic separation, photobioreactor, solid adsorbents, and ionic liquids. So far, new advancement has been discussed in previous chapters for CO_2

capture and utilization. Furthermost common cations such as pyridinium, imidazolium, and fluorine-based anions do not demonstrate biocompatibility and biodegradability [9]. Therefore, cations and anions originated from bio-organic biomolecules are potential eco-friendly options such as amino acids and choline. Amino acids are zwitterions, as they contain both anion and cations together on a single molecule structure. Many research works reported amino acid-based ionic liquids (AAILs) for CO_2 capture.

Recently, Davarpanah et al. [10] reported CO_2 immersion in organic liquids of AAILs, which were choline-based amino acids such as glycine proline, alanine, and serine. Biobased ILs were dissolved in polar aprotic DMSO solution due to their high viscosity. The choline-based amino acids showed suitable absorption capabilities as reported 'up to 0.3 mol CO_2/mol IL, with only 12.5 wt.% IL in DMSO,' which signified the better functioning of CO_2. In another study, Gao et al. [11] have reported CO_2 absorption in AAILs activated N-methyl diethanolamine (MDEA) and tetramethylammonium glycinate$[N_{1111}][Gly]$, as well as its optimization for temperature and pressure. It was observed that with the introduction of ILs or MDEA in the form of blends, viscosity rose dramatically while no remarkable effect on density was observed.

In the middle of the last decade, a hydrophilic amino acid ionic liquid $[C_2OHmim][Gly]$ was produced by Lv et al. [12] and employed to increase the performance of the monoethanolamine (MEA) based CO_2 capture process from low-pressure flue gas. The synthesized mixture's regeneration and absorption ability were improved with the increase of $[C_2OHmim][Gly]$ with excellent stability even after four regeneration cycles than standalone MEA solutions.

Thermodynamic studies are vital to investigate the effect of variables such as pressure and temperature. Recently, Li et al. [13] reported thermodynamic study on CO_2 absorption in biobased ionic liquids (ILs). Five choline-based AAILs were synthesized using different molecular weights. CO_2 solubility was found to be maximum in 30 wt% AAILs. It was reported that based on zwitterion reaction mechanism, the dissolvability of CO_2 in the aqueous solutions of amino acid was associated with a reaction equilibrium thermodynamic model (RETM) [14].

The supportive thermodynamic parameters, such as Henry's law constants, enthalpy, reaction equilibrium constants, and chemical reactions, were determined and equaled to assess the CO_2 absorption in five amino acids [14]. Authors have postulated that the CO_2 dissolution in aqueous solutions of amino acids will be supplemented by green ionic liquids and might be thermodynamically favorable for CO_2 capture. However, some of them with targeted functional groups demonstrate physical advantages for CO_2 capture. The viscosity of biobased ILs needs redressal. ILs are mostly highly viscous (being comparatively higher than conventional solvents) and cannot be treated in the same reactor set-up together with amine solvents [15]. Moreover, additional analysis of biobased ILs for direct flue gas streams is necessary to investigate the 'real-time' performance. Furthermore, models are required to understand the theory of such materials. Investigating on lab and large scales can help find the most favorable biobased ILs and help researchers to adopt the best strategy for the development of biobased ILs solvents.

13.2.2 Microalgae Roles in CO_2 Capture

The natural carbon cycle consists of four major reservoirs (i.e., atmosphere, oceans, lands, and fossil fuels). The carbon is exchanged between these reservoirs. There are three reasons to highlight the importance of the carbon cycle: (1) carbon forms the structure of all lives on the Earth, (2) cycling of carbon approximates the energy flow around the globe, and (3) utilization of fossil fuels has increased [16]. Figure 13.1 depicts the carbon cycle. Climate change due to excessive emission of CO_2 endangers the whole ecosystem. Nature has its self-recycled terrestrial and marine ecosystems for CO_2 capture. Though scientists have researched and continue to investigate several methods for carbon capture up till now, there is a gap to be filled with the most sustainable process for carbon capture. To advance the process of bio-sequestration for CO_2 utilization in microalgal cell factories has shown a favorable route for recycling of CO_2 captured hooked on biomass and value-added chemicals [18]. Microalgae-based carbon capture technology is biodegradable and highly efficient for CO_2 reduction. CO_2 fixation and transformation into maximum microalgal biomass can be accomplished in the optimum growing condition. The nutritional environment is important for cell growth and culture media. Nonetheless, stressful environment such as a change in pH, salinity, light, intensity, and temperature are the key factors, which significantly reduce the biomass product yield [14]. Its wide dispersion, high biomass output, capacity to adapt to unfavorable conditions, rapid carbon uptake and utilization, and ability to develop value-added products contribute to algae's potential for industrial use, as shown in Figure 13.2 [19].

Many research studies have been performed for CO_2 absorption using microalgae in the last decade. Al-Zuhair *et al.* [20] have reported CO_2 absorption via freshwater microalgae strains from diethanolamine (DEA) solutions. Three strains of microalga (*Pseudochlorococcum* sp., *Chlamydomonas* sp., and *Chlorella* sp.) were selected for cultivation in DEA solutions enriched with CO_2. The efficiency of these strains to employ the dissolved CO_2 was evaluated. The reported results showed a 10% DEA solution saturated with CO_2 for all strains, with growing rates of 0.352, 0.365, and 0.669 per day, corresponding to *Chlorella* sp., *Chlamydomonas* sp., and *Pseudochlorococcum* sp., respectively. These microalgae were efficient enough

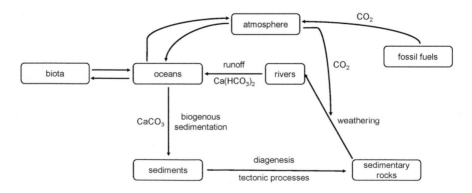

FIGURE 13.1   Carbon cycle of the Earth [17].

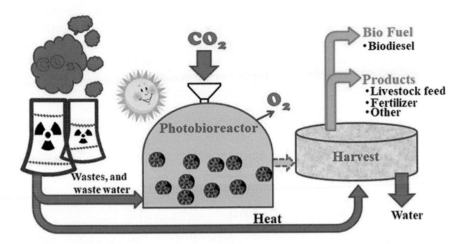

FIGURE 13.2 Products derived from algae that use CO_2 as a carbon source, their applications adopted from [19], and their ability to develop value-added products contribute to algae's potential for industrial use.

to remove the dissolved CO_2 with a drop rate of 0.135, 0.0120, and 0.0123 mole/mL/day, respectively. These observations demonstrated substantial improvement over traditional solvent rejuvenation processes for amine solutions generally used for CO_2 absorption. The utilization of microalgae for CO_2 is quite complicated because of variables, especially pH value and temperature. Piiparinen et al. [21] have recently reported on the effect of extreme pH for microalgae CO_2 capture. CO_2 was found to be absorbed in large quantity at acidic pH levels, while only a small amount of CO_2 was wasted at neutral pH. Conclusively, cryotolerant algae can be used for CO_2 capture.

Similarly, not all microalgae species are cryotolerant. Some microalgae are more efficient towards CO_2 capture; others are not or moderately efficient depending on species and their biomass composition and growth rate. Rodas-Zuluaga et al. [22] have reported the influence of CO_2 absorption on the growing rate of microalgae. *Scenedesmus* sp. exhibited more growth in carbohydrates, biomass, and protein production when CO_2 was dissolved at 20%. The study demonstrated that each microalgae strain was inflated by biomass growth and composition conditional to algal species. Additionally, this analysis presents that microalga can be an outstanding substitute towards sustainable CO_2 capture, wherever the CO_2 emitted from flue gas can be siphoned from an industrialized smokestack to the photo-bioreactors.

Microalgae can play a prominent role in reducing the gap between the separated CO_2 purity and its low content in flue gas streams. Chen and Xu [23] have reported optimal carbon capture method from flue gas-based CO_2 levels using microalgae. The major findings revealed that the elevated gas stream limits the CO_2 gap from growing optimum to flue gas concentrations. Another factor of removal of oxygen from inlet gas improved the CO_2 fixation by 2.7 times. The steady increase of CO_2

increased the C-tolerance to 30% and C-fixation by 2.5 times. Moreover, the main variable for CO_2 capture through microalgae was influenced by pH control, which is vital and could be adjusted by CO_2 treatment. Another advancement towards CO_2 capture using chemical wastewater was recently investigated by Yang *et al.* [23]. In this study, authors reported that CO_2 capture effectiveness was up to 92% while using wastewater-mediated culture with purified terephthalic acid (PTA). The application of industrial chemical wastewater successfully reduced CO_2 emissions.

Concluding, the utilization of algae for CO_2 capture in wastewater has been reported extensively. Hence, it can be concluded that using microalgae for carbon capture and fixation is a promising platform in reducing and utilizing CO_2, with process flexibility for SO_x and NO_x emissions.

13.2.3 CARBON CAPTURE VIA FUEL CELLS

The fuel cell (FC) is an electrochemical device that generates electricity with the help of a chemical reaction. Batteries are portable sources of electrical power that contain the chemicals required to generate power. Fuel Cells (FCs) do not include any chemical fuel. They merely serve as a reaction chamber for the fuel cell reaction [24]. FCs have high performance, compact size, low emissions, and are environmentally friendly devices [25]. Various kinds of FCs are considered for capturing CO_2, as shown in Figure 13.3. Microbial fuel cells (MFCs), molten carbonate fuel cells (MCFCs), and solid oxide fuel cells (SOFCs) exhibited promising outcomes in this area [26–27]. Recent studies also explain the carbon capture with the help of high-temperature fuel cells. Wang *et al.* reviewed high-temperature fuel cells (HTFCs)

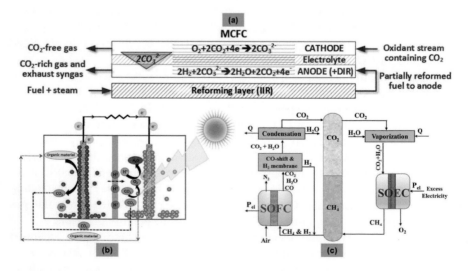

FIGURE 13.3 (a) Molten carbonate fuel cells adopted from [29]; (b) microbial fuel cells; (c) solid oxide fuel cells.

Source: **Adapted from [26].**

and their hybrid systems to capture CO_2; they also reviewed the techno-economic performance of FCs [28].

Although advances in FCs for carbon capture (CC) applications demonstrate definite growth and exciting possibilities, commercialization will need further experimentation and growth. Several challenges and observations have been reported in the literature (e.g., fuel cell module scale-up, fuel cell cost reduction, CO_2 capture energy reduction, safety and reliability, project demonstration) [28]. The key conclusions are that MCFCs are best suited for post-combustion CCS and are commonly used for carbon capture when active concentrators are needed downstream of conventional electricity. On the other hand, SOFCs are ideally suited for oxy-combustion due to their improved oxygen permeation capability [26].

13.2.4 CARBONIC ANHYDRASE FOR CARBON CAPTURE

Carbonic anhydrase (CA), a natural rapid biocatalyst stimulated by the CO_2 metabolic process in microbial cells, is a promising method that has high potential to improve the implementation and economics of carbon capture (CC) while meeting severe environmental requirements. Carbonic anhydrases can be used to facilitate the reversible hydration of the CO_2 molecule effectively [30]. The process design of biomimetic CO_2 capture is heavily reliant on selecting enzyme forms that can withstand harsh operating conditions, such as high concentration of salt, extreme temperature, and higher alkalinity, impairing enzyme performance. The absorption processes are carried out at 40°C to 60°C. The desorption temperature, in comparison, is 100°C, though this can be reduced when the unit is run under a vacuum pressure (about 0.3 bar) [31]. The sequestration of CO_2 by carbonic anhydrase is shown in Figure 13.4 [32].

Carbonic anhydrase (which mimics the use of carbonic acid as a building block for biomimetic CO_2 capture) is still in its early stages, but it has shown encouraging promise thus far. Although enzymes have many practical applications, they have limited applicability in carbon capture processes due to their high cost, low catalytic activity, poor temporal stability, high temperature sensitivity, low resistance to contaminants such as sulfur compounds, and recyclability. Additional advances are necessary to address these challenges so that significantly greater viability and a substantial amount of growth on various characteristics can be predicted [33]. This approach has already been adopted by some companies, as shown in Table 13.1 [34].

13.2.5 ARTIFICIAL INTELLIGENCE (AI) IN CARBON CAPTURE

Numerical simulation has gained popularity and significance in several scientific and engineering fields due to the rapid growth of computer technology over the last two decades. There is a growing interest in AI (artificial intelligence) technology, such as machine learning (ML) [35–36]. The basic machine learning implementation has been visualized in Figure 13.5 for different scales and sources for a carbon capture process [37].

Data from large-scale CO_2 capture facilities, such as TMC Mongstad in Norway and BD3 SaskPower in Canada, can be utilized to generate knowledge that will

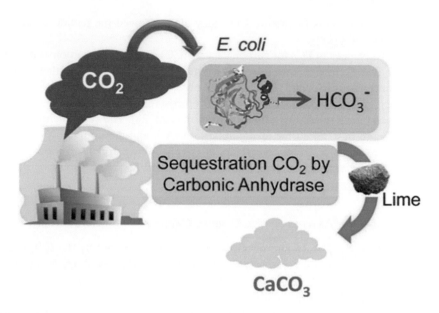

FIGURE 13.4 CO_2 capture with carbonic anhydrase.

Source: **Adapted from [32].**

TABLE 13.1
Commercial Utilization of Carbonic Anhydrase

Company	Process
Novozymes	Set up a carbon capture facility at the Amager Resource Center (ARC) in Copenhagen. Novozymes' biosolution for enzymatic carbon capture is tested as a possible solution for the future carbon capture plant.
Codexis Inc.	Jointly working with CO_2 Solutions Inc. (CSI).
Carbozyme, Inc.	The design comprises two microporous membranes made from fibrous material that are separated by a thin liquid membrane that serves as a separator.
Akermin, Inc.	Immobilization – stabilization of carbonic anhydrase for CO_2 capture from flue gas.
Sandia National Laboratories and University of New Mexico	An ultrathin liquid membrane restrained via capillary forces and catalyzed by carbonic anhydrase has been created for CO_2 separation.

improve the CO_2 capture process [38]. The AI has been used frequently to determine solubility and physical properties, gas compressibility factors, mass transfer of CO_2, and process design for carbon capture [39]. Summarizing, the use of AI in the carbon capture process has shown promising results in calculating important parameters for this process. To ensure the effective use of AI technology, AI scientists and engineers

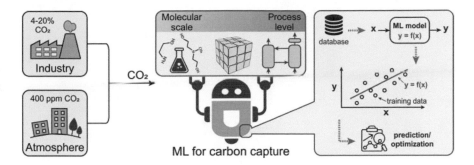

FIGURE 13.5 Representation of ML applications in carbon capture processes.

need to work together effectively. As a result of these collaborative efforts, CO_2 capture facilities can be designed more efficiently, with lower installation costs and a more efficient CO_2 collection process [40].

13.2.6 DIRECT AIR CARBON CAPTURE

Direct air capture (DAC) methods are gaining popularity among scientists, business organizations, legislators, and governments. While complete decarbonization of all sectors is required to reach the Paris Agreement objective, DAC can assist in dealing with emissions that cannot be reduced [41]. The ultimate goal of DAC technology is to remove CO_2 from the atmosphere and generate a concentrated stream of CO_2. This CO_2 is beneficial, as it can be used in a variety of ways. Many promising and developing DAC methods exist, along with the definition of DAC being so broad, as shown in Figure 13.6 [42–45]. Recently, a group of researchers has extensively reviewed the materials required for scaling up DAC technologies and the concept of learning by doing, which may reduce the costs [46]. Moreover, Fujikawa *et al.* highlighted the membrane-based direct air capture (m-DAC). They performed the process simulation of multistage m-DAC to find the energy requirement. They concluded that achieving the targeted m-DAC performance with competitive energy expense depends on the requirements in the subsequent utilization processes [47].

When evaluating DAC, some limiting factors may be considered important. The success of DAC will be determined not only by the energy and cost feasibility but also by factors affecting the region surrounding operations and the effects of the environment on the operation itself. Critical factors include land use and location, emissions from the operation, and water loss, to name just a few [48].

13.2.7 HYBRID APPROACH FOR CARBON CAPTURE

Currently, the studies have focused on testing and optimizing a single technology for carbon capture. There are few stories in the canvass about the hybrid approach for carbon capture. The hybrid approach studies integrating two or more technologies for carbon dioxide capture. Due to this, the integration of technologies overcome the

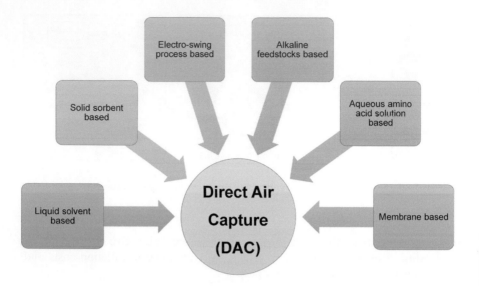

FIGURE 13.6   Direct air capture (DAC) methods.

disadvantage of one technology [49–51]. Recently, researchers reviewed the hybrid configuration using absorption, adsorption, membrane, and cryogenic for carbon dioxide capture [52]. Different arrangements (i.e., series, parallel, and integration) were tested to design the hybrid processes for CO_2 capture. Among these arrangement types, series and parallel are commonly used. But in terms of effectiveness, the integration arrangement is a good choice. More effective application arrangements should be investigated far enough by using each individual feature appropriately [49]. Figure 13.7 shows the different arrangements for the CO_2 capture hybrid processes. Moreover, some of the novel hybrid configurations for CO_2 capture have been tabulated in Table 13.2.

In contrast to stand-alone processes, hybrid approaches have better energy and CO_2 recovery results and lower capital equipment [52]. Although increased interest was attracted in recent years in CO_2 capture via hybrid processes and exciting results were reported, more studies were carried out through simulations or laboratory scales. As a result, much work remains to be done before this approach becomes commercially viable.

13.2.8   Calcium Looping for Carbon Capture

Calcium looping is used to trap carbon dioxide from flue gases from fossil fuel sources at low cost and with very high efficiency [59]. This technology use calcination/carbonation cycles of CaO-based sorbents (such as lime) to split CO_2 [60–61]. Figure 13.8 illustrates the basic schematic of calcium looping for CO_2 capture. In calcium looping, two reactors (carbonation and calcination) are used. At 873–973 K [62], the flue gas carries CO_2, CaO in the carbonator adsorbs the CO_2, and

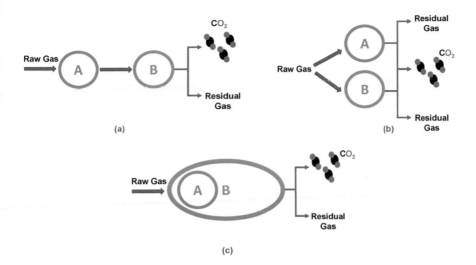

FIGURE 13.7 (a) Series arrangement, (b) parallel arrangement, and (c) integration arrangement for hybrid CO_2 capture processes.

Source: Adapted from [52].

TABLE 13.2
Hybrid Configurations for CO_2 Capture

Authors	Approach	Reference
Nakhjiri and Heydarinasab	Hybrid membrane absorption process using ethylenediamine (EDA), 2-(1-piperazinyl)-ethylamine (PZEA), and potassium sarcosinate (PS) absorbents	[53]
Scholes *et al.*	CO_2 capture using a hollow fiber membrane, MEA pilot plant	[54]
Lian *et al.*	CO_2 capture via IL-based hybrid processes	[55]
Zhang *et al.*	Hybrid electrocatalysts for efficient electrochemical CO_2 reduction	[56]
Rashidi *et al.*	MEA-MeOH and MEA-water hybrid solvents for CO_2 capture	[57]
Chen *et al.*	Integrated process having a membrane separator and nonthermal plasma reactor	[58]

$CaCO_3$ is formed. This $CaCO_3$ is then calcined and becomes CaO in the calciner during oxyfuel combustion [63]. The regenerated CaO sorbent is returned to the carbonator to continue CO_2 sorption, while the CO_2 stream is produced and processed or used [64].

Researchers Chen *et al.* recently conducted a study on the advancement in CaO-based sorbent design and reinforcement. Activation approaches, all of which significantly improved the sorbents' cyclic performance, include doping, incorporation of supports with increased Tammann temperatures, chemical pretreatment, and structural modification [65].

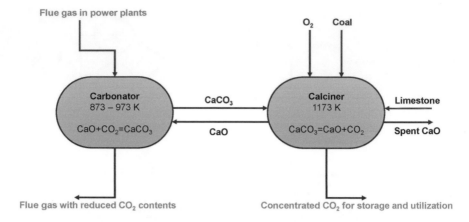

FIGURE 13.8 Schematic diagram of CO_2 capturing by calcium looping from power plants. *Source*: **Adapted from [61].**

13.2.9 OLIVINE MINERALIZATION FOR CARBON CAPTURE

Olivine is a naturally occurring magnesium-based silicate mineral used for CO_2 capture and mineralization [66]. Earth's mantle has the highest concentrations of olivine, up to 700 km in depth [67]. Olivine is also useful in capturing CO_2 from the air indirectly. The mineral can be treated with NH_4HSO_4 to produce $MgSO_4$ in the solid state. Adding NH_4OH to the solution results in the precipitation of brucite (Mg $(OH)_2$). Subsequently, it is ready to capture CO_2 [68–69]. Olivine could be a suitable and efficient alternative to traditional CO_2 capture techniques, especially environmental CO_2 [70].

13.3 PROSPECTS

It is suggested that additional modeling be done to evaluate the particular criteria designed to improve commercial usage of innovative carbon capture and utilization (CCU) alternatives, along with exciting applications and ideas for future consideration. As a result, environmental impact assessment, particularly life cycle assessment (LCA), becomes essential to certain CCU technology developments, such as the growth of circular economy and economic growth models. Thereafter, global studies will be required to help investigators narrow their focus on CCU. Proper knowledge dissemination channels should be developed to inform policymakers and industries to direct financing and prioritize funding for the growth of heading CCU contenders. In addition, research must be done to provide an understanding of the overall monetary, technical, and environmental factors, especially with regard to distribution networks and industrial base (e.g., CO_2 capture and transportation competencies, engineering plants for processing and converting CO_2) [71]. To assess the social structure and carbon capture technology based long-term goals, researchers should summarize CCU-based technologies' development status and innovation activities–based technologies in a dedicated patent database [72].

The supplementary growth and implementation of CCU technologies must be consistent with the social, eco-friendly, and monetary SDGs in future policies and financing schemes. SDGs serve as a valuable tool for assessing technology applications to avoid or minimize potential negative indirect effects by anticipating policy-making and encouraging and enhancing potential direct and indirect positive effects [73]. We need to investigate more transdisciplinary and geographically diversified links among different approaches to carbon capture and the delivery of SDGs. This may also include developing joint evaluation principles or metrics to measure and evaluate the environmental impacts of sustainable development policies across the globe in order to support international climate regulation [74].

If we may want to compare the best prospects for CO_2 capture, there are indeed many options. However, we must also consider the process requirements, materials, costs, and life cycle effectiveness in reducing CO_2 emissions. Many materials can be synthesized, combined, and simulated on the way forward to get the best practice for each requirement. Hence, it is always good to develop new materials by trying novel methods and combinations, where CO_2 capture with a sustainable outlook will always be the main concern for all.

REFERENCES

1. B. Metz, O. Davidson, H. De Coninck, M. Loos and L. Meyer, *IPCC Special Report on Carbon Dioxide Capture and Storage*. Cambridge University Press, 2005.
2. W. S. L. Lesson, "The global goals for sustainability development," May 20, 2021. [Online]. Available: https://worldslargestlesson.globalgoals.org/.
3. E. Chiu, "Economic Equity and Sustainable Development," in *Encyclopedia of Sustainability in Higher Education*, W. Leal Filho, Ed., Springer International Publishing, 2019, pp. 1–6.
4. T. M. L. Wigley, "The Paris warming targets: Emissions requirements and sea level consequences," *Climatic Change*, vol. 147, pp. 31–45, 2018.
5. V. Masson-Delmotte, P. Zhai, H. O. Pörtner, D. Roberts, J. Skea, P. R. Shukla *et al.*, "Global warming of 1.5 C," *An IPCC Special Report on the Impacts of Global Warming of*, vol. 1, pp. 1–9, 2018.
6. T. S. Schmidt and S. Sewerin, "Technology as a driver of climate and energy politics," *Nature Energy*, vol. 2, pp. 1–3, 2017.
7. B. Hare and N. Höhne, "We can limit global warming to 1.5°C if we do these things in the next ten years," *The Conversation*, Nov. 23, 2016.
8. G. Latini, M. Signorile, V. Crocellà, S. Bocchini, C. F. Pirri and S. Bordiga, "Unraveling the CO_2 reaction mechanism in bio-based amino-acid ionic liquids by operando ATR-IR spectroscopy," *Catalysis Today*, vol. 336, pp. 148–160, 2019.
9. W. Gouveia, T. F. Jorge, S. Martins, M. Meireles, M. Carolino, C. Cruz *et al.*, "Toxicity of ionic liquids prepared from biomaterials," *Chemosphere*, vol. 104, pp. 51–56, 2014.
10. E. Davarpanah, S. Hernández, G. Latini, C. F. Pirri and S. Bocchini, "Enhanced CO_2 absorption in organic solutions of biobased ionic liquids," *Advanced Sustainable Systems*, vol. 4, p. 1900067, 2020.
11. Y. Gao, F. Zhang, K. Huang, J. W. Ma, Y. T. Wu and Z. B. Zhang, "Absorption of CO_2 in amino acid ionic liquid (AAIL) activated MDEA solutions," *International Journal of Greenhouse Gas Control*, vol. 19, pp. 379–386, 2013.

12. B. Lv, Y. Shi, C. Sun, N. Liu, W. Li and S. Li, "CO_2 capture by a highly-efficient aqueous blend of monoethanolamine and a hydrophilic amino acid ionic liquid [C_2OHmim] [Gly]," *Chemical Engineering Journal*, vol. 270, pp. 372–377, 2015.

13. B. Li, Y. Chen, Z. Yang, X. Ji and X. Lu, "Thermodynamic study on carbon dioxide absorption in aqueous solutions of choline-based amino acid ionic liquids," *Separation and Purification Technology*, vol. 214, pp. 128–138, 2019.

14. Y. Y. Choi, A. K. Patel, M. E. Hong, W. S. Chang and S. J. Sim, "Microalgae bioenergy with carbon capture and storage (BECCS): An emerging sustainable bioprocess for reduced CO_2 emission and biofuel production," *Bioresource Technology Reports*, vol. 7, p. 100270, 2019.

15. Y. Park, K. Y. A. Lin, A. H. A. Park and C. Petit, "Recent advances in anhydrous solvents for CO_2 capture: Ionic liquids, switchable solvents, and nanoparticle organic hybrid materials," *Frontiers in Energy Research*, vol. 3, Oct. 1, 2015.

16. W. M. Post, T. H. Peng, W. R. Emanuel, A. W. King, V. H. Dale and D. L. DeAngelis, "The global carbon cycle," *American Scientist*, vol. 78, pp. 310–326, 1990.

17. J. H. Mathewson, "Oceanography, chemical," in *Encyclopedia of Physical Science and Technology*, 3rd ed., R. A. Meyers, Ed. Academic Press, 2003, pp. 99–115.

18. J. Singh and D. W. Dhar, "Overview of carbon capture technology: Microalgal biorefinery concept and state-of-the-art," *Frontiers in Marine Science*, vol. 6, Feb. 5, 2019.

19. S. Kondaveeti, I. M. Abu-Reesh, G. Mohanakrishna, M. Bulut and D. Pant, "Advanced routes of biological and bio-electrocatalytic carbon dioxide (CO_2) mitigation toward carbon neutrality," *Frontiers in Energy Research*, vol. 8, June 12, 2020.

20. S. Al-Zuhair, S. AlKetbi and M. Al-Marzouqi, "Regenerating diethanolamine aqueous solution for CO_2 absorption using microalgae," *Industrial Biotechnology*, vol. 12, pp. 105–108, 2016.

21. J. Piiparinen, D. Barth, N. T. Eriksen, S. Teir, K. Spilling and M. G. Wiebe, "Microalgal CO_2 capture at extreme pH values," *Algal Research*, vol. 32, pp. 321–328, 2018.

22. L. I. Rodas-Zuluaga, L. Castañeda-Hernández, E. I. Castillo-Vacas, A. Gradiz-Menjivar, I. Y. López-Pacheco, C. Castillo-Zacarías *et al.*, "Bio-capture and influence of CO_2 on the growth rate and biomass composition of the microalgae *Botryococcus braunii* and *Scenedesmus* sp," *Journal of CO_2 Utilization*, vol. 43, p. 101371, 2021.

23. Q. Yang, H. Li, D. Wang, X. Zhang, X. Guo, S. Pu *et al.*, "Utilization of chemical wastewater for CO_2 emission reduction: Purified terephthalic acid (PTA) wastewater-mediated culture of microalgae for CO_2 bio-capture," *Applied Energy*, vol. 276, p. 115502, 2020.

24. P. Breeze, "Chapter 1 – an introduction to fuel cells," in *Fuel Cells*, P. Breeze, Ed., Academic Press, 2017, pp. 1–10.

25. A. Olabi, M. A. Abdelkareem, T. Wilberforce and E. T. Sayed, "Application of graphene in energy storage device – a review," *Renewable and Sustainable Energy Reviews*, vol. 135, p. 110026, 2021.

26. M. A. Abdelkareem, M. A. Lootah, E. T. Sayed, T. Wilberforce, H. Alawadhi, B. A. A. Yousef *et al.*, "Fuel cells for carbon capture applications," *Science of the Total Environment*, vol. 769, p. 144243, 2021.

27. M. A. Abdelkareem, E. T. Sayed, H. O. Mohamed, M. Obaid, H. Rezk and K. J. Chae, "Nonprecious anodic catalysts for low-molecular-hydrocarbon fuel cells: Theoretical consideration and current progress," *Progress in Energy and Combustion Science*, vol. 77, p. 100805, 2020.

28. F. Wang, S. Deng, H. Zhang, J. Wang, J. Zhao, H. Miao *et al.*, "A comprehensive review on high-temperature fuel cells with carbon capture," *Applied Energy*, vol. 275, p. 115342, 2020.

29. M. Spinelli, D. Di Bona, M. Gatti, E. Martelli, F. Viganò and S. Consonni, "Assessing the potential of molten carbonate fuel cell-based schemes for carbon capture in natural gas-fired combined cycle power plants," *Journal of Power Sources*, vol. 448, p. 227223, 2020.

30. A. Di Fiore, V. Alterio, S. M. Monti, G. De Simone and K. D'Ambrosio, "Thermostable carbonic anhydrases in biotechnological applications," *International Journal of Molecular Sciences*, vol. 16, pp. 15456–15480, 2015.
31. M. Russo, G. Olivieri, A. Marzocchella, P. Salatino, P. Caramuscio and C. Cavaleiro, "Post-combustion carbon capture mediated by carbonic anhydrase," *Separation and Purification Technology*, vol. 107, pp. 331–339, 2013.
32. S. I. Tan, Y. L. Han, Y. J. Yu, C. Y. Chiu, Y. K. Chang, S. Ouyang et al., "Efficient carbon dioxide sequestration by using recombinant carbonic anhydrase," *Process Biochemistry*, vol. 73, pp. 38–46, 2018.
33. Z. Zhang, W. Zhang and E. Lichtfouse, *Membranes for Environmental Applications*. Springer, 2020.
34. N. Boucif, D. Roizard and E. Favre, "The carbonic anhydrase promoted carbon dioxide capture," *Membranes for Environmental Applications*, pp. 1–44, 2020.
35. W. M. S. Alabdraba, C. A. Arslan and Z. B. Mohammed, "Application of artificial neural networks ANN and adaptive neuro fuzzy inference system ANFIS models in water quality simulation of Tigris River at Baghdad City," *Transactions on Machine Learning and Artificial Intelligence*, vol. 5, p. 47, 2017.
36. Y. Kim, H. Jang, J. Kim and J. Lee, "Prediction of storage efficiency on CO_2 sequestration in deep saline aquifers using artificial neural network," *Applied Energy*, vol. 185, pp. 916–928, 2017.
37. M. Rahimi, S. M. Moosavi, B. Smit and T. A. Hatton, "Toward smart carbon capture with machine learning," *Cell Reports Physical Science*, vol. 2, p. 100396, 2021.
38. C. W. Chan, Q. Zhou, and P. Tontiwachiwuthikul, "Part 4a: Applications of knowledge-based system technology for the CO_2 capture process system," *Carbon Management*, vol. 3, pp. 69–79, 2012.
39. L. Helei, P. Tantikhajorngosol, C. Chan and P. Tontiwachiwuthikul, "Technology development and applications of artificial intelligence for post-combustion carbon dioxide capture: Critical literature review and perspectives," *International Journal of Greenhouse Gas Control*, vol. 108, p. 103307, 2021.
40. S. Sarkar, D. Sen, A. Sarkar, S. Bhattacharjee, S. Bandopadhya, S. Ghosh et al., "Modelling aspects of carbon dioxide capture technologies using porous contactors: A review," *Environmental Technology Reviews*, vol. 3, pp. 15–29, 2014.
41. M. Ozkan, "Direct air capture of CO_2: A response to meet the global climate targets," *MRS Energy & Sustainability*, pp. 1–6, 2021.
42. T. von Hippel, "Thermal removal of carbon dioxide from the atmosphere: Energy requirements and scaling issues," *Climatic Change*, vol. 148, pp. 491–501, 2018.
43. X. Shi, H. Xiao, K. Kanamori, A. Yonezu, K. S. Lackner and X. Chen, "Moisture-driven CO_2 sorbents," *Joule*, vol. 4, pp. 1823–1837, 2020.
44. S. Voskian and T. A. Hatton, "Faradaic electro-swing reactive adsorption for CO_2 capture," *Energy & Environmental Science*, vol. 12, pp. 3530–3547, 2019.
45. N. McQueen, P. Kelemen, G. Dipple, P. Renforth and J. Wilcox, "Ambient weathering of magnesium oxide for CO_2 removal from air," *Nature Communications*, vol. 11, pp. 1–10, 2020.
46. N. McQueen, K. V. Gomes, C. McCormick, K. Blumanthal, M. Pisciotta and J. Wilcox, "A review of direct air capture (DAC): Scaling up commercial technologies and innovating for the future," *Progress in Energy*, 2021. doi:10.1088/2516-1083/abf1ce.
47. S. Fujikawa, R. Selyanchyn and T. Kunitake, "A new strategy for membrane-based direct air capture," *Polymer Journal*, vol. 53, pp. 111–119, 2021.
48. M. Broehm, J. Strefler and N. Bauer, "Techno-economic review of direct air capture systems for large scale mitigation of atmospheric CO_2," Available: SSRN 2665702, 2015.
49. B. Freeman, P. Hao, R. Baker, J. Kniep, E. Chen, J. Ding et al., "Hybrid membrane-absorption CO_2 capture process," *Energy Procedia*, vol. 63, pp. 605–613, 2014.

50. M. Scholz, B. Frank, F. Stockmeier, S. Falß and M. Wessling, "Techno-economic analysis of hybrid processes for biogas upgrading," *Industrial & Engineering Chemistry Research*, vol. 52, pp. 16929–16938, 2013.

51. X. Wang and C. Song, "Carbon capture from flue gas and the atmosphere: A perspective," *Frontiers in Energy Research*, vol. 8, p. 265, 2020.

52. C. Song, Q. Liu, N. Ji, S. Deng, J. Zhao, Y. Li *et al.*, "Alternative pathways for efficient CO_2 capture by hybrid processes – a review," *Renewable and Sustainable Energy Reviews*, vol. 82, pp. 215–231, 2018.

53. A. T. Nakhjiri and A. Heydarinasab, "Computational simulation and theoretical modeling of CO_2 separation using EDA, PZEA and PS absorbents inside the hollow fiber membrane contactor," *Journal of Industrial and Engineering Chemistry*, vol. 78, pp. 106–115, 2019.

54. C. A. Scholes, S. E. Kentish and A. Qader, "Membrane gas-solvent contactor pilot plant trials for post-combustion CO_2 capture," *Separation and Purification Technology*, vol. 237, p. 116470, 2020.

55. S. Lian, C. Song, Q. Liu, E. Duan, H. Ren and Y. Kitamura, "Recent advances in ionic liquids-based hybrid processes for CO_2 capture and utilization," *Journal of Environmental Sciences*, vol. 99, pp. 281–295, 2021.

56. B. Zhang, Y. Jiang, M. Gao, T. Ma, W. Sun and H. Pan, "Recent progress on hybrid electrocatalysts for efficient electrochemical CO_2 reduction," *Nano Energy*, vol. 80, p. 105504, 2021.

57. H. Rashidi, P. Valeh-e-Sheyda and S. Sahraie, "A multiobjective experimental based optimization to the CO_2 capture process using hybrid solvents of MEA-MeOH and MEA-water," *Energy*, vol. 190, p. 116430, 2020.

58. H. Chen, Y. Mu, C. Hardacre and X. Fan, "Integration of membrane separation with nonthermal plasma catalysis: A proof-of-concept for CO_2 capture and utilization," *Industrial & Engineering Chemistry Research*, vol. 59, pp. 8202–8211, Apr. 29, 2020.

59. S. Chen, C. Qin, J. Yin, X. Zhou, S. Chen and J. Ran, "Understanding sulfation effect on the kinetics of carbonation reaction in calcium looping for CO_2 capture," *Fuel Processing Technology*, vol. 221, p. 106913, 2021.

60. J. Blamey, E. Anthony, J. Wang and P. Fennell, "The calcium looping cycle for large-scale CO_2 capture," *Progress in Energy and Combustion Science*, vol. 36, pp. 260–279, 2010.

61. J. Liu, J. Baeyens, Y. Deng, T. Tan and H. Zhang, "The chemical CO_2 capture by carbonation-decarbonation cycles," *Journal of Environmental Management*, vol. 260, p. 110054, 2020.

62. Y. Zhang, X. Gong, X. Chen, L. Yin, J. Zhang and W. Liu, "Performance of synthetic CaO-based sorbent pellets for CO_2 capture and kinetic analysis," *Fuel*, vol. 232, pp. 205–214, 2018.

63. C. Qin, D. He, Z. Zhang, L. Tan and J. Ran, "The consecutive calcination/sulfation in calcium looping for CO_2 capture: Particle modeling and behaviour investigation," *Chemical Engineering Journal*, vol. 334, pp. 2238–2249, 2018.

64. J. Sun, Y. Sun, Y. Yang, X. Tong and W. Liu, "Plastic/rubber waste-templated carbide slag pellets for regenerable CO_2 capture at elevated temperature," *Applied Energy*, vol. 242, pp. 919–930, 2019.

65. J. Chen, L. Duan and Z. Sun, "Review on the development of sorbents for calcium looping," *Energy & Fuels*, vol. 34, pp. 7806–7836, 2020.

66. S. Kwon, M. Fan, H. F. DaCosta and A. G. Russell, "Factors affecting the direct mineralization of CO_2 with olivine," *Journal of Environmental Sciences*, vol. 23, pp. 1233–1239, 2011.

67. R. Dabirian, M. Beiranvand and S. Aghahoseini, "Mineral carbonation in peridotite rock for CO_2 sequestration and a method of leakage reduction of CO_2 in the rock," *Nafta*, vol. 63, pp. 44–48, 2012.

68. E. Nduagu, "Mineral carbonation: Preparation of magnesium hydroxide [Mg $(OH)_2$] from serpentinite rock," M. Sc.(Eng.) Thesis, Heat Engineering Laboratory Faculty of Chemical Engineering, Åbo Akademi University, Finland, 2008.

69. C. M. García-Hernández, J. López-Cuevas, C. A. Gutiérrez-Chavarría and A. Flores-Valdés, "Use of mechanical activation to obtain $Mg(OH)_2$ from olivine mineral for CO_2 capture," *Boletín de la Sociedad Española de Cerámica y Vidrio*, vol. 60, pp. 163–174, 2021.

70. R. A. Dwairi, "Mineralogical and geochemical characterization of Jordanian Olivine and its ability to capture CO_2 by mineralization process," *Indonesian Journal on Geoscience*, vol. 6, 2019.

71. S. P. Philbin, "Critical analysis and evaluation of the technology pathways for carbon capture and utilization," *Clean Technologies*, vol. 2, pp. 492–512, 2020.

72. J. N. Kang, Y. M. Wei, L. C. Liu and J. W. Wang, "Observing technology reserves of carbon capture and storage via patent data: Paving the way for carbon neutral," *Technological Forecasting and Social Change*, vol. 171, p. 120933, 2021.

73. B. Olfe-Kräutlein, "Advancing CCU technologies pursuant to the SDGs: A challenge for policy making," *Frontiers in Energy Research*, vol. 8, Aug. 27, 2020.

74. M. Honegger, A. Michaelowa and J. Roy, "Potential implications of carbon dioxide removal for the sustainable development goals," *Climate policy*, vol. 21, pp. 678–698, 2021.

Index

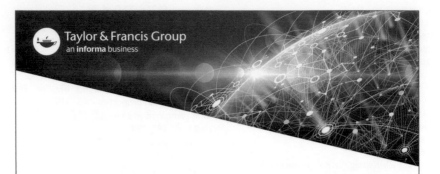